AF787212

# Complexity and Statistical Physics Approaches to Earthquakes

# Complexity and Statistical Physics Approaches to Earthquakes

Editor

**Georgios Michas**

MDPI

Basel • Beijing • Wuhan • Barcelona • Belgrade • Novi Sad • Cluj • Manchester

*Editor*
Georgios Michas
National and Kapodistrian
University of Athens
Athens
Greece

*Editorial Office*
MDPI
St. Alban-Anlage 66
4052 Basel, Switzerland

This is a reprint of articles from the Special Issue published online in the open access journal *Entropy* (ISSN 1099-4300) (available at: https://www.mdpi.com/journal/entropy/special_issues/complex_earthquake).

For citation purposes, cite each article independently as indicated on the article page online and as indicated below:

Lastname, A.A.; Lastname, B.B. Article Title. *Journal Name* **Year**, *Volume Number*, Page Range.

**ISBN 978-3-7258-0205-0 (Hbk)**
**ISBN 978-3-7258-0206-7 (PDF)**
**doi.org/10.3390/books978-3-7258-0206-7**

© 2024 by the authors. Articles in this book are Open Access and distributed under the Creative Commons Attribution (CC BY) license. The book as a whole is distributed by MDPI under the terms and conditions of the Creative Commons Attribution-NonCommercial-NoDerivs (CC BY-NC-ND) license.

# Contents

# About the Editor

**Georgios Michas**

Dr. Georgios Michas currently holds a Research Fellow position in Seismology and Geophysics at the National and Kapodistrian University of Athens, Greece. His PhD from University College London in 2016 was accompanied by a prominent Post-Doctoral Fellowship from the AXA Research Fund. He is the leading author and co-author of more than 40 journal articles, 4 book chapters and more than 60 publications in conference proceedings. His research currently focuses on solid Earth physics, regional geodynamics, fluid-induced seismicity, stochastic models of earthquake occurrence and earthquake forecasting.

*entropy*

*Editorial*

# Complexity and Statistical Physics Approaches to Earthquakes

**Georgios Michas**

Section of Geophysics-Geothermics, Department of Geology and Geoenvironment, National and Kapodistrian University of Athens, 15772 Athens, Greece; gemichas@geol.uoa.gr

**Citation:** Michas, G. Complexity and Statistical Physics Approaches to Earthquakes. *Entropy* **2024**, *26*, 59. https://doi.org/10.3390/e26010059

Received: 18 December 2023
Accepted: 8 January 2024
Published: 10 January 2024

**Copyright:** © 2024 by the author. Licensee MDPI, Basel, Switzerland. This article is an open access article distributed under the terms and conditions of the Creative Commons Attribution (CC BY) license (https://creativecommons.org/licenses/by/4.0/).

This Special Issue of *Entropy*, "Complexity and Statistical Physics Approaches to Earthquakes", sees the successful publication of 11 original scientific articles. This collection presents broad perspectives on the complexity of earthquakes and the use of statistical physics as a consistent, but also necessary, theoretical framework to unravel the complex dynamics that lead to the nucleation and evolution of the phenomenon.

Earthquakes are inherently a complex phenomenon, incorporating intermittency, hierarchy, nonlinear dynamics and interactions over a wide range of spatial and temporal scales [1–5]. However, a simple phenomenology seems to apply to their collective behavior. The most prominent property is scale-invariance. This applies to a variety of key attributes of seismicity manifested as power–law distributions [6–8], as the distribution of fault-trace lengths [9], the Gutenberg–Richter scaling relation that resembles power–law scaling in the frequency of dissipative seismic energies [7] or the Omori–Utsu relation for the power–law decay rate of aftershocks [10]. Such properties motivate the statistical physics approach to fracturing and earthquakes as a consistent and promising theoretical framework for deriving the macroscopic properties observed in fault and earthquake populations from the specification of the laws that govern friction, fluid–rock interactions, fracture nucleation, propagation and so on, at the microscopic level [11].

Since the 1980s, when concepts such as fractals, entropy and self-organized criticality (SOC) became relevant to seismicity, considerable progress has been made in the statistical physics of earthquakes. Within this context, earthquakes are considered a critical-point phenomenon undergoing continuous phase transition [8]. According to SOC, the Earth's crust spontaneously self-organizes in a dynamical stationary state to generate earthquakes with self-similar size distributions and fractal geometries [12]. Earthquakes occur on a fractal set of faults, characterized by long-range correlations and scale-invariant properties in their size and spatiotemporal organization [6–9]. Moreover, based on the maximum entropy principle, classic and generalized statistical mechanics can be used to infer the macroscopic properties of fractures and earthquakes from the specification of their microscopic constituents and their interactions [13]. Other statistical-physics-based models and analysis techniques that have been applied to understand the multiscale dynamics of earthquakes include renormalization group theory, phase diagrams, stochastic models, cellular automata models, correlation lengths, turbulence, percolation and fiber models, multifractals, damage mechanics models, random walks and wavelets and network theory, among others [2,3,7,8,14–17].

Some of these concepts and tools have been applied to the articles found in this Special Issue. This collection features original studies on regional seismicity that evolves into large and destructive earthquakes, as with the recent cases of the 2023 $M_w$ 7.8 and $M_w$ 7.6 doublet that struck the Kahramanmaraş region in East Turkey (contribution 1), the 2019 $M_w$ 7.1 Ridgecrest earthquake in California (contribution 2) and the large subduction earthquakes of magnitudes greater than 7 that occurred on the Cocos subducting plate in Mexico over the last years (contribution 3). The development of early warning systems is exceptionally important in managing such extreme seismic risks, as pointed out by Donciu et al. (contribution 4) in their uniaxial shaking table testing regarding seismic frequency response. To effectively mitigate seismic risk, pattern recognition and probabilistic forecasting of earthquake occurrence using appropriate statistical models are essential, as discussed by Varini and Rotondi (contribution 5) in relation to large earthquakes in Italy and by Anyfadi et al. (contribution 6) regarding major subduction zone aftershock sequences. In this vein, emerging machine learning approaches, such as the NESTORE algorithm applied to the seismicity of Greece by Anyfadi et al. (contribution 7), are becoming more and more valuable. Nonetheless, the theoretical comprehensiveness of fundamental empirical scaling relations in observational seismology, such as the Omori–Utsu relation of the aftershock production rate discussed by Abe et al. (contribution 8) and the development of models that can mimic the physical mechanisms of earthquakes (contribution 9), are important in better understanding earthquake interactions and evolution. In addition, complex network approaches to the physics of earthquakes, as applied to intraplate seismicity in Norway by Pavez-Orrego and Pastén (contribution 10), have been in constant development in recent times. Finally, the integrative study of other natural complex systems with earthquakes, such as solar flare fluctuations, as discussed by Morikawa and Nakamichi (contribution 11), may provide universal patterns regarding the physical behavior of such systems.

Despite the considerable progress that has been achieved over the last forty years, fundamental challenges regarding the complexity and the statistical physics of earthquakes remain wide open, with many important findings anticipated in the years to come. Not only do the exact dynamics that lead to the deformation of the Earth's brittle crust and the subsequent generation of earthquakes remain unknown, but the physical laws that govern friction, rheological and chemical processes, as well as fracture nucleation and propagation at a microscopic scale, are generally elusive and at a primal stage [18,19]. Statistical physics thus remains an expedient framework for bridging the gap between the complex microscopic laws that govern the deformation and brittle failure of solid earth materials and the macroscopic behavior of their ensemble average manifested in fault networks and regional seismicity [11,13,19,20]. Given the overwhelming amount of data that are continually collected, the constantly increasing computational power available and the new models and artificial intelligence methods that emerge, statistical physics, in synergy with seismology and other related fields branching from geology and physics, can lead to a unified framework that will provide a better understanding of the earthquake generation phenomenon, with the ultimate goal of providing efficient earthquake forecasting that can effectively mitigate risk for people and infrastructures.

**Acknowledgments:** I would like to thank all the submitting authors and reviewers for their valuable contributions, as well as *Entropy*'s editorial office for their endless support throughout the compiling of this Special Issue.

**Conflicts of Interest:** The author declares no conflict of interest.

## List of Contributions

1. Zaccagnino, D.; Telesca, L.; Tan, O.; Doglioni, C. Clustering Analysis of Seismicity in the Anatolian Region with Implications for Seismic Hazard. *Entropy* **2023**, *25*, 835.
2. Sardeli, E.; Michas, G.; Pavlou, K.; Vallianatos, F. Spatiotemporal Variations of the Frequency–Magnitude Distribution in the 2019 Mw 7.1 Ridgecrest, California, Earthquake Sequence. *Entropy* **2023**, *25*, 1612.
3. Ramírez-Rojas, A.; Flores-Márquez, E.L.; Vargas, C.A. Visibility Graph Analysis of the Seismic Activity of Three Areas of the Cocos Plate Mexican Subduction Where the Last Three Large Earthquakes ($M > 7$) Occurred in 2017 and 2022. *Entropy* **2023**, *25*, 799.
4. Donciu, C.; Serea, E.; Temneanu, M.C. Frequency Seismic Response for EEWS Testing on Uniaxial Shaking Table. *Entropy* **2023**, *25*, 655.
5. Varini, E.; Rotondi, R. Connection between Variations of the Probability Distribution of the Recurrence Time and Phases of the Seismic Activity. *Entropy* **2023**, *25*, 1441.

6. Anyfadi, E.-A.; Avgerinou, S.-E.; Michas, G.; Vallianatos, F. Universal Non-Extensive Statistical Physics Temporal Pattern of Major Subduction Zone Aftershock Sequences. *Entropy* **2022**, *24*, 1850.
7. Anyfadi, E.-A.; Gentili, S.; Brondi, P.; Vallianatos, F. Forecasting Strong Subsequent Earthquakes in Greece with the Machine Learning Algorithm NESTORE. *Entropy* **2023**, *25*, 797.
8. Abe, S.; Suzuki, N.; Tayurskii, D.A. Aftershocks and Fluctuating Diffusivity. *Entropy* **2023**, *25*, 989.
9. Motuzas, C.A.; Shcherbakov, R. Viscoelastic Slider Blocks as a Model for a Seismogenic Fault. *Entropy* **2023**, *25*, 1419.
10. Pavez-Orrego, C.; Pastén, D. Defining the Scale to Build Complex Networks with a 40-Year Norwegian Intraplate Seismicity Dataset. *Entropy* **2023**, *25*, 1284.
11. Morikawa, M.; Nakamichi, A. Solar Flare 1/f Fluctuations from Amplitude-Modulated Five-Minute Oscillation. *Entropy* **2023**, *25*, 1593.

## References


1. Keilis-Borok, V.I. The lithosphere of the earth as a nonlinear system with implications for earthquake prediction. *Rev. Geophys.* **1990**, *28*, 19–34. [CrossRef]
2. Kagan, Y.Y. Observational evidence for earthquakes as a nonlinear dynamic process. *Physical D* **1994**, *77*, 160–192. [CrossRef]
3. Turcotte, D.L.; Shcherbakov, R.; Rundle, J.B. Complexity and earthquakes. *Treatise Geophys.* **2007**, *4*, 675–700.
4. Ben-Zion, Y. Collective behavior of earthquakes and faults: Continuum-discrete transitions, progressive evolutionary changes, and different dynamic regimes. *Rev. Geophys.* **2008**, *46*, RG4006. [CrossRef]
5. Chelidze, T.; Vallianatos, F.; Telesca, L. *Complexity of Seismic Time Series: Measurement and Application*; Elsevier: Amsterdam, The Netherlands, 2018.
6. Main, I. Statistical physics, seismogenesis, and seismic hazard. *Rev. Geophys.* **1996**, *34*, 433–462. [CrossRef]
7. Turcotte, D.L. *Fractals and Chaos in Geology and Geophysics*, 2nd ed.; Cambridge University Press: Cambridge, UK, 1997.
8. Rundle, J.B.; Turcotte, D.L.; Shcherbakov, R.; Klein, W.; Sammis, C. Statistical physics approach to understanding the multiscale dynamics of earthquake fault systems. *Rev. Geophys.* **2003**, *41*, 1019. [CrossRef]
9. Bonnet, E.; Bour, O.; Odling, N.E.; Davy, P.; Main, I.; Cowie, P.; Berkowitz, B. Scaling of fracture systems in geological media. *Rev. Geophys.* **2001**, *39*, 347–383. [CrossRef]
10. Utsu, T.; Ogata, Y.; Matsu'ura, R.S. The centenary of the Omori formula for a decay law of aftershock activity. *J. Phys. Earth* **1995**, *43*, 1–33. [CrossRef]
11. Sornette, D.; Werner, M.J. Statistical physics approaches to seismicity. In *Encyclopedia of Complexity and Systems Science*; Meyers, R.A., Ed.; Springer: New York, NY, USA, 2009; pp. 7872–7891.
12. Bak, P. *How Nature Works: The Science of Self-Organized Criticality*; Copernicus: New York, NY, USA, 1996.
13. Vallianatos, F.; Papadakis, G.; Michas, G. Generalized statistical mechanics approaches to earthquakes and tectonics. *Proc. R. Soc. A* **2016**, *472*, 20160497. [CrossRef] [PubMed]
14. Sornette, D. *Critical Phenomena in Natural Sciences*, 2nd ed.; Springer: Berlin/Heidelberg, Germany, 2006.
15. Chelidze, T.; Kolesnikov, Y.; Matcharashvili, T. Seismological criticality concept and percolation model of fracture. *Geophys. J. Int.* **2006**, *164*, 125–136. [CrossRef]
16. Abe, S.; Suzuki, N. Complex earthquake networks: Hierarchical organization and assortative mixing. *Phys. Rev. E* **2006**, *74*, 026113. [CrossRef] [PubMed]
17. Michas, G.; Vallianatos, F. Scaling properties, multifractality and range of correlations in earthquake timeseries: Are earthquakes random. In *Statistical Methods and Modeling of Seismogenesis*; Limnios, N., Papadimitriou, E., Tsaklidis, G., Eds.; ISTE Wiley: London, UK, 2021; pp. 171–210.
18. Kawamura, H.; Hatano, T.; Kato, N.; Biswas, S.; Chakrabarti, B.K. Statistical physics of fracture, friction and earthquakes. *Rev. Mod. Phys.* **2012**, *84*, 839–884. [CrossRef]
19. Vallianatos, F.; Michas, G. Complexity of Fracturing in Terms of Non-Extensive Statistical Physics: From Earthquake Faults to Arctic Sea Ice Fracturing. *Entropy* **2020**, *22*, 1194. [CrossRef] [PubMed]
20. Fan, J.; Meng, J.; Ludescher, J.; Chen, X.; Ashkenazy, Y.; Kurths, J.; Havlin, S.; Schellnhuber, H.J. Statistical physics approaches to the complex Earth system. *Phys. Rep.* **2021**, *896*, 1–84. [CrossRef] [PubMed]


**Disclaimer/Publisher's Note:** The statements, opinions and data contained in all publications are solely those of the individual author(s) and contributor(s) and not of MDPI and/or the editor(s). MDPI and/or the editor(s) disclaim responsibility for any injury to people or property resulting from any ideas, methods, instructions or products referred to in the content.

*entropy*

MDPI

*Article*

# Clustering Analysis of Seismicity in the Anatolian Region with Implications for Seismic Hazard

**Davide Zaccagnino [1,*], Luciano Telesca [2], Onur Tan [3] and Carlo Doglioni [1,4]**

1 Department of Earth Sciences, Sapienza University of Rome, 00185 Roma, Italy; carlo.doglioni@uniroma1.it
2 Institute of Methodologies for Environmental Analysis, National Research Council, 85050 Tito, Italy; luciano.telesca@imaa.cnr.it
3 Department of Geophysical Engineering, Faculty of Engineering, Istanbul University-Cerrahpaşa, Istanbul 34320, Turkey; onur.tan@iuc.edu.tr
4 National Institute of Geophysics and Volcanology, 00143 Roma, Italy
* Correspondence: davide.zaccagnino@uniroma1.it

**Abstract:** The Anatolian region is one of the most seismically active tectonic settings in the world. Here, we perform a clustering analysis of Turkish seismicity using an updated version of the Turkish Homogenized Earthquake Catalogue (TURHEC), which contains the recent developments of the still ongoing Kahramanmaraş seismic sequence. We show that some statistical properties of seismic activity are related to the regional seismogenic potential. Mapping the local and global coefficients of variation of inter-event times of crustal seismicity which occurred during the last three decades, we find that territories prone to major seismic events during the last century usually host globally clustered and locally Poissonian seismic activity. We suggest that regions with seismicity associated with higher values of the global coefficient of variation of inter-event times, $C_V$, are likely to be more prone to hosting large earthquakes in the near future than other regions characterized by lower values, if their largest seismic events have the same magnitude. If our hypothesis is confirmed, clustering properties should be considered as a possible additional information source for the assessment of seismic hazard. We also find positive correlations between global clustering properties, the maximum magnitude and the seismic rate, while the b-value of the Gutenberg–Richter law is weakly correlated with them. Finally, we identify possible changes in such parameters before and during the 2023 Kahramanmaraş seismic sequence.

**Keywords:** clustering coefficients; b-value; maximum magnitude; seismogenic potential

**Citation:** Zaccagnino, D.; Telesca, L.; Tan, O.; Doglioni, C. Clustering Analysis of Seismicity in the Anatolian Region with Implications for Seismic Hazard. *Entropy* **2023**, *26*, 835. https://doi.org/10.3390/e25060835

Academic Editor: Georgios Michas

Received: 17 April 2023
Revised: 19 May 2023
Accepted: 20 May 2023
Published: 23 May 2023

**Copyright:** © 2023 by the authors. Licensee MDPI, Basel, Switzerland. This article is an open access article distributed under the terms and conditions of the Creative Commons Attribution (CC BY) license (https://creativecommons.org/licenses/by/4.0/).

## 1. Introduction

### 1.1. Current State of Knowledge

Earthquakes are the final outcome of long-lasting processes of energy accumulation in the brittle crust due to the action of tectonic forces [1]. The nucleation of seismic events starts as soon as the local differential stress overcomes friction and fracture resistance; however, the dynamics of rupture propagation and arrest, as well as the spatial and temporal evolution of seismicity, depend on both the physical properties and the detailed structural organization of the fault systems, e.g., [2]. Therefore, both on- and off-fault rheology and boundary conditions play a role in shaping seismic sequences [3]. Rheology is likely to control the mechanism of stress accumulation and drop, i.e., how tectonic strain is spatially accommodated and released via a wide range of possible seismic dynamics, e.g., [4–6]. For instance, slow slip events tend to be nucleated within weak interfaces along the shallow section of subduction zones, large megathrust events occur in the locked segments close to the trench and aseismic creep takes place where stress is continuously dissipated by spread ductile deformations. On the other hand, long-range interactions are mainly responsible for the temporal evolution of seismicity [7–9]; for this reason, statistical patterns of earthquake activity preceding major seismic events have been widely investigated, e.g., [10–13]. Stress transfer due to preceding events and strain arrangement within crustal volumes provide the ultimate conditions for the dynamic propagation of fracture during the coseismic phase and for the destabilization of fault patches during seismic sequences. Such a complex pattern of interactions leads to both long- and short-term clustering of seismicity over several spatial scales, e.g., [14]. For this reason, clustering features of seismic activity have been extensively studied using different approaches, ranging from classical statistical analysis to artificial intelligence, both in the laboratory and in real fault systems [15–18].

### 1.2. Aim of the Work

Collective parameters can be extremely useful for characterizing the clustering properties of seismicity [19–22]; moreover, more recently, it has been suggested that they can be related to the behavior of seismogenetic sources at a regional scale [23]. Therefore, they may be of interest to infer the seismogenic potential of still poorly investigated areas. In this work, we analyze Anatolian seismicity since 1990, considering the major ($M_w \geq 5.5$) events since 1905 listed in the Turkish Homogenized Earthquake Catalogue (TURHEC) [24,25]. Particular attention is devoted to the southeastern region of Turkey, recently affected by the Kahramanmaraş seismic sequence. We mostly focus on its statistical properties, in particular, long-term global clustering, and the scaling exponent of the frequency-size Gutenberg–Richter law.

## 2. Methods

In this work, we only consider seismic events contained in the TURHEC seismic catalogue which occurred from 1 January 1990 to 27 February 2023 from latitude 34° to 44° N and between longitude 25° and 46° E. In addition, seismic events are only considered if their depth is shallower than 30 km and their size is above the completeness magnitude (compare with the next paragraph). We also consider the $M_w$ 5.5+ earthquakes that occurred from 1905 to 2023 in the same region reported in the same catalogue. In our analysis, we divide the Anatolian region into rectangular contiguous areas. The number of parts is chosen to allow a reliable assessment of the statistical properties of seismicity according to the different sources of uncertainty and their variation along the catalogue. For the assessment of the b-value, a 15 × 6 grid, along longitude and latitude, respectively, is used to guarantee reliable statistical resultswhile a 30 × 15 grid is applied otherwise.

### 2.1. Catalogue Completeness

In this investigation, only earthquakes above the completeness magnitude are considered. We apply the Wiemer–Wyss method [26] and add a correction of +0.2 magnitude units, as suggested in [27]. The completeness magnitude is computed for samples of one thousand earthquakes each in order to take into account the different stages of seismic activity usually associated with variable catalogue completeness.

### 2.2. Coefficients of Variation

The global coefficient of variation of inter-event times, $C_V$, defined by [28]

$$C_V = \frac{\sigma_{\Delta T}}{\langle \Delta T \rangle} \tag{1}$$

where $\langle \Delta T \rangle$ represents the mean value of the inter-event time and $\sigma_{\Delta T}$ is its standard deviation, is applied to study the temporal clustering of seismicity. If $C_V < 1$, the dynamics is regular; in contrast, if $C_V > 1$, the dynamics is clustered. The condition $C_V = 1$ stands for a completely random Poisson process [29]. Conversely, the local coefficient of variation, $L_V$ defined by [30]

$$L_V = \frac{3}{N-1} \sum_{i=1}^{N-1} \frac{(T_i - T_{i+1})^2}{(T_i + T_{i+1})^2} \tag{2}$$

is routinely utilized for quantifying the local variability of the inter-event time series. The meaning of the values of $L_V$ is the same as $C_V$.

### 2.3. b-Value

The Tinti–Mulargia [31] and the maximum likelihood Aki–Utsu [32] methods are applied for the estimation of the b-value of the Gutenberg–Richter law. The first technique performs well in the case of limited catalogues; moreover, it takes into account the magnitude of binning, while the second technique is a standard method with widespread applicability in the case of quite large catalogues and magnitudes ranging at least over three bins. In order to utilize it, $\langle M_w \rangle$ and the threshold (minimum completeness) magnitude $M_{wc}$ are required.

The first is obtained by the definition of the arithmetic mean of the $N$ magnitudes in the catalogue, while the second is estimated using the Wiemer–Wyss method [26], with an additional correction of $+0.2$ magnitude units, as described above.

## 3. Analysis and Results

Since 1990, seismicity in Turkey has mainly taken place offshore in the Aegean Sea and along different segments of the Northern and, more recently, of the Eastern Anatolian fault systems (Figure 1).

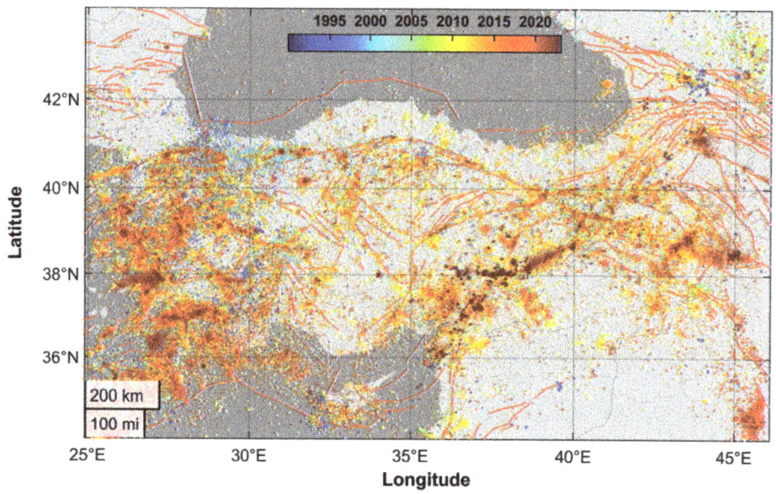


**Figure 1.** Map of seismicity in the Anatolian region. Each point represents an earthquake (TURHEC Catalogue, 1990–2023). Seismic events with epicenters located between 25 and 46° E of longitude and 34 and 44° N of latitude and a hypocenter shallower than 30 km. Red lines represent mapped active faults (data from GEM Global Active Faults database).

More than one hundred thousand earthquakes have been recorded above the completeness magnitude, whose average value has been estimated to be about $M_c \sim 2.8$, which decreased from about $\sim$3.3 in 1990 to less than 2.0 currently (compare with Figure 2) because of the increase in the number of AFAD and KOERI stations after the 1999 Izmit and Düzce earthquakes.

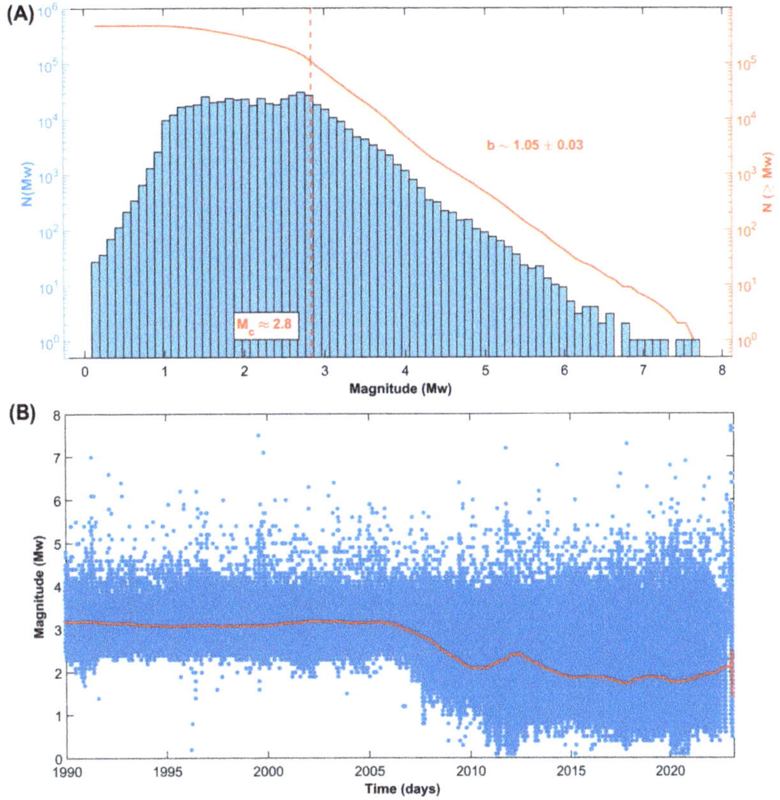


**Figure 2.** (**A**) Frequency-size distribution of the shallow Turkish seismicity (1990–2023, longitude 25–46° E and latitude 34–44° N, and hypocenters shallower than 30 km). (**B**) Catalogue completeness from 1990 to 2023. The red line represents the smoothed completeness magnitude calculated using samples of one thousand earthquakes each.

The largest seismic events of the last three decades occurred in the Kahramanmaraş region, being the 6th February 2023 $M_w$ 7.8 and 7.6 seismic doublet [33] and the $M_w$ 7.4 17 August 1999 Izmit earthquake [34]. Both areas are characterized by high values of the global coefficient of variation. This peculiarity is also shared with other zones along the western Aegean coast of Turkey, which is also prone to large seismic events (Figure 3).

Moreover, a comparative analysis of the spatial distribution of $C_V$ and $L_V$ estimated using data from the TURHEC (1990–2023) shows that the largest seismic events from 1905 to 2023 were nucleated in regions hosting globally clustered and locally Poissonian seismicity (Figure 4).

In addition, a positive correlation has been observed between the global clustering coefficient of the inter-event times and the local seismic rate, defined as the annual amount of energy nucleated by seismicity in the selected area, expressed as a moment magnitude equivalent, and the number of events. For the sake of simplicity, we use a linear fit, as the data is too scattered to apply more complex functions; however, the coefficient of variation is a positive number and the linear relationship is to be considered within the range of magnitudes constrained by the observations. The second trend shows a roughly logarithmic dependence of $C_V$ on the number of earthquakes $N$, so that, while for small subsets ($N \leq 500$) an almost linear relationship exists between the two parameters, for large datasets ($N \geq 1000$), the size effect is almost negligible (Figure 5).

Even clearer is the correlation between the global coefficient of variation of the inter-event times and the maximum magnitude observed in the catalogue (1990–2023). Surprisingly, the statistical trend is still observed considering the maximum magnitude listed in the whole TURHEC, which supports the output already reported in Figure 4. Compare with Figure 6: the blue dots represent the global coefficient of variation of seismicity from 1990 to 2023 in each investigated region as a function of the maximum observed magnitude, while the orange stars mark the same in the case of the TURHEC catalogue since 1905 ($M_w \geq 5.5$). It is worth noting that the blue points tend to be located above the dashed red fit line; in contrast, the stars (except for an outlier) are mainly below the dashed line. A possible interpretation is that regions where higher values of the coefficient of variation of the inter-event times are observed, determined by the size of the largest seismic event in catalogue, are likely to be more prone to hosting major earthquakes in the future with respect to regions characterized by lower values. So, it might just be a matter of time before the next large event. This hypothesis is consistent with what is shown in Figures 4 and 5.

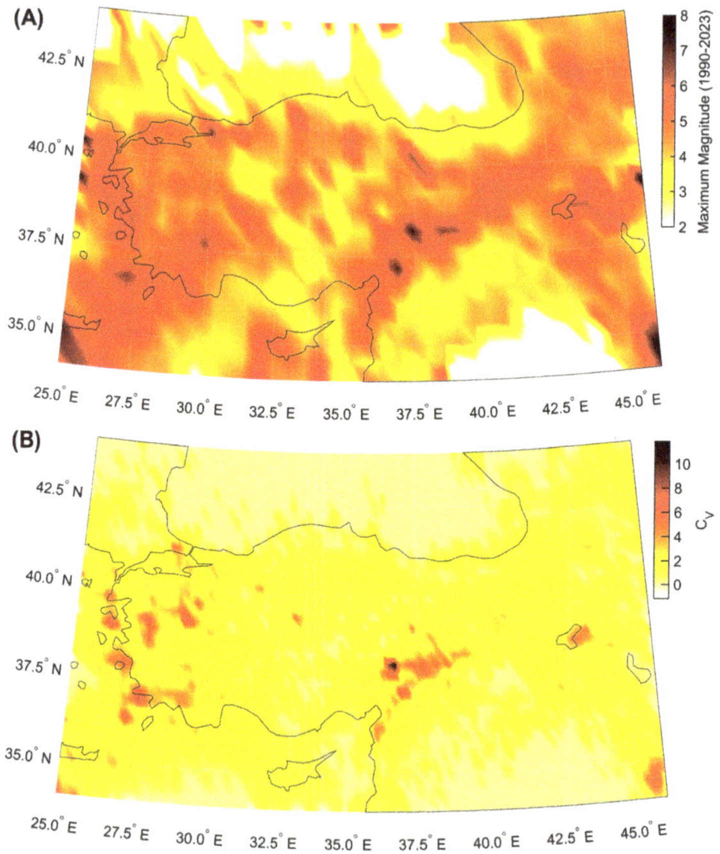


**Figure 3.** (**A**) Map of the maximum magnitude in the catalogue. (**B**) Map of the global coefficient of variation $C_V$ of inter-event times (seismic events occurring in the period 1 January 1990–27 February 2023, longitude 25–46° E, latitude 34–44° N, and hypocenters shallower than 30 km, are considered).

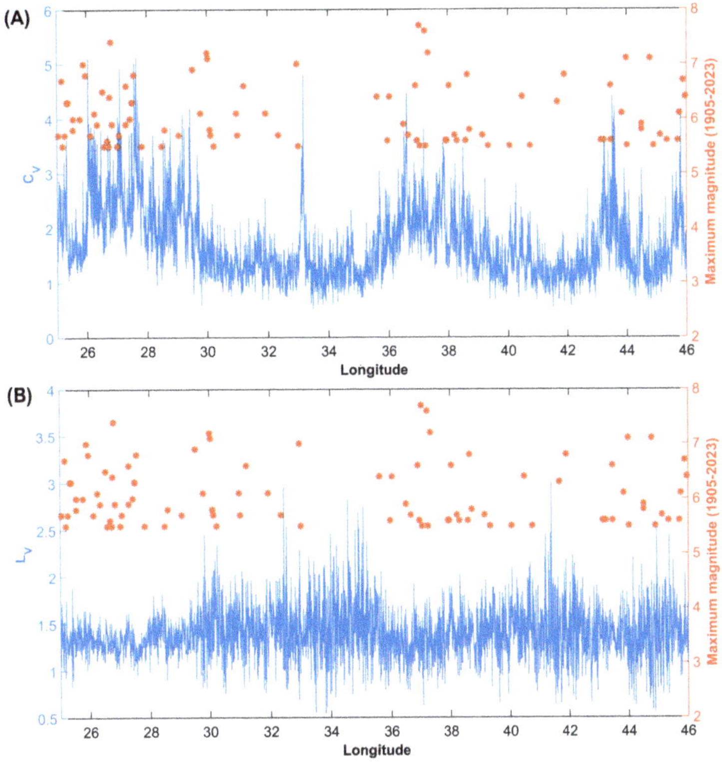


**Figure 4.** Spatial distribution of the global $C_V$ (**A**) and local $L_V$ (**B**) coefficient of variation of inter-event times. The blue line represents the $C_V$ for seismicity reported in the TURHEC catalogue from 1990 to 2023, while the orange asterisks stand for the large ($M_w$ 5.5+) recorded in Turkey since 1905.

In our study, we also investigate the spatial and temporal distribution of the b-values of the Gutenberg–Richter law. We find some regions with lower values of the scaling exponents located along the Northern Anatolian and, above all, along the Eastern Anatolian fault system and offshore of the western coasts of Turkey. A zone with an apparently low b-value is also observed close to the Karliova Triple Junction. Higher values are located in the central and western part of the country between longitude 28 and 31° E. We identify a negative correlation with the maximum magnitude in the TURHEC catalogue (shallow crustal events from 1990 to 2023). Compare with Figure 7.

A more quantitative analysis shows a negative relationship between the b-values of the Gutenberg–Richter law and the seismic rate and the maximum magnitude. However, the negative trend between the scaling exponent of the frequency-size distribution and the amount of annual nucleated energy within sub-regions of a looser grid (see the Section 2), although statistically significant, shows large residuals with respect to the linear trend. The reason is that the uncertainties of the b-values are quite small; so, the $R^2$ is extremely low, which means that the data variability cannot be explained just by the linear relationship used for fitting our data. Compare with Figure 8. The same result is found in the case of the maximum magnitudes. A possible explanation is that the b-value is investigated across the entire Anatolian region where a mixture of different tectonic settings exists and a large variation in crustal states of stress takes place and a simple linear fit is not able to take into account such local effects.

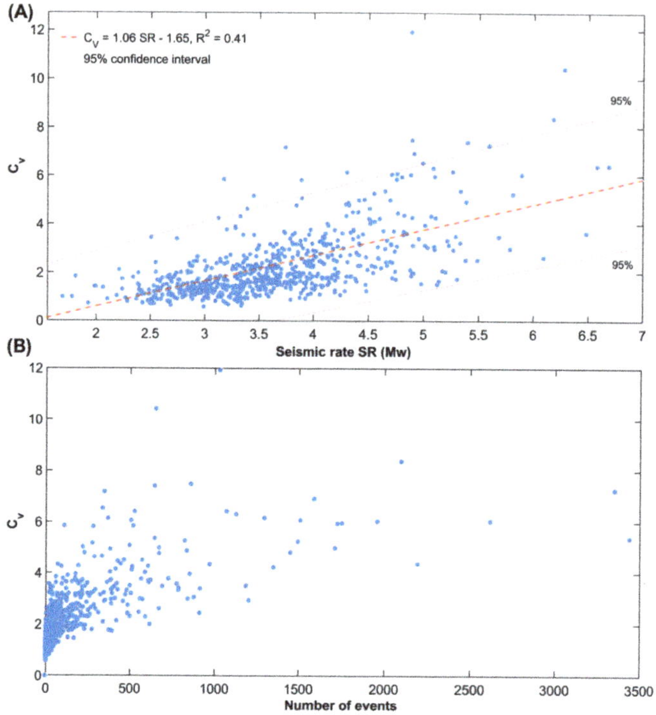


**Figure 5.** (**A**) Correlation between the global coefficient of variation $C_V$ and the annual seismic rate inferred using the regional seismic activity above the completeness magnitude from 1990 to 2023. The linear fit is represented by the dashed thick red line, while the 0.95 prediction intervals are marked by the dashed pink thin ones. (**B**) The global coefficient of variation is weakly positively related to the length of the seismic catalogue. For large seismic catalogues (≥1000 events), $C_V$ appears to be almost independent of the number of recordings.

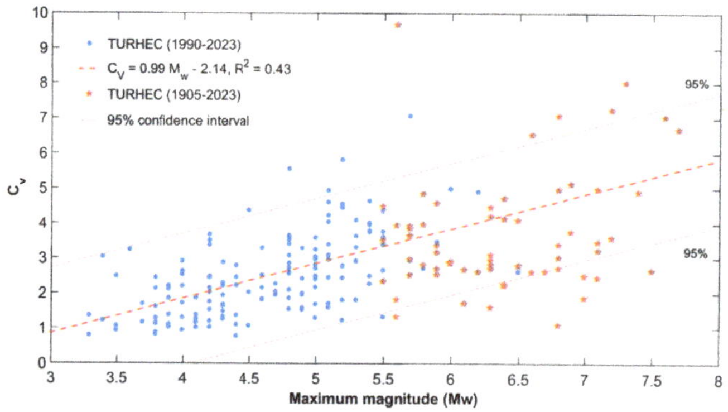


**Figure 6.** The global coefficient of variation is positively correlated with the maximum magnitude in catalogue. The blue circles represent the $C_V$ for each spatial grid element; we segmented the whole region (30 × 15) with at least one hundred seismic events in the catalogue (occurring in the period 1 January 1990–27 February 2023, longitude 25–46° E, latitude 34–44° N, and hypocenter shallower than 30 km). Orange stars stand for the largest earthquakes (≥$M_w$ 5.5) occurring since 1905 in each segment. The linear fit is represented by the dashed thick red line, while the pink thin ones mark the 0.95 prediction intervals.

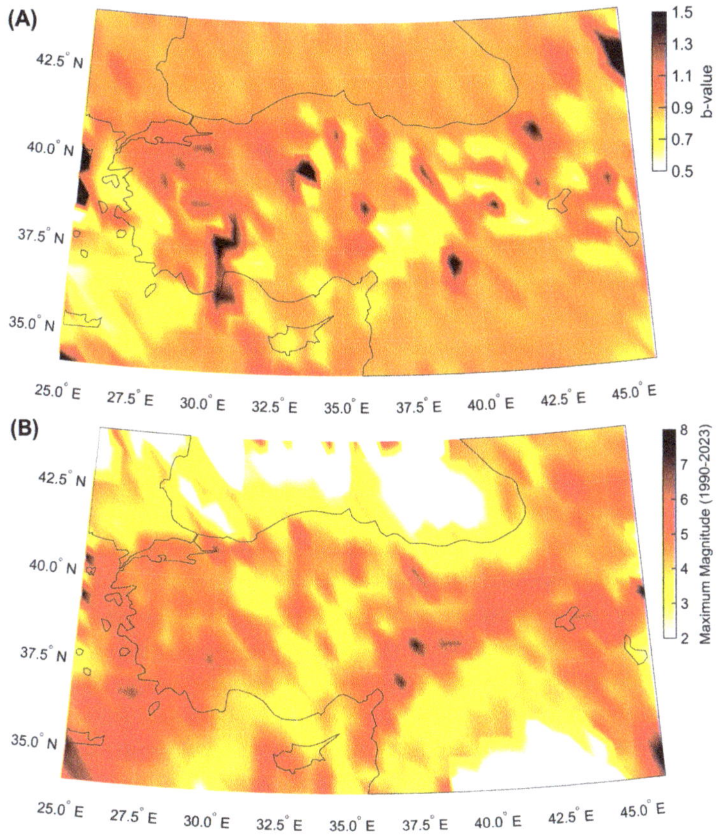


**Figure 7.** (**A**) Map of the b-values of the Gutenberg–Richter scaling law (seismic events occurring in the period 1 January 1990–27 February 2023, longitude 25–46° E, latitude 34–44° N, and hypocenter shallower than 30 km, are considered). (**B**) Map of the maximum magnitude in the catalogue.

In the second part of our investigation, we focus on the Kahramanmaraş region and the seismic sequence still ongoing there. Figure 9 represents seismicity from 1990 to 2023 (events with longitude 34–41° E, latitude 35–40° N, and hypocenter shallower than 30 km are plotted), considering the Gutenberg–Richter law (orange line, Figure 9B) and the density distribution of magnitudes (blue bars). In the lower plots (Figure 9C,D), the temporal evolution of the completeness magnitude is shown.

We analyze the spatial distribution of seismicity above the completeness magnitude since 1990 and the inter-event times (Figure 10). Although from 1990 to 2010 a decreasing trend in the duration of the inter-event intervals is observed because of the progressive lowering of the completeness magnitude due to improvements in the seismic network, a slow, but significant, acceleration in seismic activity is detected since 2014. This evolution led to the $M_w$ 6.7 Doğanyol which occurred on 24 January 2020 and culminated just after the Kahramanmaraş seismic doublet on 4 February 2023. The decrease in the inter-event times from 2014 to 2020 is mainly due to seismic events located at a depth of 10–25 km distributed along almost all the considered faulting region, without the occurrence of any sizeable spatial cluster.

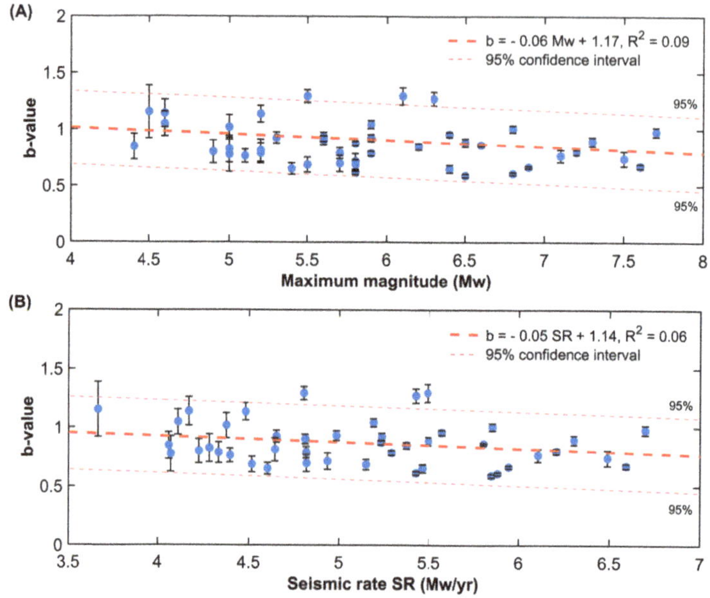


**Figure 8.** A negative trend is observed between the b-value and the annual seismic rate (**A**) and the maximum magnitude in the catalogue (**B**) in Turkey (seismic events occurring in the period 1January 1990–28 February 2023, longitude 25–46° E, latitude 34–44° N, and hypocenter shallower than 30 km, are considered). The Anatolian region is segmented using a rectangular grid (6 × 15 elements); the b-value is included in the plots above provided that at least 300 earthquakes are reported within each region. The error bars stand for the $2\sigma$ uncertainty confidence intervals and the dashed pink thin lines represent the 95% confidence intervals for the linear fit (in dashed thick red).

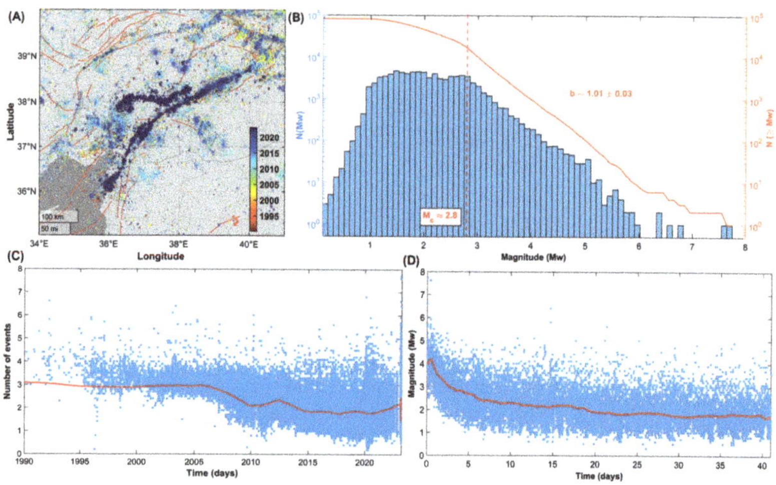


**Figure 9.** (**A**) Map of seismicity in the Kahramanmaraş region from 1990 to 2023 (events with longitude 34–41° E, latitude 35–40° N, and hypocenter shallower than 30 km, are considered). (**B**) Frequency-size distribution of seismicity. Blue bars represent the probability density function, while the orange line stands for the cumulative magnitude distribution. (**C**) Catalogue completeness from 1990 to 2023 and from 6 February 2023 to 19 March 2023 (**D**). The red line is the smoothed completeness magnitude calculated using samples of one thousand seismic events each.

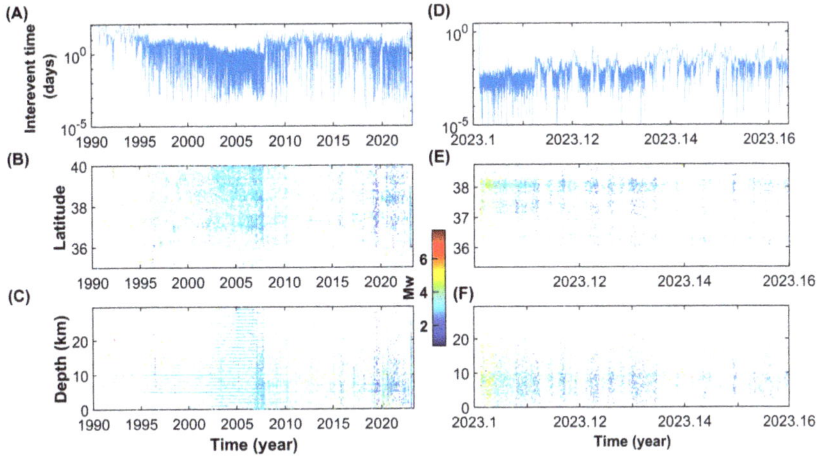


**Figure 10.** Spatial and temporal distribution of seismicity in the Kahramanmaraş region. The plots (**A–C**) show the outputs relative to the whole period of investigation (1990–2023); (**D–F**) present the results for the Kahramanmaraş seismic sequence. The upper panels represent the inter-event time, while the mid and lower plots show how the seismicity above the completeness magnitude is distributed in latitude and depth.

Jointly with the inter-event time and the global coefficient of variation, the temporal changes in the b-value of the Gutenberg–Richter law are usually estimated, providing insightful information about the dynamics preceding large seismic events. Therefore, even in this work, we report the temporal variation in the scaling exponent of the frequency-size law of earthquakes above the completeness magnitude in the investigated area. Figure 11 shows that the large 2020 and 2023 earthquakes are forewarned by a several-months-long drop in the b-value, as well as by an increase in the global coefficient of variation of the inter-event times, $C_V$ (see Figure 11A–C). The decrease in the b-value from about 1.0 to 0.4 started during the second half of 2018. A progressive increase is observed after the $M_w$ 6.7 earthquake, but its values have never returned to their previous level, with further fluctuations occurring during the Kahramanmaraş seismic sequence. The variations in the b-value are also accompanied by changes in $C_V$; in our case, an accelerated increase is observed both before the 2020 and the 2023 seismic sequences. It may suggest that seismicity tends to cluster before major events in this region. A real physical effect in the change in the b-value is likely, even more so in light of the concomitant variations in the clustering properties. Nevertheless, non-physical contributions might play a role in reducing its value, for instance, because of rapid changes in the magnitude completeness that our analysis (based on groups of several hundred events each to provide better estimates of the scaling exponent of the Gutenberg–Richter law) does not have the resolution to highlight.

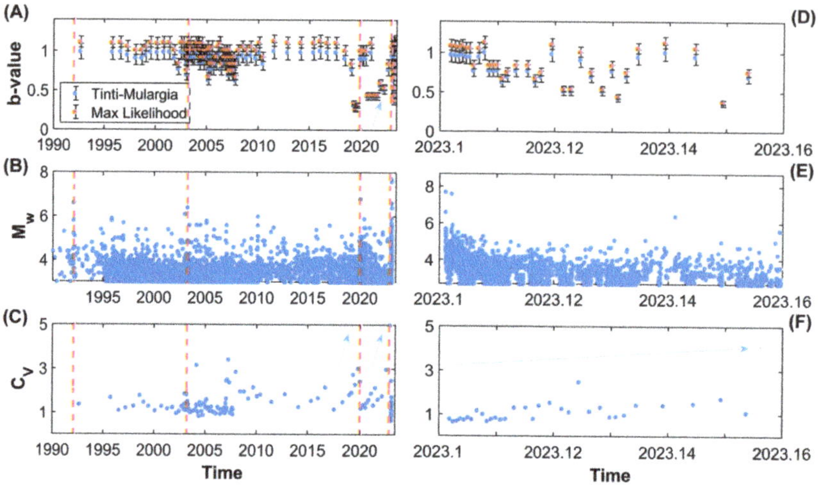


**Figure 11.** A comparison of the statistical and clustering properties of seismicity in the Kahramanmaraş region from 1990 to 2023 (**A–C**) and since 6 February 2023 (**D–F**). In the upper plots, the variations in the b-values of the Gutenberg–Richter law are reported. The b-value is estimated using two different techniques (the Tinti–Mulargia method—blue points—and Aki's maximum likelihood method—orange circles). Seismicity above $M_w$ 3.0 is shown in the central plots, while the global clustering coefficient is plotted below.

## 4. Discussion

### *4.1. Seismotectonic Context and Historical Seismicity*

The Anatolian Plate is located in a quite complex geodynamic setting, at the boundary between the African, Arabian and Eurasian Plates. Extended GNSS time surveys show that the Anatolian Plate undergoes a counter-clockwise rotation [35]. Moreover, a motion ranging between 18 and 28 mm/year is recorded by geodetic stations across the North Anatolian Fault and the Marmara Sea, e.g., [36]. Along the northern transform boundary near the Black Sea coast, as well as along the East Anatolian Fault, frequent seismic activity is recorded. Being prone to major seismic events and densely populated in some areas, great attention has been paid by the scientific community to improve the hazard assessment of this region (e.g., [37–40]). In addition, the Aegean area hosts large earthquakes. Therefore, except for a small portion of its territory, mainly located in the inner regions, Turkey is an extremely active earthquake and volcanic region [41] (Figure 1). Turkey has been hit by several large events. The 17 August 1668 North Anatolia earthquake ($M_w$ 7.8–8.0) [42] was likely the largest known. More recent sequences followed the devastating 26 December 1939 Erzincan $M_w$ 7.9 quake [43], which likely produced the further destabilization which was the reason for the 1942–1944 seismic activity, continuing with major quakes in 1949, 1951, 1957, 1966–1967, 1992 and two in 1999. The last one culminated with the 17 August 1999 $M_w$ 7.4 Izmit earthquake [44] and the following 12 November 1999 $M_w$ 7.2 Düzce event [45]. The Aegean region also nucleated large seismic events, such as the 23 July 1949 Chios [46], the 24 April 1957 $M_w$ 7.1 Ortaca, and the 30 October 2020 $M_w$ 7.0 Izmir Bay earthquake [47]. Large seismic events also occur in intraplate Turkish territories, such as in the case of the 28 March 1970 Gediz $M_w$ 7.0 earthquake [48]. More recently, south-eastern Turkey and its neighboring areas of Syria have been hit by the largest seismic events ever reported in regional instrumental catalogues. On 6 February 2023 at 01:17:35 UTC, a $M_w$ 7.8 strike-slip faulting earthquake involved the East Anatolian Fault. After nine hours, the main shock was followed by a $M_w$ 7.6 twin earthquake nucleated by the Sürgü fault at 10:24:49 UTC about 150 km to the north-west [33]. Thanks to the recent enhancement of the regional KOERI and AFAD seismic networks [49] (compare with the progressive lowering of the completeness magnitude in Figure 2) and the publication of homogeneous catalogues [25] since the occurrence of the 1999 Izmit event, it is now possible to perform an advanced statistical and clustering analysis covering more than thirty years of recordings.

### 4.2. Clustering and Scaling Properties of Turkish Seismicity and Its Regional Variability

Our analysis shows that the largest seismic events in Turkey occur in regions where seismicity is featured by locally Poissonian and globally clustered behavior (Figures 3B and 4). The maximum magnitudes in the catalogue since 1905 are also positively correlated with the global coefficient of variation, calculated using available seismicity data from events occurring during the last three decades (Figures 5 and 6). Our results are in agreement with preceding recently published research [23,50]. Moreover, the spatial mapping of the scaling exponent of the Gutenberg–Richter law also provides interesting information. This analysis is performed by taking advantage of the events above the completeness magnitude listed in the TURHEC and which occurred from 1 January 1990 to 27 February 2023 from latitude 34° to 44° N and between longitude 25° and 46° E, with hypocentral depth shallower than 30 km. We divide the Anatolian region into rectangular contiguous areas. For the assessment of the b-value, a $15 \times 6$ grid, along longitude and latitude, respectively, is utilized in order to guarantee reliable statistical results. See Figures 7 and 8. Regions where seismicity is characterized by lower b-values than the surrounding areas are identified along a large part of the East Anatolian fault system, while isolated spots are observed along the North Anatolian transcurrent boundary. The Aegean coast and sea is also characterized by low b-values. Conversely, the northwestern part of Turkey, as well as the Antalya area, host seismic activity with rather high b-values. A negative correlation between the local b-value, the seismic rate, and the maximum magnitude is observed (Figure 8).

### 4.3. Seismic Activity in the Kahramanmaraş Region

Our analysis focussing on the Kahramanmaraş region confirms the results discussed above and also highlights an anomalous drop in the b-value since 2018 (from 1.0 to $b \sim 0.4$) in the region shaken by the 2023 seismic sequence, accompanied by significant changes in the global coefficient of variation (Figure 11). The value of the scaling exponent of the Gutenberg–Richter law appears to recover its equilibrium condition ($b \approx 0.9$–$1.0$) after the occurrence of the earthquake doublet on 6 February 2023, even though fluctuations are still observed with an apparent long-term decrease. This evidence is consistent with other peculiar seismological patterns recently reported in scientific publications, e.g., [51]. Our results suggest a progressive acceleration of seismic activity in the region since 2018 with a first peak reaching to the north in January 2020, corresponding to the Doğanyol 2020 seismic sequence, which, probably, produced a further destabilization in the southern area, subsequently hit by the Kahramanmaraş events.

### 4.4. Implications and Physical Interpretation

Our research clearly shows that a relationship exists between the clustering and statistical properties of seismicity in Turkey. Large seismic events tend to occur where small to moderate activity is featured by locally Poissonian and globally clustered behavior, low b-values, and an elevated seismic rate. As suggested in [23], such a connection may arise from the mechanism of stress accumulation and release as a function of the structural complexity, fault roughness and rheological heterogeneity of fault systems, e.g., [52–55]. A mechanically weak interface is characterized by low internal friction, so it cannot hold high spatial stress concentration, producing diffuse small magnitude seismicity along the interface; conversely, strong faults enhance stress accumulation and, therefore, the probability of large seismic events increases. Frequent strain release seems to be associated with diffuse, globally Poissonian seismicity with mid-to-high b-values and a relatively low maximum magnitude; in contrast, where fault systems are completely locked, small events occur clustered in time and space, usually organized in swarms or short seismic sequences. Cascade triggering processes are ultimately responsible for larger seismic events, which play a crucial role in re-establishing the mechanical stability of the whole fault system, and also producing significant stress drop. In summary, the geophysical properties of the crustal volumes and major faults, fracturing, statistical features, and clustering of seismicity, are closely connected to each other and to the regional seismogenic potential.

## 5. Conclusions

In this study we perform a clustering analysis of seismicity in Turkey, paying special attention to the Eastern Anatolian region recently hit by the $M_w$ 7.8 and 7.6 seismic doublet, followed by widespread aftershocks. Our results suggest that large earthquakes are more likely to occur in zones characterized by globally clustered, locally Poissonian seismicity, and low b-values. A clear positive correlation is observed between $C_V$ and the annual seismic rate (Figure 5) and the maximum magnitude in catalogue (1990–2023). The effect is still observed when comparing the clustering properties with large seismic events recorded over longer time periods (1905–2023) (see Figures 4 and 6). The prediction intervals and the goodness-of-fit confirm that our conclusions are supported by statistical analysis. Regions with higher values of the global coefficient of variation of inter-event times, $C_V$, are likely to be more prone to nucleating large earthquakes in the near future than regions characterized by lower values, if their largest seismic events have the same magnitude. We think that new studies are required in order to understand to what extent such effects are common to, and statistically significant in, different tectonic regions, having already been observed in New Zealand [23]. If our hypothesis is confirmed, the clustering properties should be considered as a possible additional information source for the assessment of seismic hazard. We also highlight significant variations in both the b-values and the global coefficient of variation of inter-event time series before the largest seismic events in the Kahramanmaraş region, suggesting accelerated energy release and foreshock activity. The result is verified using two different methods for the estimation of the frequency-size scaling exponent, as shown in Figure 11.

**Author Contributions:** Conceptualization, D.Z.; methodology, D.Z. and L.T.; software, D.Z.; investigation, D.Z.; writing—original draft preparation, D.Z.; data curation, O.T.; writing—review and editing, C.D., L.T., O.T. and D.Z.; visualization, D.Z.; supervision, C.D. and L.T.; project administration, L.T. All authors have read and agreed to the published version of the manuscript.

**Funding:** This research received no external funding.

**Institutional Review Board Statement:** Not applicable

**Data Availability Statement:** The updated version of the Turkish Homogenized Earthquake Catalogue (TURHEC), first version available at https://doi.org/10.5281/zenodo.5056801, can be obtained by reasonable request from Onur Tan (onur.tan@iuc.edu.tr), last accessed on 27 February 2023 for the present work; while the catalogue of aftershocks is available on the AFAD portal (https://deprem.afad.gov.tr/event-catalog, accessed on 27 February 2023). The maps in the article were realized using the Matlab mapping toolbox and the Global Active Faults database available at https://github.com/GEMScienceTools/gem-global-active-faults (last access on 20 March 2023).

**Acknowledgments:** The authors thank the editor and four anonymous reviewers for their useful suggestions and comments.

**Conflicts of Interest:** The authors declare no conflict of interest.

## References

1. Kagan, Y.Y.; Knopoff, L. Random stress and earthquake statistics: Time dependence. *Geophys. J. Int.* **1987**, *88*, 723–731. [CrossRef]
2. Zaccagnino, D.; Telesca, L.; Doglioni, C. Scaling properties of seismicity and faulting. *Earth Planet. Sci. Lett.* **2022**, *584*, 117511. [CrossRef]
3. Dieterich, J.H.; Smith, D.E. Nonplanar faults: Mechanics of slip and off-fault damage. In *Mechanics, Structure and Evolution of Fault Zones*; Ben-Zion, Y., Sammis, C., Eds.; Springer: Basel, Switzerland, 2010; pp. 1799–1815.
4. Kagan, Y.Y.; Jackson, D.D. Spatial aftershock distribution: Effect of normal stress.*J. Geophys. Res. Solid Earth* **1998**, *103*, 24453–24467. [CrossRef]

5. Liu, Y.K.; Ross, Z.E.; Cochran, E.S.; Lapusta, N. A unified perspective of seismicity and fault coupling along the San Andreas Fault. *Sci. Adv.* **2022**, *8*, eabk1167. [CrossRef]
6. Volpe, G.; Pozzi, G.; Carminati, E.; Barchi, M.R.; Scuderi, M.M.; Tinti, E.; Aldega, L.; Marone, C.; Collettini, C. Frictional controls on the seismogenic zone: Insights from the Apenninic basement, Central Italy. *Earth. Planet. Sci. Lett.* **2022**, *583*, 117444. [CrossRef]
7. Dieterich, J.H. Earthquake simulations with time-dependent nucleation and long-range interactions. *Nonlinear Process. Geophys.* **1995**, *2*, 109–120. [CrossRef]
8. Knopoff, L.; Levshina, T.; Keilis-Borok, V.I.; Mattoni, C. Increased long-range intermediate-magnitude earthquake activity prior to strong earthquakes in California. *J. Geophys. Res. Solid Earth* **1996**, *101*, 5779–5796. [CrossRef]
9. Vallianatos, F.; Papadakis, G.; Michas, G. Generalized statistical mechanics approaches to earthquakes and tectonics. *Proc. Math. Phys. Eng. Sci.* **2016**, *472*, 20160497. [CrossRef]
10. Sornette, D.; Sammis, C.G. Complex critical exponents from renormalization group theory of earthquakes: Implications for earthquake predictions. *J. Phys. I* **1995**, *5*, 607–619. [CrossRef]
11. Mintzelas, A.; Sarlis, N.V. Minima of the fluctuations of the order parameter of seismicity and earthquake networks based on similar activity patterns. *Physica A* **2019**, *527*, 121293. [CrossRef]
12. Varotsos, P.A.; Sarlis, N.V.; Skordas, E.S. Phenomena preceding major earthquakes interconnected through a physical model. *Ann. Geophys.* **2019**, *37*, 315–324. [CrossRef]
13. Skordas, E.S.; Christopoulos, S.R.G.; Sarlis, N.V. Detrended fluctuation analysis of seismicity and order parameter fluctuations before the M7.1 Ridgecrest earthquake. *Nat. Hazards* **2020**, *100*, 697–711. [CrossRef]
14. Griffin, J.D.; Stirling, M.W.; Wang, T. Periodicity and clustering in the long-term earthquake record. *Geophys. Res. Lett.* **2020**, *47*, e2020GL089272. [CrossRef]
15. Samadi, H.R.; Kimiaefar, R.; Hajian, A. Robust Earthquake Cluster Analysis Based on K-Nearest Neighbor Search. *Pure Appl. Geophys.* **2020**, *177*, 5661–5671. [CrossRef]
16. Seydoux, L.; Balestriero, R.; Poli, P.; Hoop, M.D.; Campillo, M.; Baraniuk, R. Clustering earthquake signals and background noises in continuous seismic data with unsupervised deep learning. *Nat. Commun.* **2020**, *11*, 3972. [CrossRef]
17. Wozniakowska, P.; Eaton, D.W. Machine learning-based analysis of geological susceptibility to induced seismicity in the Montney Formation, Canada. *Geophys. Res. Lett.* **2020**, *47*, e2020GL089651. [CrossRef]
18. Yuan, R. An improved K-means clustering algorithm for global earthquake catalogs and earthquake magnitude prediction. *J. Seismol.* **2021**, *25*, 1005–1020. [CrossRef]
19. Hillers, G.; Mai, P.M.; Ben-Zion, Y.; Ampuero, J.P. Statistical properties of seismicity of fault zones at different evolutionary stages. *Geophys. J. Int.* **2007**, *169*, 515–533. [CrossRef]
20. Goh, K.I.; Barabasi, A.L. Burstiness and memory in complex systems. *Europhys. Lett.* **2008**, *81*, 48002. [CrossRef]
21. Lennartz, S.; Livina, V.N.; Bunde, A.; Havlin, S. Long-term memory in earthquakes and the distribution of interoccurrence times. *Europhys. Lett.* **2008**, *81*, 69001. [CrossRef]
22. Sardeli, E.; Michas, G.; Pavlou, K.; Vallianatos, F.; Karakonstantis, A.; Chatzopoulos, G. Complexity of Recent Earthquake Swarms in Greece in Terms of Non-Extensive Statistical Physics. *Entropy* **2023**, *25*, 667. [CrossRef]
23. Zaccagnino, D.; Telesca, L.; Doglioni, C. Global versus local clustering of seismicity: Implications with earthquake prediction. *Chaos Solitons Fractals* **2023**, *170*, 113419. [CrossRef]
24. Tan, O. Turkish Homogenized Earthquake Catalogue (TURHEC). *Nat. Hazards Earth Syst. Sci. (NHESS)* **2021**. [CrossRef]
25. Tan, O. A homogeneous earthquake catalogue for Turkey. *Nat. Hazards Earth Syst. Sci.* **2021**, *21*, 2059–2073. [CrossRef]
26. Wiemer, S.; Wyss, M. Minimum magnitude of completeness in earthquake catalogs: Examples from Alaska, the western United States, and Japan. *Bull. Seismol. Soc. Am.* **2000**, *90*, 859–869. [CrossRef]
27. Woessner, J.; Wiemer, S. Assessing the quality of earthquake catalogues: Estimating the magnitude of completeness and its uncertainty. *Bull. Seismol. Soc. Am.* **2005**, *95*, 684–698. [CrossRef]
28. Kagan, Y.Y.; Jackson, D.D. Long-term earthquake clustering. *Geophys. J. Int.* **1991**, *104*, 117–133. [CrossRef]
29. Chelidze, T.; Vallianatos, F.; Telesca, L. *Complexity of Seismic Time Series: Measurement and Application*; Elsevier: Amsterdam, The Netherlands, 2018.
30. Shinomoto, S.; Miura, K.; Koyama, S. A measure of local variation of inter-spike intervals. *Biosystems* **2005**, *79*, 67–72. [CrossRef]
31. Tinti, S.; Mulargia, F. Confidence intervals of b values for grouped magnitudes. *Bull. Seismol. Soc. Am.* **1987**, *77*, 2125–2134.
32. Aki, K. Maximum likelihood estimate of b in the formula log N = a-bM and its confidence limits. *Bull. Earthq. Res. Inst.* **1965**, *43*, 237–239.
33. Melgar, D.; Taymaz, T.; Ganas, A.; Crowell, B.; Öcalan, T.; Kahraman, M.; Tsironi, V.; Yolsal-Çevikbilen, S.; Valkaniotis, S.; Irmak, T.S.; et al. Sub- and super-shear ruptures during the 2023 Mw 7.8 and Mw 7.6 earthquake doublet in SE Türkiye. *Seismica* **2023**, *2*. [CrossRef]
34. Ozalaybey, S.; Ergin, M.; Aktar, M.; Tapirdamaz, C.; Biçmen, F.; Yörük, A. The 1999 Izmit earthquake sequence in Turkey: Seismological and tectonic aspects. *Bull. Seismol. Soc. Am.* **2002**, *92*, 376–386. [CrossRef]
35. Mcclusky, S.; Balassanian, S.; Barka, A.; Demir, C.; Ergintav, S.; Georgiev, I.; Gurkan, O.; Hamburger, M.; Hurst, K.; Kahle, H.; et al. Global Positioning System constraints on plate kinematics and dynamics in the eastern Mediterranean and Caucasus. *J. Geophys. Res. Solid Earth* **2000**, *105*, 5695–5719. [CrossRef]

36. Reilinger, R.; McClusky, S.; Vernant, P.; Lawrence, S.; Ergintav, S.; Cakmak, R.; Ozener, H.; Kadirov, F.; Guliev, I.; Stepanyan, R.; et al. GPS constraints on continental deformation in the Africa-Arabia-Eurasia continental collision zone and implications for the dynamics of plate interactions. *J. Geophys. Res. Solid Earth* **2006**, *111*. [CrossRef]
37. Akinci, A.; Malagnini, L.; Herrmann, R.B.; Gok, R.; Sørensen, M.B. Ground motion scaling in the Marmara region, Turkey. *Geophys. J. Int.* **2006**, *166*, 635–651. [CrossRef]
38. Ansal, A.; Akinci, A.; Cultrera, G.; Erdik, M.; Pessina, V.; Tönük, G.; Ameri, G. Loss estimation in Istanbul based on deterministic earthquake scenarios of the Marmara Sea region (Turkey). *Soil Dyn. Earthq. Eng.* **2009**, *29*, 699–709. [CrossRef]
39. Murru, M.; Akinci, A.; Falcone, G.; Pucci, S.; Console, R.; Parsons, T. M $\geq$ 7 earthquake rupture forecast and time-dependent probability for the Sea of Marmara region, Turkey. *J. Geophys. Res. Solid Earth* **2016**, *121*, 2679–2707. [CrossRef]
40. Lange, D.; Kopp, H.; Royer, J.Y.; Henry, P.; Çakir, Z.; Petersen, F.; Sakic, P.; Ballu, V.; Bialas, J.; Özeren, M.S.; et al. Interseismic strain build-up on the submarine North Anatolian Fault offshore Istanbul. *Nat. Commun.* **2019**, *10*, 3006. [CrossRef]
41. Barka, A.; Reilinger, R. Active Tectonics of the Eastern Mediterranean Region: Deduced from GPS, Neotectonic and Seismicity Data. 1997. Available online: http://hdl.handle.net/2122/1520 (accessed on 9 April 2023).
42. Hartleb, R.D.; Dolan, J.F.; Akyüz, H.S.; Yerli, B. A 2000-year-long paleoseismologic record of earthquakes along the central North Anatolian Fault, from trenches at Alayurt, Turkey. *Bull. Seismol. Soc. Am.* **2003**, *93*, 1935–1954. [CrossRef]
43. Emre, Ö.; Kondo, H.; Özalp, S.; Elmaci, H. Fault geometry, segmentation and slip distribution associated with the 1939 Erzincan earthquake rupture along the North Anatolian fault, Turkey. *Geol. Soc. Spec. Publ.* **2021**, *501*, 23–70. [CrossRef]
44. Barka, A. The 17 august 1999 Izmit earthquake. *Science* **1999**, *285*, 1858–1859. [CrossRef]
45. Burgmann, R.; Ayhan, M.E.; Fielding, E.J.; Wright, T.J.; McClusky, S.; Aktug, B.; Demir, C.; Lenk, O.; Tu, A. Deformation during the 12 November 1999 Duzce, Turkey, earthquake, from GPS and InSAR data. *Bull. Seismol. Soc. Am.* **2002**, *92*, 161–171. [CrossRef]
46. Melis, N.S.; Okal, E.A.; Synolakis, C.E.; Kalogeras, I.S.; Kânoglu, U. The Chios, Greece earthquake of 23 july 1949: Seismological reassessment and tsunami investigations. *Pure Appl. Geophys.* **2020**, *177*, 1295–1313. [CrossRef]
47. Jamalreyhani, M.; Büyükakpınar, P.; Cesca, S.; Dahm, T.; Sudhaus, H.; Rezapour, M.; Isken, M.P.; Asayesh, B.M.; Heimann, S. Seismicity related to the eastern sector of Anatolian escape tectonic: The example of the 24 January 2020 Mw 6.77 Elazig-Sivrice earthquake. *Solid Earth Discuss.* **2020**, 1–22. [CrossRef]
48. Ambraseys, N.N.; Tchalenko, J.S. Seismotectonic aspects of the Gediz, Turkey, earthquake of March 1970. *Geophys. J. Int.* **1972**, *30*, 229–252. [CrossRef]
49. Tan, O.; Tapirdamaz, M.C.; Yörük, A. The earthquake catalogues for Turkey. *Turk. J. Earth Sci.* **2008**, *17*, 405–418.
50. Zaccagnino, D.; Telesca, L.; Doglioni, C. Variable seismic responsiveness to stress perturbations along the shallow section of subduction zones: The role of different slip modes and implications for the stability of fault segments. *Front. Earth Sci.* **2022**, *10*, 989697. [CrossRef]
51. Kwiatek, G.; Martínez-Garzón, P.; Becker, D.; Dresen, G.; Cotton, F.; Beroza, G.; Acarel, D.; Ergintav, S.; Bohnhoff, M. Months-long preparation of the 2023 MW 7.8 Kahramanmaraş earthquake, Türkiye. *Preprint* **2023**. [CrossRef]
52. Romanet, P.; Sato, D.S.; Ando, R. Curvature, a mechanical link between the geometrical complexities of a fault: Application to bends, kinks and rough faults. *Geophys. J. Int.* **2020**, *223*, 211–232. [CrossRef]
53. Zaccagnino, D.; Doglioni, C. The impact of faulting complexity and type on earthquake rupture dynamics. *Commun. Earth Environ.* **2022**, *3*, 258. [CrossRef]
54. Cochran, E.S.; Page, M.T.; van der Elst, N.J.; Ross, Z.E.; Trugman, D.T. Fault Roughness at Seismogenic Depths and Links to Earthquake Behavior. *Seism. Rec.* **2023**, *3*, 37–47. [CrossRef]
55. Goebel, T.H.W.; Brodsky, E.E.; Dresen, G. Fault roughness promotes earthquake-like aftershock clustering in the lab. *Geophys. Res. Lett.* **2023**, *50*, e2022GL101241. [CrossRef]


**Disclaimer/Publisher's Note:** The statements, opinions and data contained in all publications are solely those of the individual author(s) and contributor(s) and not of MDPI and/or the editor(s). MDPI and/or the editor(s) disclaim responsibility for any injury to people or property resulting from any ideas, methods, instructions or products referred to in the content.

*entropy*

*Article*

# Spatiotemporal Variations of the Frequency–Magnitude Distribution in the 2019 $M_w$ 7.1 Ridgecrest, California, Earthquake Sequence

**Eirini Sardeli [1], Georgios Michas [1,*], Kyriaki Pavlou [1] and Filippos Vallianatos [1,2]**

1 Section of Geophysics-Geothermics, Department of Geology and Geoenvironment, National and Kapodistrian University of Athens, 15772 Athens, Greece; eirsard@geol.uoa.gr (E.S.); kpavlou@geol.uoa.gr (K.P.); fvallian@geol.uoa.gr (F.V.)
2 Institute of Physics of Earth's Interior and Geohazards, UNESCO Chair on Solid Earth Physics and Geohazards Risk Reduction, Hellenic Mediterranean University Research & Innovation Center, 73133 Chania, Greece
* Correspondence: gemichas@geol.uoa.gr

**Abstract:** Significant seismic activity has been witnessed in the area of Ridgecrest (Southern California) over the past 40 years, with the largest being the $M_w$ 5.8 event on 20 September 1995. In July 2019, a strong earthquake of $M_w$ 7.1, preceded by a $M_w$ 6.4 foreshock, impacted Ridgecrest. The mainshock triggered thousands of aftershocks that were thoroughly documented along the activated faults. In this study, we analyzed the spatiotemporal variations of the frequency–magnitude distribution in the area of Ridgecrest using the fragment–asperity model derived within the framework of non-extensive statistical physics (NESP), which is well-suited for investigating complex dynamic systems with scale-invariant properties, multi-fractality, and long-range interactions. Analysis was performed for the entire duration, as well as within various time windows during 1981–2022, in order to estimate the $q_M$ parameter and to investigate how these variations are related to the dynamic evolution of seismic activity. In addition, we analyzed the spatiotemporal $q_M$ value distributions along the activated fault zone during 1981–2019 and during each month after the occurrence of the $M_w$ 7.1 Ridgecrest earthquake. The results indicate a significant increase in the $q_M$ parameter when large-magnitude earthquakes occur, suggesting the system's transition in an out-of-equilibrium phase and its preparation for seismic energy release.

**Keywords:** Ridgecrest; earthquake sequence; frequency–magnitude distribution; fragment–asperity model; Tsallis entropy; non-extensive statistical physics; complexity

**Citation:** Sardeli, E.; Michas, G.; Pavlou, K.; Vallianatos, F. Spatiotemporal Variations of the Frequency–Magnitude Distribution in the 2019 $M_w$ 7.1 Ridgecrest, California, Earthquake Sequence. *Entropy* **2023**, *25*, 1612. https://doi.org/10.3390/e25121612

Academic Editor: Roberto Zivieri

Received: 1 November 2023
Revised: 27 November 2023
Accepted: 29 November 2023
Published: 1 December 2023

**Copyright:** © 2023 by the authors. Licensee MDPI, Basel, Switzerland. This article is an open access article distributed under the terms and conditions of the Creative Commons Attribution (CC BY) license (https://creativecommons.org/licenses/by/4.0/).

## 1. Introduction

The 2019 Ridgecrest earthquake sequence took place in the eastern California shear zone, near the town of Ridgecrest and southwest of Searles Valley. The sequence initially evolved as a series of foreshocks, with the largest one of magnitude $M_w$ 6.4 on 4 July 2019, preceding a strong mainshock of magnitude $M_w$ 7.1 that occurred two days later, on 6 July 2019 (3:19:53 UTC). The $M_w$ 7.1 event was accompanied by thousands of aftershocks during the following months (Figure 1) [1–3]. The spatial distribution of the two events of strong magnitudes, $M_w$ 6.4 and $M_w$ 7.1, as well as the thousands of aftershocks, revealed the activation of two main strike-slip fault zones: a previously unnoticed NE–SW left-lateral strike-slip fault zone associated with the $M_w$ 6.4 seismic event and a NW–SE right-lateral strike-slip Little Lake fault zone associated with the $M_w$ 7.1 mainshock [1,4–6]. The activated area is situated near the Airport Lake and Little Lake fault zones, both of which have a lengthy history of seismic activity [7]. More specifically, the activated fault zone displays widespread orthogonal faulting over multiple length scales, characterized by intricate geometric patterns [1]. The largest scale is approximately 55 km in a northwest-striking direction, intersected orthogonally by a fault roughly 15 km in length [1,4]. During the $M_w$ 7.1 earthquake, the larger of these structures was the main one that ruptured, whereas, the $M_w$ 6.4 event affected the smaller structure [1,4].

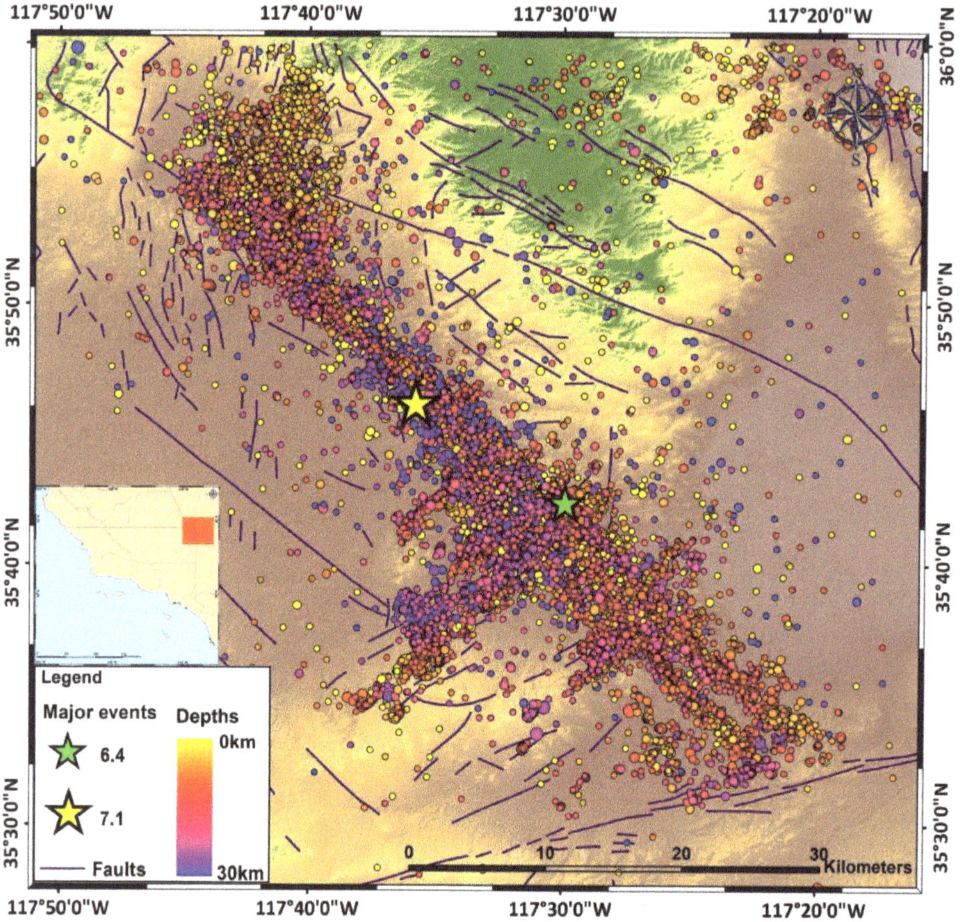


**Figure 1.** The spatial distribution of the 2019 Ridgecrest earthquake sequence for 38,452 events that occurred during the period between 4 July 2019 and 31 December 2019. The green and yellow stars indicate the foreshock and mainshock of magnitudes 6.4 and 7.1, respectively. Regional faults are marked with solid purple lines (https://koordinates.com/layer/701-california-faults/, accessed on 26 June 2023), and the seismic events are colored according to depth.

The rupture processes of the $M_w$ 6.4 and the $M_w$ 7.1 strong events have also been studied using geodetic and seismic data, revealing a complex interaction between multiple fault segments and branches, as well as the spatial and temporal variations of slip, stress drop, rupture speed, and directivity [2,4,5,8]. The 2019 Ridgecrest earthquake sequence has provided an exceptional occasion to investigate the physics of faulting and earthquake interactions in a complex fault system.

Over the past forty years, seismic activity in the Ridgecrest area has been characterized by swarms of earthquakes, with some lasting for over a year, and some notable moderate-magnitude events, such as the 1982 $M_L$ 4.9 Indian Wells Valley event and the 1995–1996 Ridgecrest sequence, including three $M_w$ 5+ earthquakes [9]. The 1995 earthquake sequence began on 17 August 1995, with an earthquake of magnitude $M_w$ 5.4, followed by a $M_w$ 5.8 event on 20 September 1995 and a $M_w$ 5.2 event on 7 January 1996. The 1995 events occurred on normal and strike-slip NW- and NE-trending faults [9], showing similar complexity to the 2019 events.

In the present study, Ridgecrest's seismicity is being investigated in terms of the frequency–magnitude distribution (FMD), which is an inherent component of the regional level of seismic activity and a fundamental part of seismic hazard assessments. The analysis of the FMD is performed with the fragment–asperity (F–A) model, initially developed by Sotolongo-Costa and Posadas [10] within the framework of non-extensive statistical physics (NESP) [11]. Using the F–A model, we analyzed the spatiotemporal variations of the derived entropic index $q_M$, which is used as an index for the physical state of the studied region. The F–A model is used to calculate the seismic energy distribution function (EDF) utilizing fragment size distribution, providing an EDF that includes the Gutenberg–Richter (G–R) scaling relation as a specific case [12,13].

The $q_M$ parameter is herein estimated for the entire duration and also for various time windows, during the period 1981–2022. The variations of $q_M$ are examined to identify patterns associated with the evolution of the seismic activity and the results are subject to discussion. Additionally, we estimated the spatiotemporal variations in $q_M$ values derived from the F–A model for the Ridgecrest earthquake sequence. We analyzed the background seismicity from 1981 to 2019 (before the foreshock of $M_w$ 6.4 on 4 July 2019) to estimate the background $q_M$ values during this period, and then for the period after the foreshock of $M_w$ 6.4, before the occurrence of the $M_w$ 7.1 mainshock, and finally, after the $M_w$ 7.1 mainshock on 6 July 2019 to November 2019 for each month of aftershock activity.

## 2. Methodological Approach Based on NESP

*Non-Extensive Fragment–Asperity Model for Seismic Energies*

In 2004, Sotolongo-Costa and Posadas [10] introduced the fragment–asperity (F–A) model of earthquake interactions, outlining the earthquake generation mechanism based on the small-scale processes within fault zones. This general model [10], developed within the non-extensive statistical physics (NESP) framework, considers the interaction between two rough profiles (fault blocks) and the fragments that occupy the space between them, caused by the local fracturing of tectonic plates. The fragments can have a significant impact on the earthquake triggering process. As stress between two fault surfaces rises, these rough fragments act both as roller bearings, expediting the slipping process, whereas when a fragment is displaced or an asperity brakes, the fault planes slip and seismic energy releases. Since fragments result from the violent fractioning between fault planes, it is anticipated that there will be long-range interactions between all parts of existent fragments. This implies that the size distribution function of the fragments is more appropriately treated using the NESP formalism.

NESP, introduced by Tsallis [14–16], is proposed as a possible generalization of Boltzmann–Gibbs (BG) statistical physics and has found wide applications in various non-linear dynamical systems [11]. Furthermore, the non-additive Tsallis entropy $Sq$, includes the parameter $q$, which quantifies the non-extensivity of a system. In the limit, where $q = 1$, $Sq$ approaches the BG entropy. Even though $Sq$ and $S_{BG}$ have several common characteristics, such as non-negativity, expansibility, and concavity, there is a characteristic dissimilarity between the two entropies. The BG entropy is additive, signifying that the entropy of a combined system is the sum of the entropies of its individual parts, whereas the Tsallis entropy $Sq$ is non-additive. In addition, the equilibrium phase of various short-range interacting systems (e.g., Hamiltonians) is well approximated with BG statistical physics, whereas various non-linear long-range interacting systems are better described with NESP [16,17].

Following Sotolongo-Costa and Posadas [10], the non-additive entropy $Sq$, in terms of the probability $p(\sigma)$ of finding a fragment of area $\sigma$, is presented as follows:

$$S_q = k\frac{1 - \int p^q(\sigma)d\sigma}{q-1} \tag{1}$$

where $q$ is the non-extensive parameter. Silva et al. [18] introduced the fragment size distribution function as:

$$p(\sigma) = \left[1 - \frac{(1-q)}{(2-q)}(\sigma - \sigma_q)\right]^{\frac{1}{1-q}} \quad (2)$$

Moreover, Silva et al. [18] introduced a scaling law between the released relative energy $E$ and the volume of fragments $r$ with the relationship $E \sim r^3$, in agreement with the standard seismic moment theory [19]. The proportionality between the released energy $E$ and $r^3$ can then be expressed as:

$$\sigma - \sigma_q = \left(\frac{E}{A}\right)^{\frac{2}{3}} \quad (3)$$

where $\sigma$ scales with $r^2$, and $A$ is proportional to the volumetric energy density.

Using Equations (2) and (3), the energy distribution function (EDF) of earthquakes is derived as:

$$p(E) = \frac{C_1 E^{\frac{1}{3}}}{\left[1 + C_2 E^{2/3}\right]^{1/q-1}} \quad (4)$$

with $C_1 = \frac{2}{3A^{2/3}}$ and $C_2 = -\frac{(1-q)}{(2-q)A^{2/3}}$

Telesca [20] further used the relation $M = 2/3\log E$ to derive the cumulative magnitude distribution:

$$\log\left(\frac{N(>M)}{N}\right) = \frac{2-q_M}{1-q_M}\log\left[1 - \left(\frac{1-q_M}{2-q_M}\right)\left(\frac{10^M}{A^{2/3}}\right)\right] \quad (5)$$

Furthermore, by considering the threshold magnitude ($Mc$), Telesca [13] introduced a modified function that links the cumulative number of earthquakes with magnitude, expressed as:

$$\log\left(\frac{N(>M)}{N}\right) = \frac{2-q_M}{1-q_M}\log\left(\frac{1 - \left(\frac{1-q_M}{2-q_M}\right)\left(\frac{10^M}{A^{2/3}}\right)}{1 - \left(\frac{1-q_M}{2-q_M}\right)\left(\frac{10^{Mc}}{A^{2/3}}\right)}\right) \quad (6)$$

Temporal variations and an increase in $q_M$ suggest that the seismic area's physical state is departing from equilibrium [21–23]. The fitting procedure of Equation (6) to the observed distribution, is the Levenberg–Marquardt non-linear least-square method [24,25] which is used to solve non-linear least squares problems. As mentioned in [26], this approach is widely known for its effectiveness in achieving high precision and swiftly converging to the best possible solution. The F–A model has found utility in diverse applications, including regional and local seismic activity, as well as volcanic seismicity [21,22,27,28].

According to Telesca [13], the maximum likelihood $q_M$ values are linked to the maximum likelihood Gutenberg–Richter $b$ value, as follows:

$$b = \frac{2-q_M}{q_M-1} \quad (7)$$

which is equivalent to the relation derived by [23].

As it is commonly known, the computation of $b$ value using the maximum likelihood approach [29] is highly sensitive to the initial choice of minimum earthquake magnitude $Mc$ in the seismic catalogue. On the other hand, $q_M$ value estimation is relatively stable irrespective of the choice of $Mc$ [30].

## 3. Seismological Data

In this work, we used the waveform relocation catalogue for Southern California [31] provided by the Southern California Earthquake Data Center (SCEDC), which expands from 1981 to 2022. A total of 103,706 earthquakes, which occurred in the period between 3 January 1981 and 31 March 2022, within a defined boundary of 117.2° E~118.0° E longitude and 35.4° N~36.0° N latitude, were considered. The depth distribution of seismicity varies from −1.41 to 30.8 km, whereas the magnitudes range between −1.02 and 7.1.

The research area has experienced thousands of small events, with some of the largest in the magnitude range of 4 to 7. On 4 July 2019, a $M_W$ 6.4 earthquake occurred, preceded by a series of small events in the prior hour. The largest earthquake occurred 34 h later, on 6 June, with a magnitude of $M_W$ 7.1. Eleven months later, on 4 June 2020, a $M_W$ 5.53 aftershock took place to the east of Ridgecrest (Figure 1). Over the past 40 years, 11 other $M > 5$ earthquakes have occurred in this area. The largest one was an earthquake of magnitude $M_W$ 5.8 on 20 September 1995, about 13 km to the west–northwest of Ridgecrest [31]. In addition, according to the catalogue, an earthquake of magnitude $M_W$ 5.2 occurred on 1 October 1982.

Within the study area, there were a total of 8924 events in 1995, with 2722 of them occurring in September of that year (Figure 2a). In 2019, there was a seismic burst of 38,530 events associated with the Ridgecrest earthquake sequence, with 6 of them over a magnitude of $M$ 5. Figure 2b illustrates the temporal progression of the earthquake magnitudes in the Ridgecrest area, as well as the cumulative seismic moment release throughout the 41-year observation period. The equation $\log Mo = 1.5 \times M + 9.1$ was used to approximate the seismic moment release (in Nm) in relation to earthquake magnitudes [32].

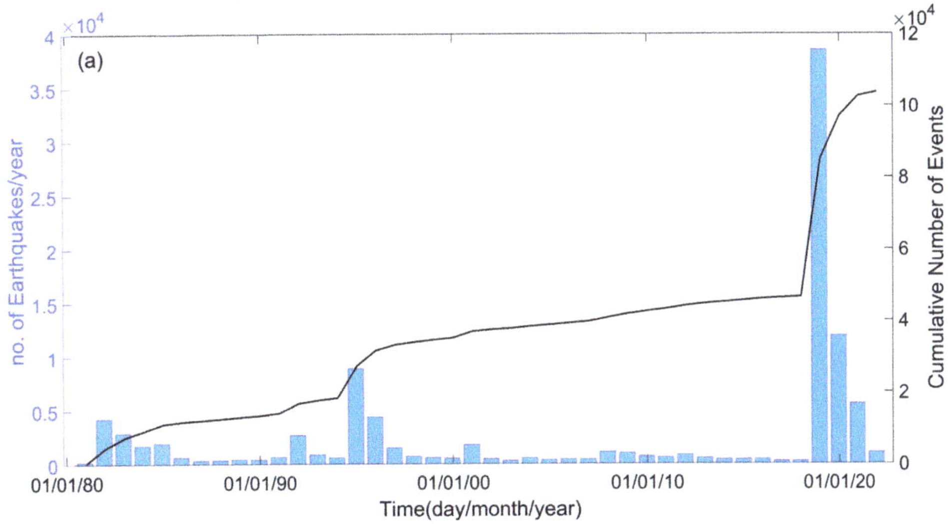


**Figure 2.** *Cont.*

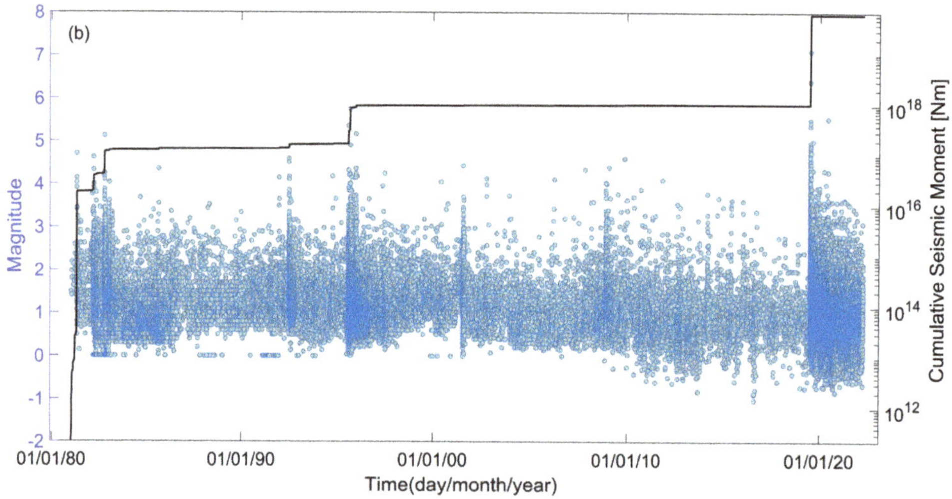


**Figure 2.** (**a**) Histogram showing the annual number of events during the period between January 1981 and April 2022. The black line illustrates the cumulative number of seismic events, N, in the research area. (**b**) Magnitude distribution per day versus time. The black line shows the cumulative seismic moment release.

## 4. Results

### *4.1. Frequency–Magnitude Distribution during 1981–2022*

The F–A model, as discussed in the preceding section, is applied to the normalized cumulative magnitude distribution for our dataset from 1981 to 2022, for the entire magnitude range. This model describes quite well the observed magnitude distribution, while fitting Equation (6) to the observed data provides the values $q_M$ and $A$ using a non-linear squares algorithm. The results of this analysis are presented in Figure 3. The F–A model fits well the data for the values of $q_M = 1.52 \pm 0.01$ and $A = 27.92 \pm 8.04$.

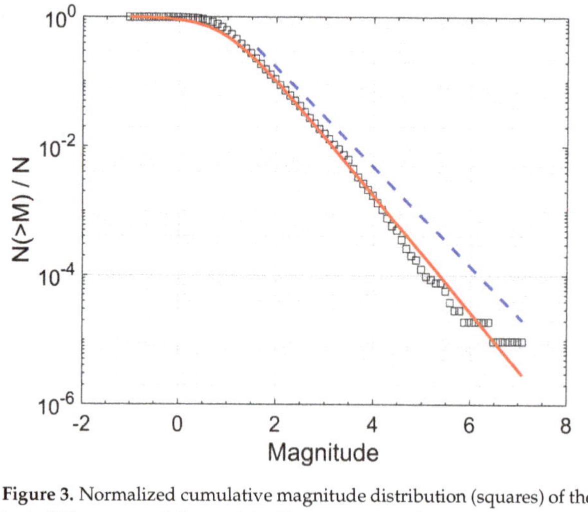


**Figure 3.** Normalized cumulative magnitude distribution (squares) of the 1981–2022 earthquake activity in Ridgecrest and the model of Equation (6) (red solid line) for the values of $q_M = 1.52 \pm 0.01$ and $A = 27.92 \pm 8.04$. The blue dashed line illustrates the Gutenberg–Richter relation for $b = 0.771 \pm 0.005$ and $\alpha = 5.561 \pm 0.077$.

One of the most renowned empirical scaling relations in seismology is the Gutenberg–Richter (G-R) [33] relation, which expresses a power-law dependence between the number of earthquakes $N$ and the released seismic energies $E$. In terms of earthquake magnitudes $M$, it is expressed as $N(>M) = 10^{a-bM}$, where $N(>M)$ is the number of earthquakes greater than a threshold magnitude $M$, and $a$, $b$ are positive scaling parameters. The $b$ value is usually calculated using the maximum likelihood method [29], as subsequently amended by [34], as: $b = (1/\overline{M} - M_0)\log e$, where $\overline{M}$ is the observed mean magnitude and $M_0$ is the minimum magnitude. Therefore, for comparison, a maximum likelihood fit that corresponds to the G–R relation is also plotted in Figure 3, for the values of $M_0 = 1.6$, $b = 0.771 \pm 0.005$, and $\alpha = 5.561 \pm 0.077$.

### 4.2. Variations of $q_M$ Values with Time

The F–A model is initially applied to the seismic catalogue for the entire period from 1981 to 2022. A notable aspect in this analysis is to examine the temporal variations of the $q_M$ parameter and how these variations are related to the dynamic evolution of earthquake activity, which can offer valuable insights into the physical processes of earthquake generation. Initially, we divided the data into various time intervals using a sliding window approach. Subsequently, we calculated the $q_M$ values within time intervals containing 1000 events, with a 500-event overlap, resulting in a 50% overlap between consecutive windows. In the statistical analysis we used, in one case, all the seismic events (blue solid line in Figure 4), while in the other case, we focused on events with $M \geq Mc$ (red solid line in Figure 4). In the latter case, within each temporal window consisting of 1000 events, we applied the maximum curvature method [35] along with an additional +0.2 correction, to determine the magnitude of completeness ($Mc$). The findings of this analysis are displayed in Figure 4, showing the temporal variations of the $q_M$ values along with their standard deviations. The parameter $q_M$ varies between 1.3 and 1.6 during the studied period. Furthermore, as observed in Figure 4, the $q_M$ value estimation is relatively stable in each temporal window, irrespective of the selection of $Mc$, an important aspect for reliable analysis of the FMD.

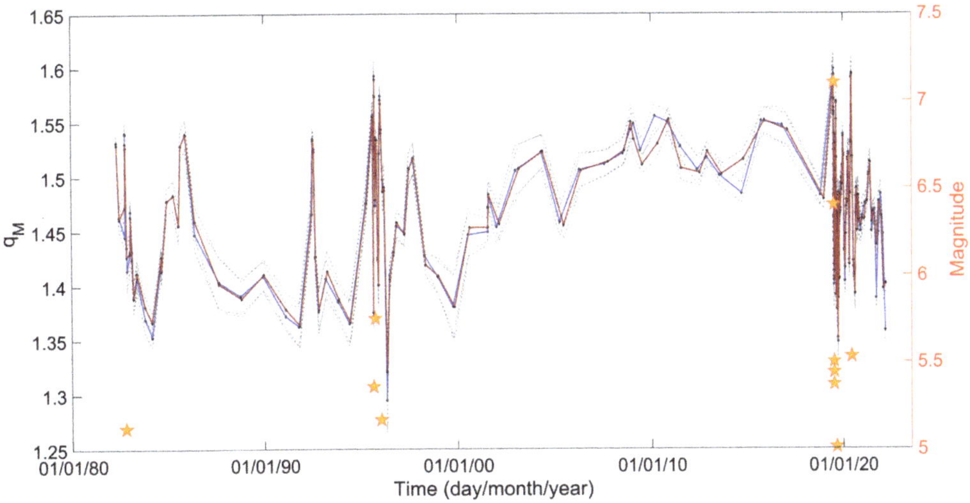


**Figure 4.** Temporal evolution of the $q_M$ value for all the seismic events (blue solid line) and for $M \geq Mc$ (red solid line), along with their corresponding standard deviations (gray dashed lines), calculated in successive time intervals with a 50% overlap covering the period from 1981 to 2022. Stars indicate earthquakes of a magnitude greater than 5.

The $q_M$ values exhibit an increase during periods characterized by higher-magnitude earthquakes and decrease during more seismically quiet intervals. In Figure 4, we can observe that the largest magnitude events of 1982, 1995, and 2019 induced variations in the values of $q_M$. The occurrence of such significant events, which are illustrated with stars in Figure 4, causes the increase in the $q_M$ value. We remind that when $q_M$ approaches unity, the system reaches equilibrium and transitions into BG statistical physics. Conversely, as $q_M$ increases, the system deviates from equilibrium and this instability may cause larger magnitude events to occur.

In Figure 5, we zoom in into the variations of $q_M$ values from 2019 to 2020, during the period of the Ridgecrest earthquake sequence. The parameter $q_M$ exhibits a notable increase, reaching its peak ($q_M$ = 1.6) on 7 July 2019 and on 4 June 2020, when the $M_w$ 7.1 and the $M_w$ 5.53 events occurred. After the strong events, the $q_M$ parameter starts decreasing rapidly.

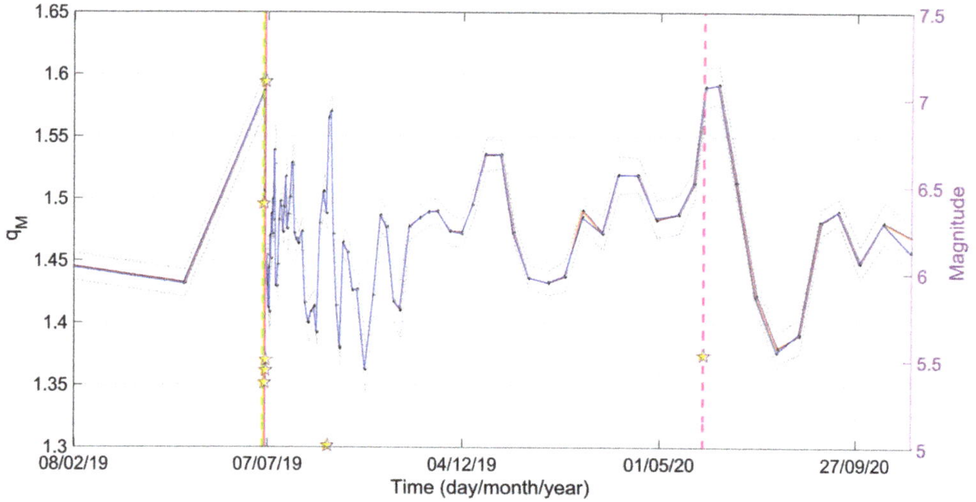


**Figure 5.** Temporal evolution of the $q_M$ value for all the seismic events (blue solid line) and for $M \geq Mc$ (red solid line), along with their corresponding standard deviations (gray dashed lines), calculated in successive time intervals with a 50% overlap. The stars indicate earthquakes of a magnitude greater than 5 during the period from 2019 to 2020. The $M_w$ 6.4 on 4 July 2019, the $M_w$ 7.1 on 6 July 2019, and the $M_w$ 5.53 on 4 June 2020, seismic events are represented with green dashed, orange solid and pink dashed lines, respectively.

To better resolve the correlations between the released seismic energy and $q_M$ value variations, we perform a cross-correlation analysis [36] designed to quantify the statistical confidence between two datasets, the $q_M$ value and the seismic moment release ($Mo$). This analysis is used to delineate the strength of correlations and the time lag between $q_M$ and $Mo$ for the period 1981–2022. Cross-correlation analysis is based on two discrete datasets in time, which are then normalized by subtracting the mean value so that the estimated correlation coefficients range between $-1$ and 1 [36,37]. The normalized cross-correlation close to zero suggests no correlation, while maximum positive or negative peaks may indicate correlated or anti-correlated signals, respectively. The next step is the use of surrogate reshuffling tests which allow dynamic testing against statistical confidence intervals of anticipated spurious correlations [36]. These tests determine the confidence curves of the estimated normalized cross-correlation. In particular, surrogate datasets of the original series are determined using Monte Carlo methods [36,37]. The reshuffling procedure removes any correlation from the original time series while maintaining their spectral amplitudes in order to enforce the same cyclic autocorrelation [38]. In this test, 10,000 surrogates were computed, and the main peak was observed at cross-correlations greater than 99.9% confidence curves. The cross-correlation between $q_M$ and *Mo* for the time period 1981–2022 is shown in Figure 6 along with the 95%, 99%, and 99.9% confidence curves. We observe nine positive peaks exceeding the 95% confidence curve. The highest peak, with an amplitude of 0.71, occurs with a statistical significance of 99.9% of being non-coincidental. The results of the analysis indicate that the surrogate tests place confidence greater than 99.9% which shows that the $q_M$ value in each temporal window and the corresponding cumulative seismic energy *Mo* are positively correlated.

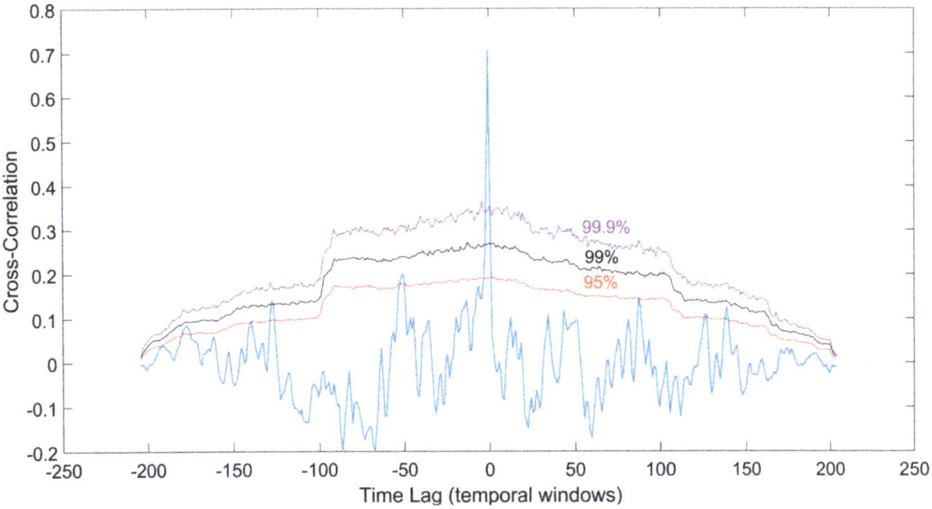


**Figure 6.** The cross-correlation between the $q_M$ values and cumulative seismic energy *Mo* for the period 1981–2022 and the corresponding 95%, 99%, and 99.9% confidence curves (are labeled with text).

### 4.3. Spatiotemporal Distributions of $q_M$ Values

To investigate the spatiotemporal distributions of $q_M$ values within the activated fault zone based on the F–A model, we proceeded to assess the spatial distribution of the parameter $q_M$ for different time windows. According to the model, the observation of $q_M$ values in space reflects the scale of interactions between fault planes and the fragments that occupy the space between them. Furthermore, an increase in $q_M$ signifies that the physical state moves away from equilibrium in a statistical physics context.

Initially, we divided the dataset into temporal segments. The first one covers the period from 1981 to 2019, before the foreshock of magnitude $M_w$ 6.4 on 4 July 2019 (Figure 7a), while the second, covers the period starting from the foreshock of $M_w$ 6.4 to the mainshock of $M_w$ 7.1 on 6 July 2019 (Figure 7b). To estimate the spatial $q_M$ values, we employed the nearest neighbor search method, where for each seismic event, the 200 nearest neighboring events within the dataset were identified. Then, we applied the F–A model to calculate the parameter $q_M$, using Equation (6). Since $q_M$ remains relatively stable regardless of the choice of completeness magnitude, as it was previously discussed, we opted not to incorporate *Mc* in our calculations. By executing this analysis for each seismic event, we obtained a detailed and comprehensive overview of the $q_M$ parameter in the studied area.

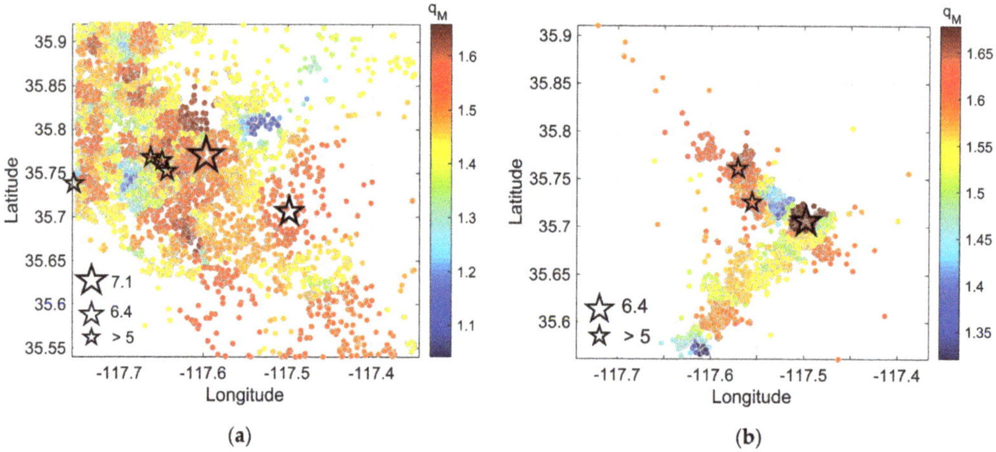


**Figure 7.** Spatial distribution of $q_M$ values along the 2019 Ridgecrest earthquake fault zone in various time windows, within the studied region defined by coordinates 117.35° E~117.75° E, 35.54° N~35.92° N. (**a**) During the period 1981–2019 (before the $M_w$ 6.4 foreshock), (**b**) from the $M_w$ 6.4 foreshock to the $M_w$ 7.1 mainshock. Larger stars represent the seismic events of $M_w$ 6.4 and $M_w$ 7.1, while other stars indicate earthquakes with magnitudes greater than 5, respectively.

An inspection of Figure 7 indicates that the $q_M$ value ranges from 1.1 to 1.7, supporting subadditivity. In Figure 7a, we observe that the seismicity background is characterized by relatively high $q_M$ values in the areas where the $M_w$ 6.4 foreshock and the $M_w$ 7.1 mainshock took place. In Figure 7b, which covers the period from the $M_w$ 6.4 foreshock to the $M_w$ 7.1 mainshock on 6 July 2019, we can observe the high $q_M$ values in the area where the $M_w$ 6.4 occurred. Furthermore, we can observe that the $M_w$ 7.1 mainshock, located to the NW of the $M_w$ 6.4 event, occurred in a high $q_M$ value region (Figure 7b).

Moreover, we analyzed the aftershock sequence of the $M_w$ 7.1 Ridgecrest mainshock from July to November 2019 based on the F–A model. We divided the dataset into four monthly segments, covering the periods from 6 July 2019 (including the $M_w$ 7.1 mainshock) to 6 August 2019, from 7 August 2019 to 7 September 2019, from 8 September 2019 to 8 October 2019, and from 9 October 2019 to 9 November 2019.

Two-dimensional spatial analysis (Figure 8) shows that the parameter $q_M$ varies from 1.2 to 1.7 and is high at the locations where the strongest earthquakes occurred. In particular, the black star in Figure 8a indicates the seismic event of $M_w$ 7.1, with the $q_M$ value reaching as high as 1.7, the highest value in the area. The presence of numerous substantial events with a magnitude of $M > 4$ results in an increase in $q_M$ in the studied area, in which we have depicted the two largest ones to the north of the $M_w$ 7.1 event (Figure 8a). However, in the same area where the $M_w$ 7.1 event occurred (Figure 8b), a decrease from 1.7 to 1.45 was observed after 1 month, suggesting stress relaxation in the area of the mainshock. Moreover, in Figure 8b, we observe that the two largest earthquakes during August–September 2019 coincide with the high $q_M$ value regions to the north of the active fault zone. Similar observations are made in Figure 8c,d, where the highest magnitude aftershocks (black stars) coincide with high $q_M$ value areas, particularly in the south and central parts of the activated zone.

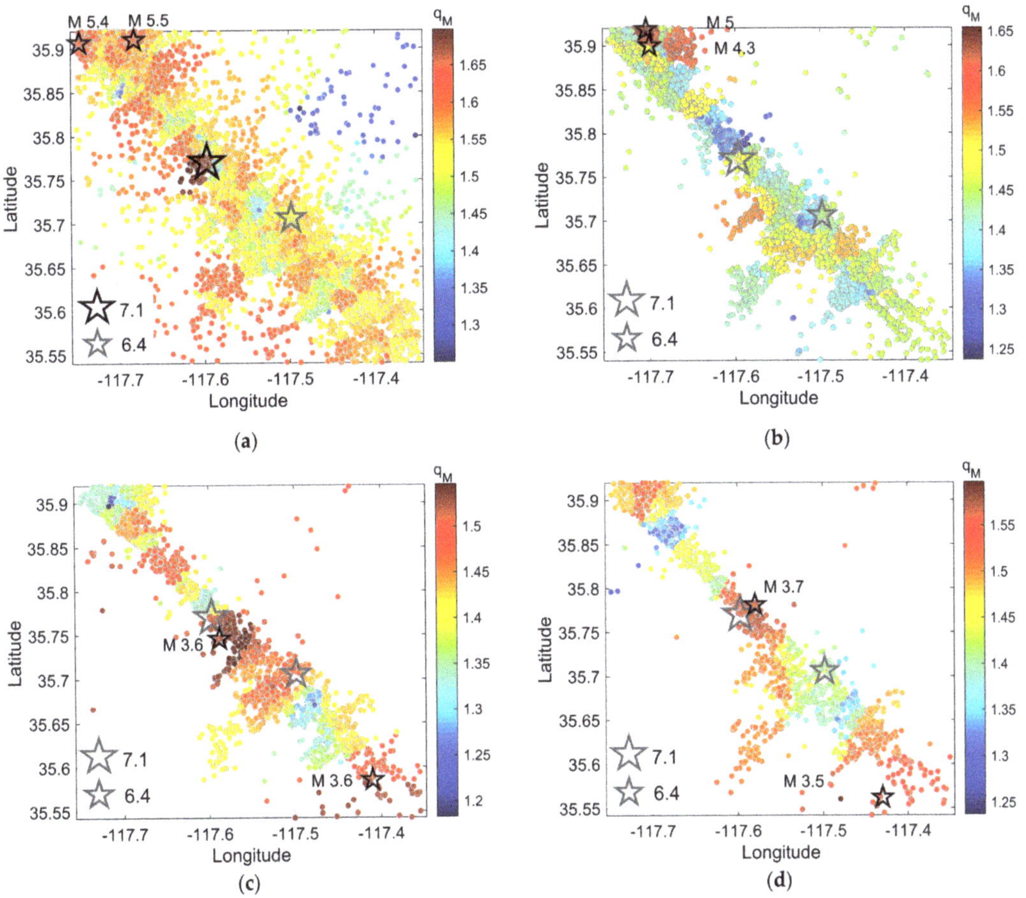


**Figure 8.** The $q_M$ values along the 2019 activated fault zone. (**a**) From 6 July to 6 August 2019, in the area 117.35° E~117.75° E, 35.54° N~35.92° N. (**b**) The same as (**a**), but for the period from 7 August to 7 September 2019, (**c**) from 8 September to 8 October 2019, (**d**) from 9 October to 9 November 2019. Black stars indicate the strongest events within each month. The gray stars illustrate the locations of the $M_w$ 6.4 and $M_w$ 7.1 events.

## 5. Discussion

In the present work, we used the F–A model, developed within the framework of NESP, to study the temporal and spatial variations of the $q_M$ parameter over the period from 1981 to 2022 in the area of Ridgecrest. The remarkable consistency observed between the F–A model (Equation (6)) and the earthquake magnitude distributions highlights the effectiveness of the F–A model. The parameter $q_M$ informs about the scale of interactions between the fault planes and the fragments that occupy the space between them. When $q_M$ is low ($\approx$1), it indicates the presence of short-ranged spatial correlations and physical states that are near equilibrium [26,39–41]. As $q_M$ increases, it signifies a departure from equilibrium in the physical state, suggesting a non-equilibrium state where more earthquakes occur [26]. Regarding the Ridgecrest seismicity during the period 1981–2022, the values obtained for $q_M$ with time vary between 1.3 and 1.6. The observed increase in the entropic index $q_M$ implies that the system is moving further away from an equilibrium state and is in a preparatory process for seismic energy release. In our study, an increase in the $q_M$ parameter can be observed when the major earthquakes of 1982, 1995, 2019, and

2020 occur, with a significant positive correlation between $q_M$ values and seismic moment release throughout the studied period.

Additional studies into the analysis of $q_M$ variations in various cases and within different seismotectonic settings have previously been conducted. In particular, according to [22], the $q_M$ parameter showed a significant increase on 9 April 1994, signaling the onset of a transitional phase leading up to the 1995 Kobe earthquake. Furthermore, the $q_M$ parameter exhibited variations, with an increase observed in the days prior to the strong earthquake of $M_L$ 5.8 in the L'Aquila area (central Italy) [26]. It should be noticed that in [41], a sharp increase in $q_M$ was observed a few days before the occurrence of the significant $M_w$ 6.4 event in the southwest segment of the Hellenic Arc. Similar research suggests a possible association between $q_M$ and seismicity patterns [40] in the South Pacific coast of Mexico. Moreover, seismic activity in the Hellenic region from 1976 to 2009 was investigated using the method of NESP along with the G–R relation by [42], which concludes that the $q_M$ parameter can be viewed as a distinctive parameter that characterizes the seismic history of a specific region. Previous studies indicate that the NESP approach seems to be a suitable method for analyzing the spatiotemporal patterns of seismicity, as also demonstrated by [28] for the spatial variability of $q_M$ within the Yellowstone Park volcanic region.

Non-extensivity is incorporated in the F–A model as a fundamental statistical component for deriving a cumulative magnitude distribution, of which the Gutenberg–Richter (G–R) relation can be regarded as a specific case [13]. In addition, it is a widely acknowledged and nearly universally observed phenomenon that the stress alterations induced by significant earthquakes have a substantial impact on seismic activity in surrounding areas [43,44]. As suggested by several case studies, mainshock-induced stress changes are therefore anticipated to consistently influence $b$ values [35,45–49]. The Ridgecrest earthquake sequence of $M_w$ 7.1 in California in July 2019 offered an opportunity for [50] to assess both the temporal and spatial variations of the $b$ value and its forecasting skills. Hence, a comparison can be made between the fluctuations in the values of $q_M$ and $b$. According to [50], the $b$ values were substantially lower after the $M_w$ 6.4 event compared to the background $b$ value, whereas after the mainshock of $M_w$ 7.1, the $b$ value increased within the first week. A decreasing $b$ value inside the seismogenic volume has been observed to correspond with increased effective stress levels before significant shocks [51]. Our results for the temporal analysis show that the parameter $q_M$ is higher when the strong events in 2019 occur, and then it decreases over time, in accordance with the results of [50].

Furthermore, the spatiotemporal patterns of variations in $b$ values provide additional insights into the prospective location of forthcoming significant events. According to the findings of [50], the $M_w$ 7.1 event took place near the area of the steepest $b$ value decrease. In our study, we can observe that the $M_w$ 7.1 event occurs in the area where the parameter $q_M$ is increased. The spatiotemporal $q_M$ value distributions, based on the F–A model for each month after the mainshock of $M_w$ 7.1 can further be compared with the study by [52], in which the aftershock sequence is investigated in terms of the spatiotemporal $b$ value distributions within the three-dimensional fault zone. The findings indicate that $b$ values were initially homogeneous throughout the spatial area, with a low level of $b$ value immediately following the mainshock. However, within 3 months, a rapid increase occurred, reaching a level that is considered typical for California during the interseismic period. As for the parameter $q_M$, when the $M_w$ 7.1 occurred, a high value of 1.7 was found in the epicentral area, which is indicated by a black star (Figure 8a). The next month, as we can see in Figure 8b, there was a decrease in the same region (gray star) from 1.7 to 1.45, while an increase in the $q_M$ parameter was observed in the areas where strong aftershocks occurred. Furthermore, the same pattern is observed in the case of the $M_w$ 6.4 foreshock, where the parameter $q_M$ decreased from 1.7 (Figure 7b) to 1.4 (Figure 8b) within 2 months.

## 6. Conclusions

In the present work, the Ridgecrest earthquake sequence is studied using non-extensive statistical physics (NESP) and the fragment–asperity (F–A) model. Within the context of the F–A model, we calculated the non-extensive parameter $q_M$ and its spatiotemporal variations during 1981–2022, an analysis that informs about the physical state of the studied area. To study the temporal variations of $q_M$ values, we used the entire period and a sliding time window method. Notably, the results show a significant increase in the $q_M$ parameter, which coincides with the occurrence of the strongest earthquakes. Furthermore, it seems that $q_M$ fluctuations over time are a valuable indicator of a seismic area's physical condition, suggesting different dynamic regimes that can decipher the physical mechanisms leading to a significant seismic event. In addition, we analyzed the seismic events for the spatiotemporal $q_M$ value distributions along the activated fault zone during 1981–2019 and for each month separately after the $M_w$ 7.1 Ridgecrest earthquake. The results show that $q_M$ values exhibit significant increases in areas where the higher-magnitude events occur, and after the mainshock, $q_M$ values decrease over time, highlighting the stress relaxation process in the activated area.

**Author Contributions:** Conceptualization, F.V. and G.M.; methodology, E.S., G.M., K.P. and F.V.; software, E.S., G.M. and K.P.; validation, E.S., G.M., K.P. and F.V.; formal analysis, E.S., G.M. and K.P.; investigation, E.S. and G.M.; writing—original draft preparation, E.S.; writing—review and editing, E.S., G.M., K.P. and F.V.; supervision, F.V. and G.M. All authors have read and agreed to the published version of the manuscript.

**Funding:** This research received no external funding.

**Institutional Review Board Statement:** Not applicable.

**Data Availability Statement:** Data are openly available at the Southern California Earthquake Data (SCEDC) (https://scedc.caltech.edu/eq-catalogs/altcatalogs.html), accessed on 20 June 2023. The relocated earthquake catalogue is based on the methods described in [31] (https://scedc.caltech.edu/data/alt-2011-dd-hauksson-yang-shearer.html)—last accessed on 20 June 2023.

**Conflicts of Interest:** The authors declare no conflict of interest.

## References

1. Ross, Z.E.; Idini, B.; Jia, Z.; Stephenson, O.L.; Zhong, M.; Wang, X.; Zhan, Z.; Simons, M.; Fielding, E.J.; Yun, S.-H.; et al. Hierarchical Interlocked Orthogonal Faulting in the 2019 Ridgecrest Earthquake Sequence. *Science* **2019**, *366*, 346–351. [CrossRef]
2. Feng, W.; Samsonov, S.; Qiu, Q.; Wang, Y.; Zhang, P.; Li, T.; Zheng, W. Orthogonal Fault Rupture and Rapid Postseismic Deformation Following 2019 Ridgecrest, California, Earthquake Sequence Revealed from Geodetic Observations. *Geophys. Res. Lett.* **2020**, *47*, e2019GL086888. [CrossRef]
3. Marchetti, D.; De Santis, A.; Campuzano, S.A.; Soldani, M.; Piscini, A.; Sabbagh, D.; Cianchini, G.; Perrone, L.; Orlando, M. Swarm Satellite Magnetic Field Data Analysis Prior to 2019 Mw = 7.1 Ridgecrest (California, USA) Earthquake. *Geosciences* **2020**, *10*, 502. [CrossRef]
4. DuRoss, C.B.; Gold, R.D.; Dawson, T.E.; Scharer, K.M.; Kendrick, K.J.; Akciz, S.O.; Angster, S.J.; Bachhuber, J.; Bacon, S.; Bennett, S.E.K.; et al. Surface Displacement Distributions for the July 2019 Ridgecrest, California, Earthquake Ruptures. *Bull. Seismol. Soc. Am.* **2020**, *110*, 1400–1418. [CrossRef]
5. Wang, K.; Dreger, D.S.; Tinti, E.; Bürgmann, R.; Taira, T. Rupture Process of the 2019 Ridgecrest, California Mw 6.4 Foreshock and Mw 7.1 Earthquake Constrained by Seismic and Geodetic Data. *Bull. Seismol. Soc. Am.* **2020**, *110*, 1603–1626. [CrossRef]
6. Fialko, Y. Estimation of Absolute Stress in the Hypocentral Region of the 2019 Ridgecrest, California, Earthquakes. *JGR Solid Earth* **2021**, *126*, e2021JB022000. [CrossRef]
7. Jin, Z.; Fialko, Y. Finite Slip Models of the 2019 Ridgecrest Earthquake Sequence Constrained by Space Geodetic Data and Aftershock Locations. *Bull. Seismol. Soc. Am.* **2020**, *110*, 1660–1679. [CrossRef]
8. Wang, K.; Bürgmann, R. Co- and Early Postseismic Deformation Due to the 2019 Ridgecrest Earthquake Sequence Constrained by Sentinel-1 and COSMO-SkyMed SAR Data. *Seismol. Res. Lett.* **2020**, *91*, 1998–2009. [CrossRef]
9. Hauksson, E.; Hutton, K.; Kanamori, H.; Jones, L.; Mori, J.; Hough, S.; Roquemore, G. Preliminary Report on the 1995 Ridgecrest Earthquake Sequence in Eastern California. *Seismol. Res. Lett.* **1995**, *66*, 54–60. [CrossRef]
10. Sotolongo-Costa, O.; Posadas, A. Fragment-Asperity Interaction Model for Earthquakes. *Phys. Rev. Lett.* **2004**, *92*, 048501. [CrossRef]
11. Tsallis, C. Nonadditive Entropy and Nonextensive Statistical Mechanics—An Overview after 20 Years. *Braz. J. Phys.* **2009**, *39*, 337–356. [CrossRef]
12. Sarlis, N.V.; Skordas, E.S.; Varotsos, P.A. Nonextensivity and Natural Time: The Case of Seismicity. *Phys. Rev. E* **2010**, *82*, 021110. [CrossRef]

13. Telesca, L. Maximum Likelihood Estimation of the Nonextensive Parameters of the Earthquake Cumulative Magnitude Distribution. *Bull. Seismol. Soc. Am.* **2012**, *102*, 886–891. [CrossRef]
14. Tsallis, C. Possible Generalization of Boltzmann-Gibbs Statistics. *J. Stat. Phys.* **1988**, *52*, 479–487. [CrossRef]
15. Tsallis, C.I. Nonextensive Statistical Mechanics and Thermodynamics: Historical Background and Present Status. In *Nonextensive Statistical Mechanics and Its Applications*; Abe, S., Okamoto, Y., Eds.; Lecture Notes in Physics; Springer: Berlin/Heidelberg, Germany, 2001; Volume 560, pp. 3–98. [CrossRef]
16. Tsallis, C. *Introduction to Nonextensive Statistical Mechanics: Approaching a Complex World*; Springer: Berlin/Heidelberg, Germany, 2009.
17. Tirnakli, U.; Borges, E.P. The Standard Map: From Boltzmann-Gibbs Statistics to Tsallis Statistics. *Sci. Rep.* **2016**, *6*, 23644. [CrossRef] [PubMed]
18. Silva, R.; França, G.S.; Vilar, C.S.; Alcaniz, J.S. Nonextensive Models for Earthquakes. *Phys. Rev. E* **2006**, *73*, 026102. [CrossRef] [PubMed]
19. Lay, T.; Wallace, T. *Modern Global Seismology*; Academic Press, Inc.: Cambridge, MA, USA, 1995; 521p.
20. Telesca, L. Tsallis-Based Nonextensive Analysis of the Southern California Seismicity. *Entropy* **2011**, *13*, 1267–1280. [CrossRef]
21. Michas, G.; Vallianatos, F.; Sammonds, P. Non-Extensivity and Long-Range Correlations in the Earthquake Activity at the West Corinth Rift (Greece). *Nonlin. Process. Geophys.* **2013**, *20*, 713–724. [CrossRef]
22. Papadakis, G.; Vallianatos, F.; Sammonds, P. A Nonextensive Statistical Physics Analysis of the 1995 Kobe, Japan Earthquake. *Pure Appl. Geophys.* **2015**, *172*, 1923–1931. [CrossRef]
23. Vallianatos, F.; Papadakis, G.; Michas, G. Generalized Statistical Mechanics Approaches to Earthquakes and Tectonics. *Proc. R. Soc. A* **2016**, *472*, 20160497. [CrossRef]
24. Levenberg, K. A Method for the Solution of Certain Non-Linear Problems in Least Squares. *Quart. Appl. Math.* **1944**, *2*, 164–168. [CrossRef]
25. Marquardt, D.W. An Algorithm for Least-Squares Estimation of Nonlinear Parameters. *J. Soc. Ind. Appl. Math.* **1963**, *11*, 431–441. [CrossRef]
26. Telesca, L. A Non-Extensive Approach in Investigating the Seismicity of L'Aquila Area (Central Italy), Struck by the 6 April 2009 Earthquake ( ML = 5.8). *Terra Nova* **2010**, *22*, 87–93. [CrossRef]
27. Antonopoulos, C.G.; Michas, G.; Vallianatos, F.; Bountis, T. Evidence of q -Exponential Statistics in Greek Seismicity. *Phys. A Stat. Mech. Its Appl.* **2014**, *409*, 71–77. [CrossRef]
28. Chochlaki, K.; Michas, G.; Vallianatos, F. Complexity of the Yellowstone Park Volcanic Field Seismicity in Terms of Tsallis Entropy. *Entropy* **2018**, *20*, 721. [CrossRef]
29. Aki, K. Maximum Likelihood Estimate of b in the Formula log10N=a-bm and Its Confidence Limits. *Bull. Earthq. Res.* **1965**, *43*, 237–239.
30. Michas, G. Generalized Statistical Mechanics Description of Fault and Earthquake populations in Corinth Rift (Greece). Ph.D. Thesis, University College London, London, UK, 2016.
31. Hauksson, E.; Yang, W.; Shearer, P.M. Waveform Relocated Earthquake Catalog for Southern California (1981 to June 2011). *Bull. Seismol. Soc. Am.* **2012**, *102*, 2239–2244. [CrossRef]
32. Hanks, T.C.; Kanamori, H. A Moment Magnitude Scale. *J. Geophys. Res.* **1979**, *84*, 2348–2350. [CrossRef]
33. Gutenberg, B.; Richter, C.F. Frequency of Earthquakes in California. *Bull. Seismol. Soc. Am.* **1994**, *34*, 185–188. [CrossRef]
34. Utsu, T. Estimation of parameter values in the formula for the magnitude-frequency relation of earthquake occurrence. *Zisin* **1978**, *31*, 367–382. [CrossRef]
35. Wiemer, S.; Wyss, M. Minimum Magnitude of Completeness in Earthquake Catalogs: Examples from Alaska, the Western United States, and Japan. *Bull. Seismol. Soc. Am.* **2000**, *90*, 859–869. [CrossRef]
36. Schultz, R.; Telesca, L. The Cross-Correlation and Reshuffling Tests in Discerning Induced Seismicity. *Pure Appl. Geophys.* **2018**, *175*, 3395–3401. [CrossRef]
37. Michas, G.; Pavlou, K.; Vallianatos, F.; Drakatos, G. Correlation Between Seismicity and Water Level Fluctuations in the Polyphyto Dam, North Greece. *Pure Appl. Geophys.* **2020**, *177*, 3851–3870. [CrossRef]
38. Little, M.A.; McSharry, P.E.; Moroz, I.M.; Roberts, S.J. Testing the Assumptions of Linear Prediction Analysis in Normal Vowels. *J. Acoust. Soc. Am.* **2006**, *119*, 549–558. [CrossRef] [PubMed]
39. Telesca, L. Analysis of Italian Seismicity by Using a Nonextensive Approach. *Tectonophysics* **2010**, *494*, 155–162. [CrossRef]
40. Valverde-Esparza, S.M.; Ramírez-Rojas, A.; Flores-Márquez, E.L.; Telesca, L. Non-Extensivity Analysis of Seismicity within Four Subduction Regions in Mexico. *Acta Geophys.* **2012**, *60*, 833–845. [CrossRef]
41. Vallianatos, F.; Michas, G.; Papadakis, G. Non-Extensive and Natural Time Analysis of Seismicity before the Mw6.4, October 12, 2013 Earthquake in the South West Segment of the Hellenic Arc. *Phys. A Stat. Mech. Its Appl.* **2014**, *414*, 163–173. [CrossRef]
42. Michas, G.; Papadakis, G.; Vallianatos, F. A Non-Extensive Approach in Investigating Greek Seismicity. *Geosociety* **2013**, *47*, 1177. [CrossRef]
43. Stein, R.S. The Role of Stress Transfer in Earthquake Occurrence. *Nature* **1999**, *402*, 605–609. [CrossRef]
44. Ebel, J.E.; Bonjer, K.-P.; Oncescu, M.C. Paleoseismicity: Seismicity Evidence for Past Large Earthquakes. *Seismol. Res. Lett.* **2000**, *71*, 283–294. [CrossRef]
45. Enescu, B.; Ito, K. Values of b and p: Their variations and relation to physical processes for earthquakes in Japan. *Ann. Disas. Prev. Res. Inst. Kyoto Univ.* **2003**, *46*, 709–719.

46. Gulia, L.; Wiemer, S. Real-Time Discrimination of Earthquake Foreshocks and Aftershocks. *Nature* **2019**, *574*, 193–199. [CrossRef] [PubMed]
47. Lombardi, A.M. Anomalies and Transient Variations of b -Value in Italy during the Major Earthquake Sequences: What Truth Is There to This? *Geophys. J. Int.* **2022**, *232*, 1545–1555. [CrossRef]
48. Wiemer, S.; Katsumata, K. Spatial Variability of Seismicity Parameters in Aftershock Zones. *J. Geophys. Res.* **1999**, *104*, 13135–13151. [CrossRef]
49. Ramírez-Rojas, A.; Flores-Márquez, E.L. Nonlinear Statistical Features of the Seismicity in the Subduction Zone of Tehuantepec Isthmus, Southern México. *Entropy* **2022**, *24*, 480. [CrossRef]
50. Gulia, L.; Wiemer, S.; Vannucci, G. Pseudoprospective Evaluation of the Foreshock Traffic-Light System in Ridgecrest and Implications for Aftershock Hazard Assessment. *Seismol. Res. Lett.* **2020**, *91*, 2828–2842. [CrossRef]
51. Kanamori, H. The Nature of Seismicity Patterns Before Large Earthquakes. In *Maurice Ewing Series;* Simpson, D.W., Richards, P.G., Eds.; American Geophysical Union: Washington, DC, USA, 2013; pp. 1–19. [CrossRef]
52. Huang, K.; Tang, L.; Feng, W. Spatiotemporal Distributions of b Values Following the 2019 Mw 7.1 Ridgecrest, California, Earthquake Sequence. *Pure Appl. Geophys.* **2023**, *180*, 2529–2542. [CrossRef]


**Disclaimer/Publisher's Note:** The statements, opinions and data contained in all publications are solely those of the individual author(s) and contributor(s) and not of MDPI and/or the editor(s). MDPI and/or the editor(s) disclaim responsibility for any injury to people or property resulting from any ideas, methods, instructions or products referred to in the content.

*entropy* MDPI

*Article*

# Visibility Graph Analysis of the Seismic Activity of Three Areas of the Cocos Plate Mexican Subduction Where the Last Three Large Earthquakes ($M > 7$) Occurred in 2017 and 2022

**Alejandro Ramírez-Rojas [1,*], Elsa Leticia Flores-Márquez [2,*] and Carlos Alejandro Vargas [1]**

1 Departamento de Ciencias Básicas, Universidad Autónoma Metropolitana, Ciudad de Mexico 02128, Mexico; cvargas@azc.uam.mx
2 Instituto de Geofísica, Universidad Nacional Autónoma de Mexico, Ciudad Universitaria, Ciudad de Mexico 04510, Mexico
* Correspondence: arr@azc.uam.mx (A.R.-R.); leticia@igeofisica.unam.mx (E.L.F.-M.)

**Abstract:** The understanding of the dynamical behavior of seismic phenomena is currently an open problem, mainly because seismic series can be considered to be produced by phenomena exhibiting dynamic phase transitions; that is, with some complexity. For this purpose, the Middle America Trench in central Mexico is considered a natural laboratory for examining subduction because of its heterogenous natural structure. In this study, the Visibility Graph method was applied to study the seismic activity of three regions within the Cocos plate: the Tehuantepec Isthmus, the Flat slab and Michoacan, each one with a different level of seismicity. The method maps time series into graphs, and it is possible to connect the topological properties of the graph with the dynamical features underlying the time series. The seismicity analyzed was monitored in the three areas studied between 2010 and 2022. At the Flat Slab and Tehuantepec Isthmus, two intense earthquakes occurred on 7 and 19 September 2017, respectively, and, on 19 September 2022, another one occurred at Michoacan. The aim of this study was to determine the dynamical features and the possible differences between the three areas by applying the following method. First, the time evolution of the *a*- and *b*-values in the Gutenberg–Richter law was analyzed, followed by the relationship between the seismic properties and topological features using the VG method, the *k*–*M* slope and the characterization of the temporal correlations from the $\gamma$-exponent of the power law distribution, $P(k) \sim k^{-\gamma}$, and its relationship with the Hurst parameter, which allowed us to identify the correlation and persistence of each zone.

**Keywords:** visibility graph; seismic activity; subduction; tectonics; *k*–*M* slope

**Citation:** Ramírez-Rojas, A.; Flores-Márquez, E.L.; Vargas, C.A. Visibility Graph Analysis of the Seismic Activity of Three Areas of the Cocos Plate Mexican Subduction Where the Last Three Large Earthquakes ($M > 7$) Occurred in 2017 and 2022. *Entropy* **2023**, *25*, 799. https://doi.org/10.3390/e25050799

Academic Editor: Georgios Michas

Received: 9 April 2023
Revised: 8 May 2023
Accepted: 10 May 2023
Published: 15 May 2023

**Copyright:** © 2023 by the authors. Licensee MDPI, Basel, Switzerland. This article is an open access article distributed under the terms and conditions of the Creative Commons Attribution (CC BY) license (https://creativecommons.org/licenses/by/4.0/).

## 1. Introduction

The dynamics of tectonic plates are considered to be one of the most important topics in geosciences, not only to improve scientific knowledge, but also to ensure the safety of small towns and big cities located in seismic zones considered to be high-risk. One of the main mechanisms of earthquake generation is convection in the interior of the Earth [1,2]. The convective processes involve nonlinear interactions which are far from equilibrium processes. Seismic processes originate from the interaction between plates in relative movement due to inner convection and strong correlations in time and space, as well as released energy or magnitudes ([3] and references therein). Although the vast majority of earthquakes occur in the interplate regions, the physical mechanism of the anthropogenic triggering of large earthquakes on active faults is connected with mining, artificial reservoir impoundment, geothermal operations, oil and gas field production and hydraulic fracturing (i.e., fracking) [4]. While almost all earthquakes occur at the plate boundaries, some anomalous events, called intraplate earthquakes, occur far from the boundaries [5]. The seismic mechanism can be considered to be a critical phenomenon exhibiting dynamic phase transitions [6], where a mainshock represents the new phase. In this context, seismic properties, such as the magnitude of an earthquake and the energy released during fracture processes, behave dynamically, showing short- and long-range temporal correlations, and these characteristics are governed by fractional Brownian processes and fractal/multifractal properties [3]. In fact, the complexity of the systems in the real world arises mainly from non-linear processes; thus, when the variability of a measurable parameter is registered in time, the system can be represented by a time series that can maintain some features of the underlying complex processes. In nonlinear signals analysis methods, there has been a trend of merging nonlinear time series analysis based on complex network theory.

In recent years, complex network theory has received important attention as a powerful tool to analyze complex time series [7–19]. In 2008, Lacasa et al. [16] introduced the Visibility Graph (VG) method, whose aim is to map a time series into a complex network or a graph. Under these transformations, it is expected that the topological properties of the network inherit the dynamical properties of the original time series and vice versa. The transformation of a time series into a graph entails building a geometric structure with two classes of sets, nodes and edges, where the nodes are the time series values and the edges determine the connectivity between nodes. The topological properties on a graph are defined by imposing conditions defining the connection between pairs of nodes by edges. In the case of the VG method, the connections between pairs of nodes are given by the reciprocal "visibility" between the values of the series [16], such that the connectivity between nodes can be defined. Visibility graph analysis has been applied as a powerful statistical method in extracting characteristic features of the time series, such as the periodicity, fractality, chaoticity and non-linearity [20]. For instance, some applications have been found in various research fields, such as medicine [21,22], economy and finances [23–26], seismology [27,28], oceanography [29] and meteorology [30,31], among others. Nowadays, the seismic activity in the world is continuously monitored by many seismic monitoring station networks, which have increased the amount of available seismicity data.

Among other methods, VG has allowed a deeper focus on the presence of dynamic features in the temporal distribution of seismic sequences [27]; a merit of the VG method is its ability to capture non-trivial correlations in non-stationary time series without introducing elaborate algorithms [7,32]. In addition, VGA has been shown to be a very useful tool to reveal the complex characteristics of seismic processes [20,27,28]. Because of its simple implementation, as well as its wide range of applicability, the VG algorithm has become popular in contrast with other methods. Telesca and Lovallo [20] were the first authors to apply the VG method to analyzing seismic sequences between 2005 and 2010 in Italy. These authors found that the VG method showed a collapse effect on the distributions of the degrees of connectivity of the seismic series with an increase in the magnitude threshold, which suggests that the properties of the distribution of the degree of connectivity can be independent of the magnitude of the threshold. In 2013, Telesca et al. [28] analyzed the seismic sequences monitored from 2005 to 2012 in five areas of the subduction zone in Mexico. In this investigation, an empirical link was found between the $b$-value of the Gutenberg–Richter law and the slope of the fitted line in the $k$–$M$ plot by using the least-squares method. This outlined the relationship between $k$ and $M$ and was called the $k$–$M$ slope. Other studies focused on the relationship between the $b$ value and the $k$–$M$ slope have been carried out, for example, in three dominant seismic areas of northern Iran [33], in seismogenic zones of Alaska, the Aleutian subduction zone [34] in Taiwan and Italy [35] and, recently, seismicity from Song Tranh 2 hydropower (Vietnam) was analyzed [20]. On the other hand, synthetic seismicity has also been generated experimentally in the laboratory through a system based on a stick–slip mechanical process, where two rough surfaces with different degrees of roughness move each other with different relative velocities [36], giving rise to avalanches that emulate earthquakes, and numerically by using the Olami–Feder–Christiansen numerical model [37]. In these cases, it has also been shown that the $b$-value is closely related to the $k$–$M$ slope, suggesting that this relationship might be characterized by a kind of universality [20].

In addition, another dynamical property obtained from the VG method is the temporal correlation, which can be estimated by the $k$-degree distribution function. Kundu et al. [7] investigated the temporal correlations of two kinds of time series, a sequence of magnitudes and the inter-event times for three different types of seismicity: regular earthquakes, earthquake swarms and tectonic tremors. Their results were obtained by assessing the exponents of the $k$-degree distributions for the inter-event series, which showed a correlation similar to fractional Brownian motion. Additionally, they studied the time series of three different categories of earthquakes, from which they were able to distinguish topological features from the graph associated by using a visibility graph. As Telesca et al. [27] rightly point out, in all cited studies, the investigation of the relationship between the seismological and topological properties of seismicity has mainly been focused on tectonic or natural seismicity until now, because these regions still represent an open problem in the world.

The subduction areas located at Tehuantepec Isthmus, Chiapas (TCh), Flat Slab (FS) and Michoacan (Mi) belong to the Cocos plate, and each one has its own structural characteristics; their most important features are described in the next section. The Mexican subduction zone is a complex dynamical system with significant variation in slab geometry along the strike from northeast to southwest [38]. Some of these differences are the dip angle and subduction velocity, which are features that can determine the local seismic activity. The motivation for this work is because three large earthquakes occurred within these three zones recently: the first one (M8.2) struck on 7 September 2017, in the Tehuantepec Isthmus, which was considered an unusual event; the second one (M7.2) was a deadly event that occurred in the Flat Slab on 19 September 2017; and the third (M7.2) occurred on 19 September 2022, within the Michoacan zone. The coincidences in terms of month and day have drawn attention within the population, especially considering that two earthquakes, M8.2 and M7.9, occurred on 19 and 20 September 1985, respectively, causing huge losses of human life and damages in Mexico City. The aim of this paper is to study the seismic activity of the three zones by analyzing the catalogs from 2010 to 2022 using the VG method. First, the analysis is focused to identify the yearly variability of the $a$- and $b$-values, the seismic parameters of the Gutenberg–Richter Law, followed by the characterization of the three zones using the $k$–$M$ slope's behavior and, finally, the estimation of the temporal correlation from the $\gamma$-exponent associated with each $k$-degree distribution computed yearly and with the catalogs that comprise the whole period.

## 2. Tectonic Cocos Plate Settings

The Mexican subduction zone is a complex system characterized by variability in the shape of the subducted plate. Many authors have conducted significant studies into the crustal anisotropy and state of stress of the Mexican subduction zone ([39] and references therein). While in Jalisco, the Rivera plate subducts with a steep angle, in Michoacán, Guerrero and western Oaxaca, the Cocos plate becomes almost subhorizontal, changing again to a steep angle in the Tehuantepec Isthmus. This subducting activity also gave rise to the Trans-Mexican Volcanic Belt, an active continental volcanic arc that spans across Mexico, developed within an extensional tectonics setting. At the same time, there is new evidence of the geometry of the subducted slab being potentially due to subduction tearing or break-off. Therefore, the subducted slab is not laterally continuous but abruptly changes due to break-offs. Carciumaru et al. (2020) [39] presented a seismological study of the Earth's crust using three different methods: azimuthal anisotropy based on ambient noise, shear-wave splitting of tectonic tremors and moment tensor inversions of the M8.2 earthquakes of 7 September 2017 in Tehuantepec, Mexico. In their work, they identify two slab tearings: in the Michoacan–Guerrero border and in Oaxaca, near the Tehuantepec fracture Transform/Ridge, the trench has an inflection point that slightly changes its trend to the southeast. At the Michoacan–Guerrero border, the subducted slab is subhorizontal, whereas in central Oaxaca the plate is characterized by northeast convergence. In this paper, we focus our seismic analysis on three regions of the subduction plate: Michoacan (Mi), Flat Slab (FS) and Tehuantepec Isthmus, Chiapas (TCh), because these regions better represent the slab geometry of the Mexican subduction zone and are examples of how the subducted plate presents variability in its shape and dynamics. These three regions are analyzed in terms of complexity measures in order to show their differences and to identify the characteristic values of the subduction regimes.

As was indicated by Carciumaru et al. in [39], the slab tearing at the Michoacan–Guerrero border constitutes an important change in the subducting slab. This tearing marks two different subduction regimes: in the north-west (Michoacan), where the subduction is perpendicular, the coast follows a constant slip and is less pronounced than in Jalisco, but not as flat as in Guerrero; and at the FS, inland at the southeast of the center of the Volcanic Belt, where the subduction shows a subhorizontal behavior. This flat slab area, referring to the subhorizontal shallow dipping lower plate, is delimited by considering the depths of the Mohorovicic interface between 40 and 60 km [40], a type of ellipse that circumscribes (or encloses) the earthquakes that occur in the area, and also where the intraslab earthquakes occur. Some authors [41–46] have already introduced the north Cocos tearing. However, the variability of the Cocos subducting plate has been well established by the bathymetric evidence of the Orozco and O'Gorman fractures and the Tehuantepec Transform/Ridge, and because the dip of the subducting plate varies from steep (in the Jalisco–Colima region) to flat (in central Mexico and at the southeast), there is a steep subduction in Chiapas (see Figure 8 in [39]). The Middle American Trench (MAT) also presents changes in velocity regimes and large dip variations along the strike [47]. Singh et al. [48] consider that this variability is precisely what explains the difference in the damage pattern in Mexico City, comparing the 2017 earthquakes and the disastrous interplate earthquake of 1985 (M8.0). The intraslab earthquakes occur closer to Mexico City, at greater depth, and involve higher stress drops than their interplate counterparts. Accordingly, the ground motion is relatively enriched at high frequencies as compared with that during interplate earthquakes, and damage is dominated by site effects [48].

The Michoacan region is characterized by high seismic activity, not only due to subduction events but also because of the existence of crustal faults in the interior. Of particular interest are the large earthquakes that have occurred in the region, such as the earthquakes on 19 and 20 September 1985, measuring M8.1 and M7.5, respectively, which occurred on a segment of the subduction zone known as the Michoacan gap [49]. This region is also responsible for intraplate earthquakes, such as those that occurred on 7 April 1845 and 19 June 1858 [50]. Moreover, the existence of crustal fault systems, such as those of Morelia–Acambay and la Paloma, have produced several earthquakes that have affected populations. The research carried out by [51] concludes that there are at least 316 seismically active faults in the region, derived from an extensional tectonics setting resulting from subduction. Additionally, the volcanic activity in the area is very important (as an example, the emergence of volcanoes: Jorullo in 1759 and Paricutín in 1943) and, as a consequence, the existence of associated seismic swarms.

After the Tehuantepec earthquake (M8.2), a particular interest arose in the Tehuantepec, Chiapas area. The sequence of earthquakes that followed after this earthquake occurred in two regions: one region is located at the Isthmus of Tehuantepec (coincident with the inferred location of the Tehuantepec Transform/Ridge [52] where slab tearing was reported [44–46]), and the other is considered as the aftershock zone (see Figure 1). According to [53], the Cocos plate convergence rate and directions at both sides (6.43 and 7.2 cm/year) of the Tehuantepec ridge are different. Mandujano-Velazquez et al. [54] and Keppie and Moran-Zenteno [55] investigated spreading rate changes, leading them to propose the existence of a microplate. This microplate would have been bound by the Tehuantepec Ridge and by a pseudo-transform fault.

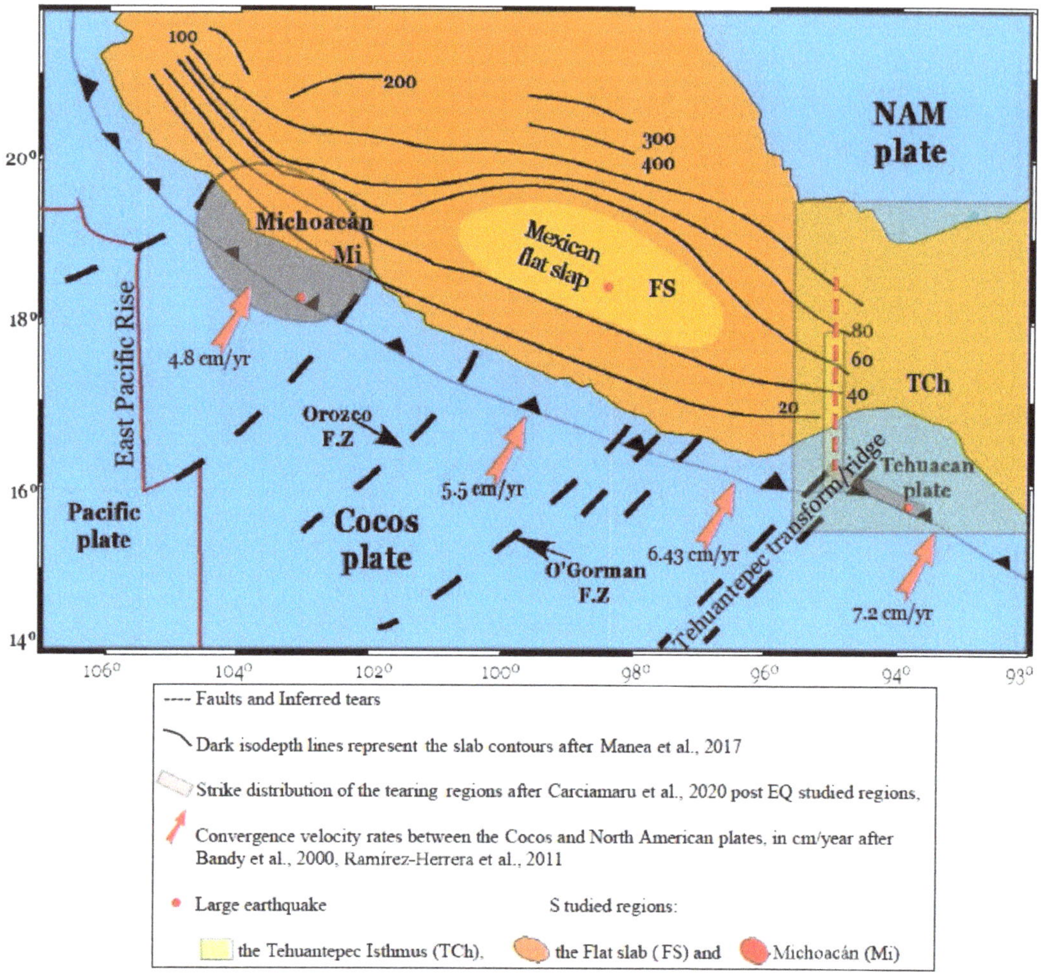


**Figure 1.** Mexican map. The three studied regions of the subduction Cocos Plate are depicted; also, the main tectonic features [39–41,47] and the larger earthquakes are shown.

## 3. Data Sets and Seismic Catalogs

The analyzed data sets, corresponding to the three seismic areas, were obtained from the National Seismic Service (SSN) website of the Universidad Nacional Autónoma de México (UNAM) (www.ssn.unam.mx (accessed on 10 November 2022) DOI:10.21766/SSNMX/EC/MX). The chosen periods for each monitored sequence were from 1 January 2012 to 10 November 2022. Within the selected areas, three large earthquakes occurred with $M > 7.5$: two in September 2017 and one in 2022. The first one, with an epicenter located within the Tehuantepec Gulf, struck on 7 September 2017 (M8.2) and was considered an unusual earthquake because it was in the intraplate. The second one, a deadly mainshock, had an epicenter in the Flat Slab area, a horizontal plate, located beneath the central area of Mexico and occurred on 19 September 2017 (M7.6) and the third and latest, whose epicenter was located in Michoacan State, struck on 19 September 2022 (M7.6). The areas where the three mainshocks occurred are indicated on the map shown in Figure 1. The catalogs were selected by considering the epicenters located in the areas for Mi, for FS and for the TCh.

The seismic activity of the three regions is shown in Figure 2, where the cumulative rate growth of the number of earthquakes monitored at Mi, FS and TCh can be observed. FS registered the minimum number of earthquakes at around 5000 over 13 years (Figure 2a in blue), whilst in Mi, around 15,000 earthquakes were registered (Figure 2a in red), and in TCh the number of earthquakes was approximately 75,000 (see Figure 2b) during the same period.

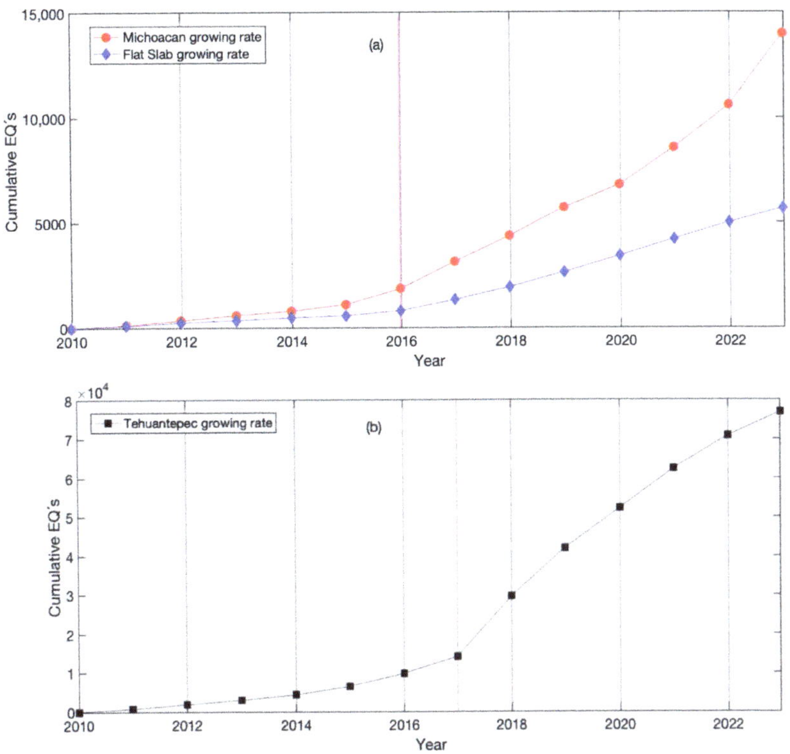


**Figure 2.** Cumulative number of earthquakes (EQs) per year; comparison between the three seismic regions. (**a**) The number of earthquakes in the Mi (red) and FS (blue) regions and (**b**) in the Tehuantepec Isthmus. The vertical line indicates the year when the activity increased: in 2016 at the Michoacan and Flat Slab and in 2017 at Tehuantepec.

## 4. Methods

### 4.1. Gutenberg–Richter Law

The Gutenberg–Richter (GR) law describes an empirical relationship between the frequency and magnitude ($M$) of earthquakes in a specific region [56] following the distribution:

$$log_{10}(N) = a - bM \tag{1}$$

where $a$ and $b$ values are constants that characterize the seismic region, and $N$ is the number of earthquakes with a magnitude $\geq M$. It has been observed that both values, $a$ and $b$, depend on the studied region and time. Changes in the $b$-value have been observed that are related to the spatial location of the analyzed area and to the time span observed. Additionally, changes in the $b$-value are inversely related to changes in stress [52]. In addition, in [57], a linear relationship between the $a$-value and $b$-value parameters of the Gutenberg–Richter (GR) law was shown, following $a$~$4b$.

## 4.2. *Visibility Graph*

This method considers a sequence of N events of a variable monitored evenly by a dynamical system. The VG means that any two events are connected by a right-line segment if they can see each other so that such a segment is not broken by any other intermediate value of the sequence. The VG maps each event as a node in a graph; therefore, each node connects with the other nodes based on the mutual visibility condition defined with the corresponding data heights. The visibility condition is expressed mathematically as follows: let two arbitrary data values (*ta*, *Ma*) and (*tb*, *Mb*), where *ta* < *tb*. Both points are visible to each other if any other data (*tc*, *Mc*) placed between them satisfy the visibility condition, as is shown in Figure 3, that is:

$$\frac{M_b - M_c}{t_b - t_c} < \frac{M_b - M_a}{t_b - t_a} \tag{2}$$

Lacasa et al. [16] showed that for this visibility condition, the associated graph of the time series is always (a) connected, (b) undirected and (c) invariant under affine transformations of the time series, including rescaling of both axes and horizontal and vertical translations. When the events represent earthquakes of magnitudes *M* such as seismic catalogs, the transformation is applied such that each earthquake of magnitude $M_i$ represents a node in the graph, and the number of earthquakes that satisfy Equation (2) indicates the connectivity between earthquakes counting the *k*-degree (*k*-degree is defined as the number of direct connections between the *i*-vertex with the other ones; see Figure 3), and can also be associated with the temporal correlations of the data set in connection with the *k*-degree distribution of the nodes, as will be described in the next subsection.

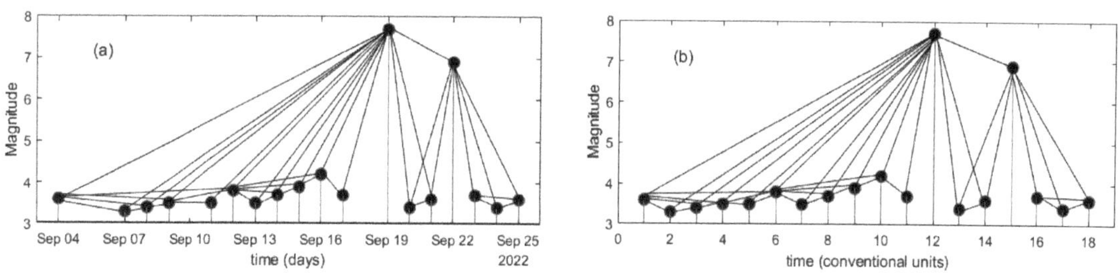


**Figure 3.** Example of the VG method of a seismic catalog. The black points are the magnitudes, and the black lines show the visibility condition. (**a**) VG for a seismic catalog as point processes; (**b**) the same catalog drawn as a discrete process where it can be seen that the connectivity does not change.

## 4.3. *k-Degree Distribution*

In complex networks, the dynamical characteristics of the underlying system can be quantified by different measures; however, in the particular case of VG, one of the most interesting and important measures is the so-called vertex degree or *k*-degree: the number of direct connections between the *i*-vertex and the other ones. Because the graph is connected, each node has at least $k = 1$ degree for the first and the last nodes and $k = 2$ degrees for the other ones. In several works where VG has been applied to analyze time series, the connectivity has highlighted topological properties in the graph that are associated with the processes underlying the time series. This information is mainly given by the *k*-degree, as well as their distribution function, *P*(*k*). For different kinds of processes, *P*(*k*) displays specific behaviors; for periodic signals, VG is transformed as a concatenation of a finite number of motifs because the basic period is an integer multiple of the sampling rate. On the other hand, the opposite extreme case is white noise. For this process, the *k*-degree distribution behaves as an exponential distribution. For fractal time series, the *k*-degree distribution is generally scale-free following the power law:

$$P(k) \sim k^{-\gamma} \tag{3}$$

where the $\gamma$-exponent is related to the Hurst exponent $H$ of the underlying time series as $\gamma = 3 - 2H$ for fractional Brownian motion and $\gamma = 5 - 2H$ for fractional Gaussian noise, so that it is possible to estimate a measure of the persistence and correlation, as well as the relationship with the $\beta$-spectral exponent from the power spectrum scaled as $f^{-\beta}$.

## 5. Results

### 5.1. Gutenberg–Richter Parameters

The local seismic activity is usually characterized by the Gutenberg–Richter law (Equation (1)), where the completeness magnitude "$M_c$", as well as the "$a$" and "$b$" parameters, are estimated. In this study, we employed the maximum likelihood method in the estimation of the $b$-value [27]. First, the $a$- and $b$-values for the entire catalogs of the three areas are listed in Table 1.

**Table 1.** Seismic parameters of the entire catalogs.

| Region | $a$ | $b$ | $Mc$ | $N\ (M \geq Mc)$ |
|---|---|---|---|---|
| Mi | 5.43 | 0.85 | 3.7 | 4357 |
| FS | 5.53 | 0.92 | 3.5 | 3232 |
| TCh | 7.83 | 1.12 | 3.6 | 62,571 |

It has been shown that these parameters can change over time [58], showing possible fluctuations in the seismic activity of each zone. We analyzed the seismic activity of three different regions: Michoacan (MI), Flat Slab (FS) and the Tehuantepec Isthmus (TCh), as is shown in Figure 1. The period analyzed was from 2010 to 2022, in which the numbers of earthquakes registered were 14,003 in Mi, 5679 in the FS and 76,889 in TCh. Figure 3 indicates the growing rate of the occurrence of earthquakes. The temporal variability of the $b$-values calculated with the Gutenberg–Richter parameters for the activity of Mi, FS and TCh are shown in Figure 4. The $b$-value changed each year and displays an irregular pattern; nevertheless, in 2017, the $b$-values of FS and TCh attained low values, and after that, such values increased. Regarding the latest main shock at Mi in 2022, the respective $b$-value also decreased at a low.

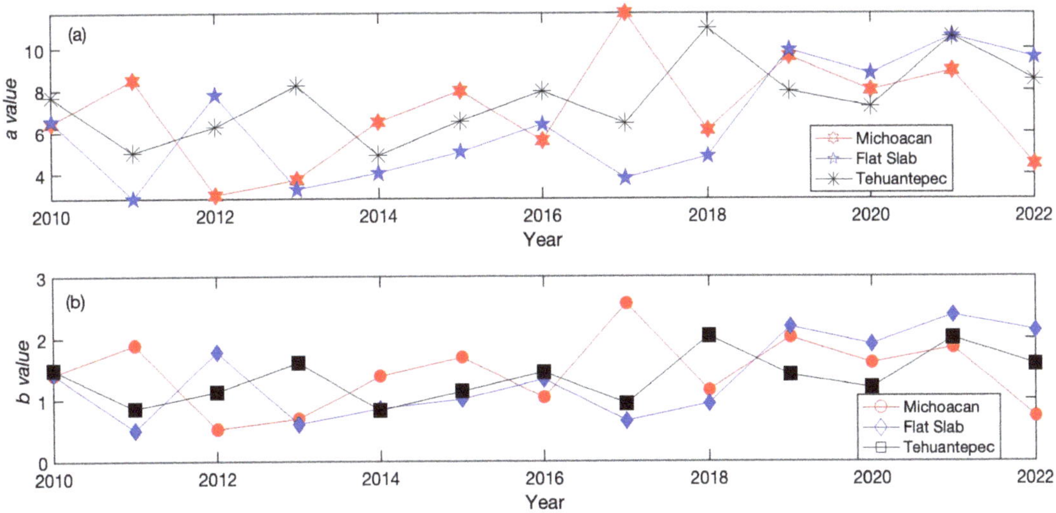


**Figure 4.** *Cont.*

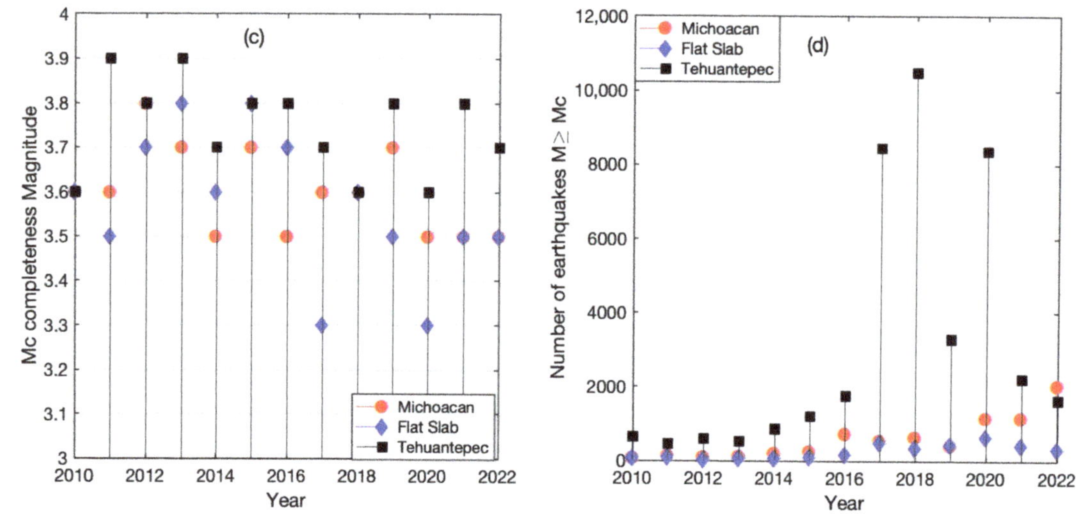


**Figure 4.** Yearly behavior of the seismic parameters (**a**) $a$-values, (**b**) $b$-values of the Gutenberg–Richter law, (**c**) completeness magnitude and (**d**) number of earthquakes with $M \geq M_c$.

The relationship between the $a$- and $b$-values of the Gutenberg–Richter law is depicted in Figure 5 and is in accordance with the result reported by [57]. It is important to point out that the expected relation $a{\sim}4b$ is fulfilled in Mi and FS in accordance with [57]; however, for the TCh catalog, the relationship is given as $a = 4.52b$. This difference could be due to the post-seismic activity of the M8.2 main shock, which could be considered aftershocks and outside the tearing region of the Tehuantepec transform/ridge [39].

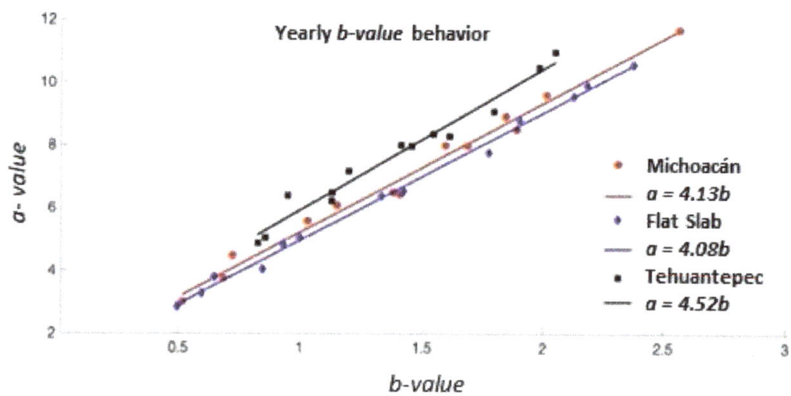


**Figure 5.** Relationship between the Gutenberg–Richter $a$-value vs. $b$-value.

*5.2. k-Degree vs. Magnitude Relationship*

In order to characterize the topological and seismic properties between the graph and the seismic activity, Telesca et al. [36] introduced the $k$–$M$ plane. The $k$–$M$ slope is obtained as the mean square fitting of the right line in the $k$–$M$ plane. The fact that the $k$–$M$ slope is positive indicates a positive correlation; thus, the larger magnitude is then higher than the connectivity degree. The $k$–$M$ slopes were calculated for the entire catalogs (including all magnitudes). In Figure 6a–c the $k$–$M$ plane is shown for the catalogs of the three regions from the completeness magnitude threshold ($M \geq Mc$). The differences in the *k–M* slope values are summarized in Table 2; in the third column, the reported values are from whole catalogs ($M < Mc$). The fitting in the *k–M* plot for the three regions can be considered linear when ($M > Mc$); however, for ($M < Mc$), the behavior is dominated by low seismicity and differs from linear fit, and is not reliable.

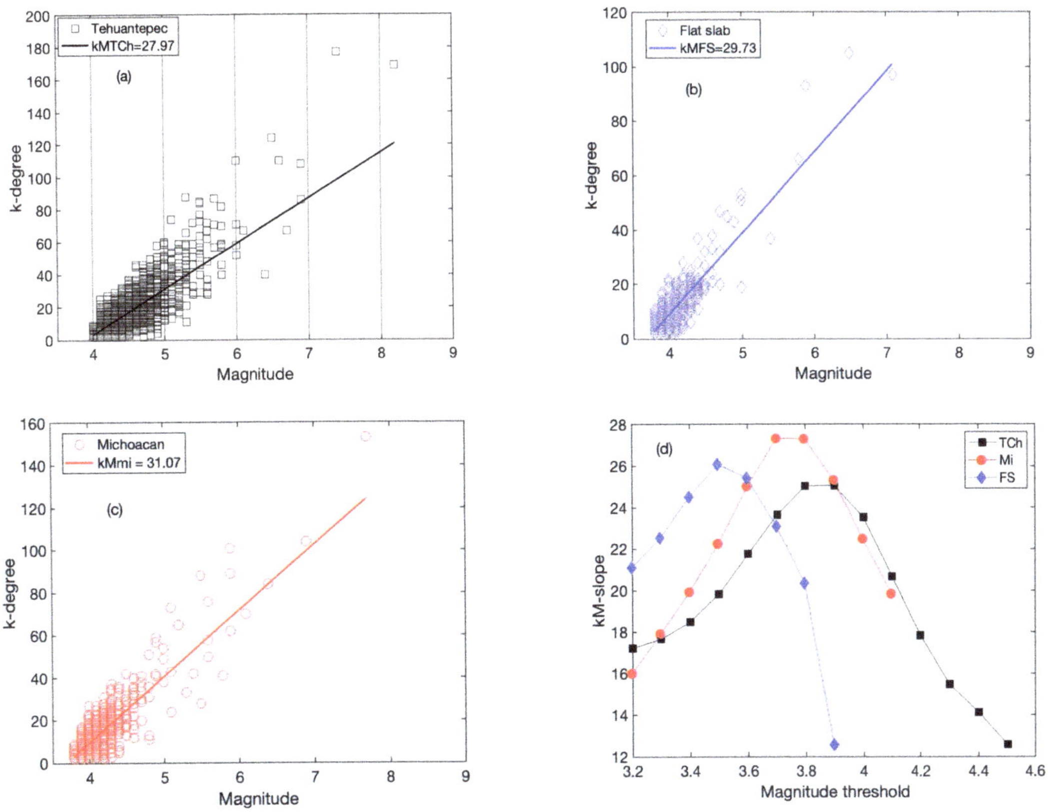


**Figure 6.** *k–M* plane of (**a**) TCh, (**b**) FS and (**c**) Mi. For the three cases, the calculation was estimated by using the completeness magnitude. (**d**) Variability of the *k–M* slope as a function of the magnitude threshold. The maximum values coincide with the completeness magnitude.

**Table 2.** Comparison of *k–M* slope between the entire catalog and with $M \geq Mc$.

| Region | *k–M* Slope with $M \geq Mc$ | *k–M* Slope Entire Catalog |
|---|---|---|
| Mi | 31.07 | 15.94 |
| FS | 29.73 | 9.78 |
| TCh | 27.97 | 17.45 |

In Figure 6d, the variation of *k–M* slope is plotted versus the threshold magnitudes for the three catalogs. The first threshold magnitude, $M_{th}$, is 3.2, and then increases by steps of 0.1; the maximum threshold magnitude depends on each catalog. In all cases, the *k–M* slope attains a maximum $M_{th} \cong M_C$, after which the number of earthquakes becomes very small.

On the other hand, the *k*–*M* slope variability as a function of year is depicted in Figure 7 for the three catalogs. The *k*–*M* slopes values show an increasing trend for the three regions. FS and TCh stopped showing this trend in 2017, but Mi continued increasing until 2022.

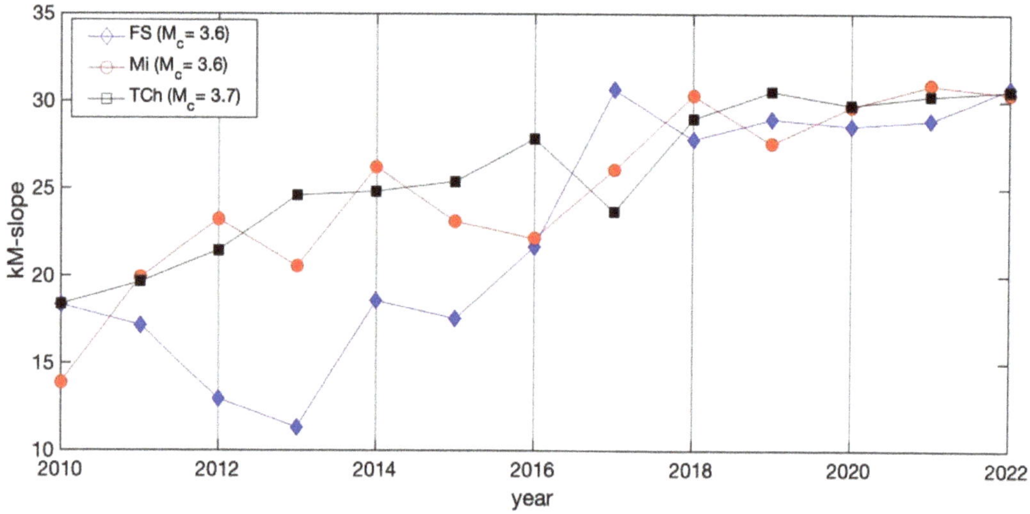


**Figure 7.** Yearly *k*–*M* slope variation of the three studied regions.

### 5.3. *P*(*k*) Distribution

When estimating the correlation of seismic activity, this can be performed using the sequence of magnitude of earthquakes by applying the VG method to seismology, as has been reported in [20,34]. To do so, the seismic catalog is considered as a magnitude-discrete time series, rather than a continuous process, where the earthquakes are events with specified occurrence time. In fact, the catalogs are magnitude time series marked as temporal point processes that are described by the sum of Dirac's delta centered on the occurrence time with amplitude proportional to the magnitude of the event; see Figure 3a, which shows a sequence of earthquakes in the catalog analyzed. In Figure 3b, the representation as a magnitude-discrete time series is displayed where the connectivity remains invariant. Next, the time series of magnitudes can be identified as a fractional Brownian motion so that the relationship between $\gamma$ and $H$ is valid [30]. The *k*-degree distribution (Equation (3)) of the visibility graph is assessed for the sequences of magnitudes of earthquakes monitored in the three regions. In [30], the authors showed the relationship between g-exponents in the *P*(*k*) power law, and the Hurst exponent was given by $\gamma = 3 - 2H$. It is well known that the significance of $H$ is determined by its deviations from the value of 0.5, which indicates randomness. The *k*-degree distribution *P*(*k*) vs. *k* in a log–log plane of each region was estimated for each one of the three regions, choosing the sequences as (a) the whole catalogs and (b) for the $M \geq M_c$, displayed in Figure 8a and b, respectively. The *P*(*k*) distributions display a behavior associated with correlated processes which are characterized by means of $H$. In Table 3, the $\gamma$-exponent values are summarized and in Figure 8a,b, the $R^2$ adjust indicates the goodness of the $\gamma$-exponent fitted. It is observed that the $\gamma$-exponent of the Tehuantepec Isthmus is the largest, whilst for FS, it is the lowest in both cases.

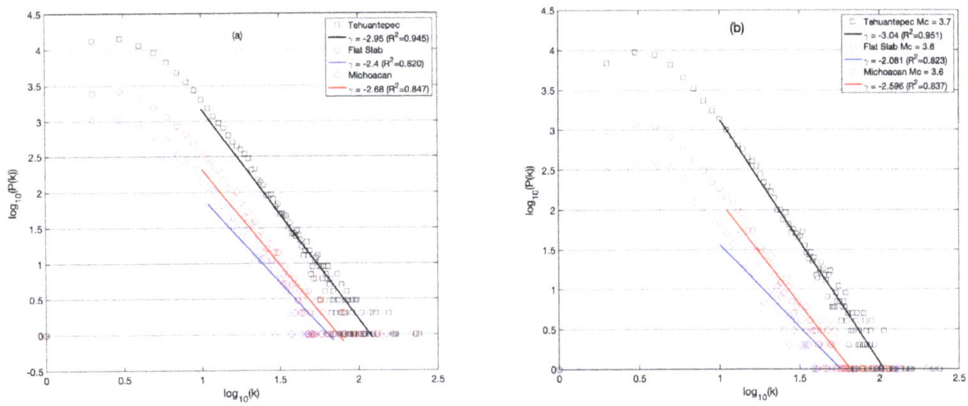


**Figure 8.** Distribution $P(k)$ vs. $k$-degree in $\log_{10}$–$\log_{10}$ scale of the catalogs from 2010 to 2022. (**a**) Distributions of $P(k)$ considering whole catalogs and (**b**) distributions of $P(k)$ estimated with $M \geq M_c$ catalogs.

**Table 3.** Comparison of $\gamma$ between the whole catalog and with $M > Mc$.

| Region | $\gamma$ Whole | $\gamma$ $M > Mc$ Catalog |
|---|---|---|
| Mi | 2.88 | 2.58 |
| FS | 2.4 | 2.08 |
| TCh | 2.95 | 3.04 |

### 5.4. Correlation Measure

Next, the $k$-degree distribution of the visibility graph was constructed (Equation (3)). The correlation of the seismic activity is directly related to the $\gamma$-exponent and $H$. A process is characterized as persistent if $(0.5 < H < 1)$, antipersistent when $(0 < H < 0.5)$ and uncorrelated or random if $H = 0.5$. In accordance with the relationship $\gamma = 3 - 2H$, in Figure 9, the yearly variability of the $H$-values is displayed.

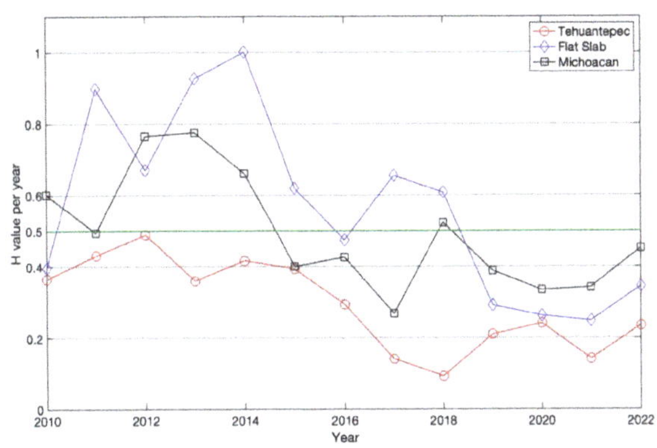


**Figure 9.** Hurst exponent behavior, obtained from relationship $\gamma = 3 - 2H$ for the three catalogs. The green line indicates the value $H = 0.5$, which indicates uncorrelation, which is the border between persistence and antipersistence.

## 5.5. Three-Dimensional Plot: *k–M* Slope–*b*-Value–*H*-Exponent

In order to describe a generalized behavior, in this study we introduce a 3D plot (Figure 10a) in which the topological, seismic and dynamical properties of the three regions are identified.

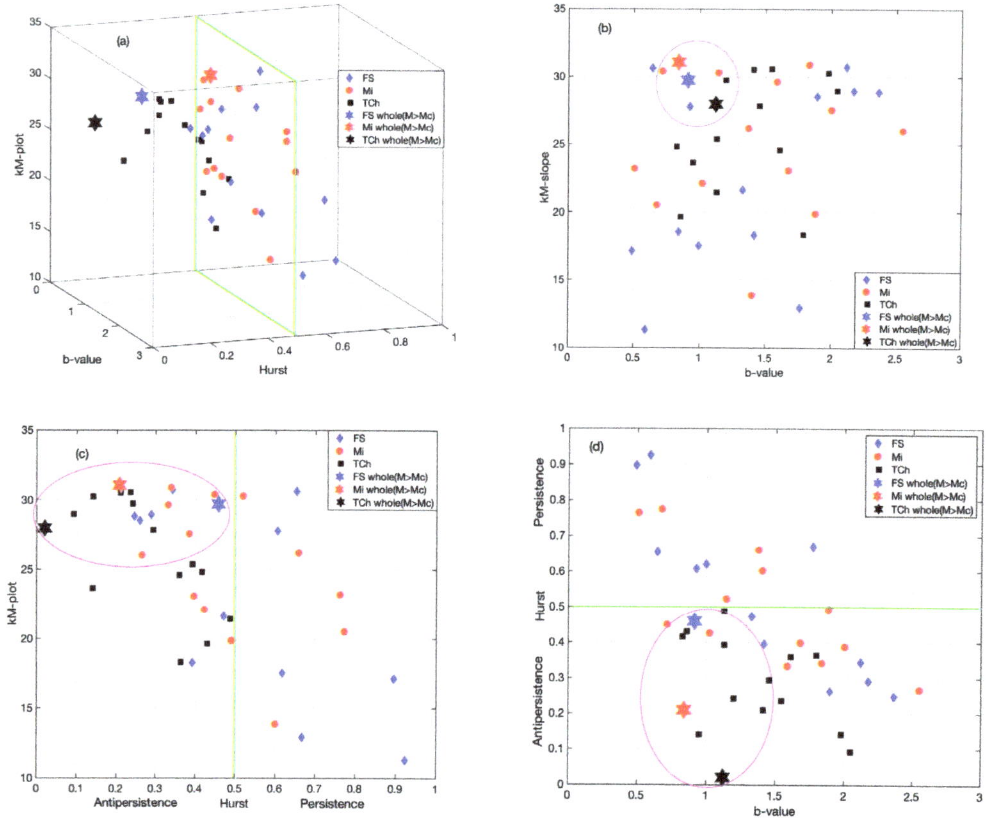


**Figure 10.** Each small point represents the yearly variation from 2010 to 2022 of the three seismic sequences: FS (blue), Mi (red), TCh (black) and the stars corresponding with the same parameters for the whole catalog with $M \geq M_c$. (**a**) Three-dimensional plot of topological (*k–M* slope), seismic (*b*-value) and dynamical (*H*) properties. (**b**) *b*-value versus *k–M* slope. (**c**) *H* versus *k–M* slope. (**d**) *H* versus *b*-value.

The 3D plot, as shown in Figure 10a, condenses the information between the yearly variability and the whole catalogs with the $\boldsymbol{M \geq M_c}$ of the three studied parameters: the topological properties represented by the *k–M* slope; the seismic features in terms of the *b*-value and the dynamical characteristic with the Hurst exponent (H). In addition, to represent the possible relationship between pairs of the three parameters, Figure 10b–d are the projections in three planes. Figure 10b, *k–M* slopes vs. *b*-value, shows a messy relationship between these parameters. Figure 10c, *k–M* slope vs. Hurst exponent, shows a poor linear fit in the sense of least squares with a negative slope. Finally, Hurst vs. *b*-value in Figure 10d also depicts some type of linearity with a negative slope. Regarding the parameters calculated for the entire catalogs with $\boldsymbol{M \geq M_c}$ (stars) inside the purple ellipses, it can be seen that the yearly variability displays dispersion; however, a clear clusterization is observed when the entire catalogs are considered as marked by the purple ellipses.

## 6. Discussion

In this work, the study of three regions of the subduction Cocos plate was performed using the VG method. First, a study of the temporal behavior of the seismic parameters, the *a*- and *b*-values in the Gutenberg–Richter law, was performed. As has been reported in different works, both *a*- and *b*-values in the Gutenberg–Richter law change with time and region. When both seismic parameters are evaluated yearly, variability is found in both of them. The yearly variability of the *b*-value for the three regions (Mi, FS and TCh) is displayed in Figure 4, where it can be observed that FS and Mi attained the minimal value ($b = 0.5$) in 2011 and 2012, respectively, while in 2014 ($b = 0.8$), the lowest value was registered at TCh. This variation is mainly associated with the stress field as is discussed in [58 and references therein]. In Figure 4b, it can also be observed that the *b*-value decreased when the strong earthquakes occurred in 2017 at TCh and FS and also at Mi in 2022. In addition, in times when the number of seisms is low, the *b*-values seem to increase and remain so. Over the years, the behavior of the mean *b*-value is approximately 1.2 until 2016, but from 2017, this parameter increases. Concerning the relationship between *a* and *b*-values, Perez-Oregon et al. [57] obtained an analytical deduction of the positive correlation between parameters *a* and *b*, where $a = (4.01 \pm 0.02M)b + logC$ (where $M$ is the magnitude and C is a constant), which is corroborated by the present analysis of the three zones. In Figure 5, the yearly relationship between both parameters for the three zones is shown; the relationship in [57] is fulfilled for FS and Mi, $a \approx 4b$, whilst for TCh, $a \approx 4.5b$, it is slightly different than expected.

This difference in TCh is most likely due to the sequence of earthquakes following the M8.2 earthquake that occurred on 7 September 2017. The sequences occurred in two regions: one that is coincident with the inferred location of the Tehuantepec Transform/Ridge [52,54,55] where the slab tearing [44–46] is reported, and the other considered to be the aftershock zone [39].

On the other hand, the relationship between the topological properties and the seismic activity is identified for the three areas considered. As was introduced in [28], the *k*–*M* slope is the parameter that connects the topological properties and the seismic variability, which is calculated by the best linear fitting in the *k*–*M* plane, as is shown in Figure 6a–c. In this representation, the slope of the *k*–*M* relationship conveys information about the earthquake productivity of the seismic areas; thus, a measure of the correlation between the connectivity and magnitude is given by the *k*–*M* slope [28]. All the *k*–*M* plots show an increasing trend in the degree with the increase of the magnitude, as was shown by [36], which means that the *k*–*M* slope is a measure of the level of correlation. In addition, for the higher-magnitude events ($M > 7$), the relationship with the higher *k*-degree becomes almost one-to-one, as is shown in Figure 6a–c, where the *k*-degree is maximum. In the three studied cases, it can be observed that the behavior indicates that the correlations are positive. The numerical value of the *k*–*M* slope is a measure of the level of correlation and consequently of the degree of connectivity.

In Figure 6d, the *k*–*M* slope versus the threshold magnitudes is plotted for the three catalogs. The *k*–*M* slope attains maxima when $M_{th} \cong M_C$, and afterward, the number of earthquakes becomes very small; therefore, it is not statistically representative.

Figure 7 shows the yearly behavior of the *k*–*M* slope, with $M \geq M_c$ for each case. It can be observed that the connectivity displays an increasing trend; nevertheless, FS and TCh stopped showing this trend in 2017, when the larger earthquakes occurred in September in each region. On the other hand, the connectivity of Mi kept increasing until 2022. This increasing r suggests that the connectivity could be associated with stress accumulation, and after the release, the variability of the connectivity is very low.

The distributions of $P(k)$ vs. *k*-degree in logarithmic scale are shown in Figure 8a,b, where the $\gamma$-values are shown in the box inside for each catalog.

The $H$ values are obtained from the $\gamma$-values. Figure 9 depicts the yearly $H$ variability; from this behavior, it can be observed that all processes analyzed show short- and long-range temporal correlations, whose characteristics are governed by processes similar to fractional Brownian motion. The case of TCh activity shows antipersistence because $H < 0.5$ along the analyzed period. This behavior indicates that the dominant fluctuations come from low-magnitude seismicity over the 13 years studied. On the other hand, FS and Mi are persistent ($H > 0.5$); in the case of the seismicity of Mi, this can be observed from 2010 to 2014, and for FS, the persistence is between 2010 and 2017. After these periods, the behavior of Mi and FS changed to antipersistent processes from 2015 to 2022 and from 2018 to 2022, respectively. These behaviors suggest that the seismic activity in each region has different dynamics, whose conditions are established by the stress fields between the Cocos and North American plates, as illustrated in Figure 1 and has been reported in other studies [39,47,52,54,55].

Finally, a 3D plot (Figure 10a) is introduced with the aim to concentrate the information on the topological ($k$–$M$ slope), seismic ($b$-value) and dynamical ($H$) properties of the three studied seismic sequences: FS, Mi and TCh. This plot contains the behavior of the yearly variability and, also, the value of the parameters, taking into account the entire catalogs with $M \geq M_c$. In this representation, we can perceive a dispersion of these properties from 2010 to 2022, but the same parameters estimated for the complete catalogs are located close to each other. When the 3D is projected, we obtain Figure 10b–d. Figure 10b conveys information about the earthquake productivity of the seismic areas. In terms of the dynamical properties of the sequences, the $k$–$M$ slope vs. $b$-value displays a clusterization for $k$–$M$ values above 14 and $b$-values ranging between 0.5 and 2.0; thus, it seems to be strongly related to the $b$-value of the Gutenberg–Richter law.

Figure 10c represents the projection of the $k$–$M$ slope vs. $H$; it can be seen that TCh indicates antipersistence with high connectivity, while Mi also displays high connectivity while it is persistent or antipersistent and finally FS shows a mainly low level of connectivity when it is antipersistent or persistent but displays high connectivity only in some cases of persistence. Figure 10d displays the relationship between $H$ and $b$-value; the main clusterization is located in the antipersistent area with $b$-values ranging between 0.5 and 2.0.

From these results, we can suggest that the VG properties of seismicity can resemble the seismological properties given by the parameters of the Gutenberg–Richter law. As already mentioned by [36], since the VG method takes into account the magnitudes of the events, this could suggest a way to analyze the statistical properties of seismicity more generally than the Gutenberg–Richter law.

As previously mentioned, Figure 10a, representing a 3D plot, was introduced with the aim of condensing the information between the yearly variability and the whole catalogs with $\boldsymbol{M} \geq \boldsymbol{M_c}$ of the three parameters studied: the topological properties represented by the $k$–$M$ slope; the seismic features in terms of the $b$-value; and the dynamical characteristic of the Hurst exponent (H). The projections of data in three planes are also shown. When observing Figure 10c,d a poor linear fit can be seen in the sense of least squares with negative slopes. The parameters calculated for the whole catalogs with $\boldsymbol{M} \geq \boldsymbol{M_c}$ (stars) show a clear clusterization.

## 7. Conclusions

The aim of this study was to identify features and their possible differences between the three seismic areas of Mi, FS and TCh, belonging to the subducted Cocos plate and the North American plate. The selection of these zones was motivated because three large earthquakes ($M > 7$) have occurred in this area in the last five years: two of them on 7 and 19 September 2017 at TCh and FS, respectively, and the third in Mi on 19 September 2022. Our finding reveals the variability of the $b$-value in the GR law, as the $b$-value estimated yearly for each zone displayed large fluctuations, ranging from 0.5 to 2.5 in Mi and FS, and from 0.8 to 2 in TCh. It is worth noting that the $b$-value attains a minimum value in the years when the main shocks occurred, although it cannot be considered a generalization. The $a$-value in the GR law is related to the $b$-value as $a = 4b$, which is fulfilled for Mi and FS; nevertheless, for TCh, $a = 4.5b$ was found, which indicates a difference in the seismic activity at Tehuantepec Isthmus. Regarding the seismic parameters, our findings suggest that the seismic activity at Tehauntepec Isthmus displays important differences with respect to Mi and FS. In terms of the topological and seismic properties, the numerical value of the $k$–$M$ slope is a measure of the level of correlation and consequently of the degree of connectivity. TCh shows the lowest value in comparison with Mi and FS. The $k$–$M$ slope values show an increasing trend for the three regions. FS and TCh stopped demonstrating this trend in 2017, after which their behavior was almost horizontal and parallel, but the Mi trend continued to increase until 2022. The $k$-degree distribution $P(k)$ allows us to evaluate the temporal correlation of each sequence of magnitudes. TCh remains antipersistent during the period analyzed, while Mi and FS remain persistent, becoming antipersistent behavior, which suggests that the seismic activity of Mi and FS changed during the period analyzed. Finally, the 3D plot of $k$–$M$ slope–$b$-value–$H$ was introduced to represent the joint seismic, topological and dynamical properties of the three studied zones. The projections of data in the three planes of Figure 10c,d show a poor linear fit in the sense of least squares with negative slopes. The entire catalogs' calculated parameters, for $M \geq M_c$ (stars), show clear clusterization.

**Author Contributions:** Conceptualization, A.R.-R., E.L.F.-M. and C.A.V.; methodology, A.R.-R. and E.L.F.-M.; software, E.L.F.-M. and A.R.-R.; validation, formal analysis, A.R.-R. and E.L.F.-M.; investigation, A.R.-R. and E.L.F.-M.; resources, A.R.-R. and E.L.F.-M.; data curation, A.R.-R. and E.L.F.-M.; writing—original draft preparation, A.R.-R., E.L.F.-M. and C.A.V.; writing—review and editing, A.R.-R., E.L.F.-M. and C.A.V. All authors have read and agreed to the published version of the manuscript.

**Funding:** This research received no external funding.

**Institutional Review Board Statement:** Not applicable.

**Data Availability Statement:** Not applicable.

**Acknowledgments:** E.L.F.-M. thanks the Institute of Geophysics, Universidad Nacional Autónoma de México. A.R.R. and C.A.V. thank the Departamento de Ciencias Básicas, Universidad Autónoma Metropolitana unidad Azcapotzalco.

**Conflicts of Interest:** The authors declare no conflict of interest.

## References


1. Mallard, C.; Coltice, N.; Seton, M.; Müller, R.D.; Tackley, P.J. Subduction controls the distribution and fragmentation of Earth's tectonic plates. *Nature* **2016**, *535*, 140–143. [CrossRef]
2. Coltice, N.; Husson, L.; Faccenna, C.; Arnould, M. What drives tectonic plates? *Sci. Adv.* **2019**, *5*, eaax4295. [CrossRef]
3. Sigalotti, L.D.G.; Ramírez-Rojas, A.; Vargas, C.A. Tsallis $q$-Statistics in Seismology. *Entropy* **2023**, *25*, 408. [CrossRef] [PubMed]
4. Mulargia, F.; Bizzarri, A. Anthropogenic Triggering of Large Earthquakes. *Sci. Rep.* **2014**, *4*, srep06100. [CrossRef]
5. Da Silva, S.L.E.F.; Julià, J.; Bezerra, F.H.R. Deviatoric Moment Tensor Solutions from Spectral Amplitudes in Surface Network Recordings: Case Study in São Caetano, Pernambuco, Brazil. *Bull. Seism. Soc. Am.* **2017**, *107*, 1495–1511. [CrossRef]
6. Holliday, J.R.; Rundle, J.B.; Turcotte, D.L.; Klein, W.; Tiampo, K.F.; Donnellan, A. Space-Time Clustering and Correlations of Major Earthquakes. *Phys. Rev. Lett.* **2006**, *97*, 238501. [CrossRef]
7. Kundu, S.; Opris, A.; Yukutake, Y.; Hatano, T. Extracting Correlations in Earthquake Time Series Using Visibility Graph Analysis. *Front. Phys.* **2021**, *9*, 656310. [CrossRef]
8. Albert, R.; Barabási, A.-L. Statistical mechanics of complex networks. *Rev. Mod. Phys.* **2002**, *74*, 47–97. [CrossRef]
9. Newman, M.E.J. The Structure and Function of Complex Networks. *SIAM Rev.* **2003**, *45*, 167–256. [CrossRef]
10. Abe, S.; Suzuki, N. Scale-free network of earthquakes. *Europhys. Lett.* **2004**, *65*, 581–586. [CrossRef]
11. Baiesi, M.; Paczuski, M. Scale-free networks of earthquakes and aftershocks. *Phys. Rev. E* **2004**, *69*, 066106. [CrossRef]
12. Newman, M.; Barabási, A.L.; Watts, D.J. *The Structure and Dynamics of Networks*; Princeton University Press: Princeton, NJ, USA, 2006.
13. Boccaletti, S.; Latora, V.; Moreno, Y.; Chavez, M.; Hwang, D.-U. Complex networks: Structure and dynamics. *Phys. Rep.* **2006**, *424*, 175–308. [CrossRef]
14. Zhang, J.; Small, M. Complex Network from Pseudoperiodic Time Series: Topology versus Dynamics. *Phys. Rev. Lett.* **2006**, *96*, 238701. [CrossRef] [PubMed]
15. Yang, Y.; Yang, H. Complex network-based time series analysis. *Phys. A Stat. Mech. Appl.* **2008**, *387*, 1381–1386. [CrossRef]
16. Lacasa, L.; Luque, B.; Ballesteros, F.; Luque, J.; Nuño, J.C. From time series to complex networks: The visibility graph. *Proc. Natl. Acad. Sci. USA* **2008**, *105*, 4972–4975. [CrossRef]

17. Donner, R.V.; Zou, Y.; Donges, J.F.; Marwan, N.; Kurths, J. Recurrence networks—A novel paradigm for nonlinear time series analysis. *New J. Phys.* **2010**, *12*, 033025. [CrossRef]
18. Hope, S.; Kundu, S.; Roy, C.; Manna, S.S.; Hansen, A. Network topology of the desert rose. *Front. Phys.* **2015**, *3*, 72. [CrossRef]
19. Gao, Z.-K.; Small, M.; Kurths, J. Complex network analysis of time series. *EPL Europhys. Lett.* **2016**, *116*, 50001. [CrossRef]
20. Telesca, L.; Lovallo, M. Analysis of seismic sequences by using the method of visibility graph. *EPL Europhys. Lett.* **2012**, *97*, 50002. [CrossRef]
21. Zhu, G.; Li, Y.; Wen, P.P. An Efficient Visibility Graph Similarity Algorithm and Its Application on Sleep Stages Classification. In *International Conference on Brain Informatics*; Springer: Berlin/Heidelberg, Germany, 2012; pp. 185–195.
22. Yu, M.; Hillebrand, A.; Gouw, A.A.; Stam, C.J. Horizontal visibility graph transfer entropy (HVG-TE): A novel metric to char-acterize directed connectivity in large-scale brain networks. *NeuroImage* **2017**, *156*, 249–264. [CrossRef]
23. Zhang, R.; Ashuri, B.; Shyr, Y.; Deng, Y. Forecasting Construction Cost Index based on visibility graph: A network approach. *Phys. A Stat. Mech. Appl.* **2018**, *493*, 239–252. [CrossRef]
24. Long, Y. Visibility graph network analysis of gold price time series. *Phys. A Stat. Mech. Appl.* **2013**, *392*, 3374–3384. [CrossRef]
25. Dai, P.-F.; Xiong, X.; Zhou, W.-X. Visibility graph analysis of economy policy uncertainty indices. *Phys. A Stat. Mech. Appl.* **2019**, *531*, 121748, ISSN 0378-4371. [CrossRef]
26. Goncalves, B.A.; Atman, A.P.F. Visibility graph combined with information theory: An estimator of stock market efficiency. *J. Netw. Theory Financ.* **2017**, *3*, 1–15. [CrossRef]
27. Telesca, L.; Thai, A.T.; Lovallo, M.; Cao, D.T. Visibility Graph Analysis of Reservoir-Triggered Seismicity: The Case of Song Tranh 2 Hydropower, Vietnam. *Entropy* **2022**, *24*, 1620. [CrossRef]
28. Telesca, L.; Lovallo, M.; Ramirez-Rojas, A.; Flores-Marquez, L. Investigating the time dynamics of seismicity by using the visi-bility graph approach: Application to seismicity of Mexican subduction zone. *Phys. A Stat. Mech. Appl.* **2013**, *392*, 6571–6577. [CrossRef]
29. Telesca, L.; Lovallo, M.; Pierini, J.O. Visibility graph approach to the analysis of ocean tidal records. *Chaos Solitons Fractals* **2012**, *45*, 1086–1091. [CrossRef]
30. Elsner, J.B.; Jagger, T.H.; Fogarty, E.A. Visibility network of United States hurricanes. *Geophys. Res. Lett.* **2009**, *36*, L16702. [CrossRef]
31. Pierini, J.; Lovallo, M.; Telesca, L. Visibility graph analysis of wind speed records measured in central Argentina. *Phys. A Stat. Mech. Appl.* **2012**, *391*, 5041–5048. [CrossRef]
32. Lacasa, L.; Luque, B.; Luque, J.; Nuño, J.C. The visibility graph: A new method for estimating the Hurst exponent of fractional Brownian motion. *EPL Europhys. Lett.* **2009**, *86*, 30001. [CrossRef]
33. Khoshnevis, N.; Taborda, R.; Azizzadeh-Roodpish, S.; Telesca, L. Analysis of the 2005–2016 Earthquake Sequence in Northern Iran Using the Visibility Graph Method. *Pure Appl. Geophys.* **2017**, *174*, 4003–4019. [CrossRef]
34. Azizzadeh-Roodpish, S.; Cramer, C.H. Visibility Graph Analysis of Alaska Crustal and Aleutian Subduction Zone Seismicity: An Investigation of the Correlation between *b* Value and *k–M* Slope. *Pure Appl. Geophys.* **2018**, *175*, 4241–4252. [CrossRef]
35. Telesca, L.; Chen, C.-C.; Lovallo, M. Investigating the Relationship Between Seismological and Topological Properties of Seismicity in Italy and Taiwan. *Pure Appl. Geophys.* **2020**, *177*, 4119–4126. [CrossRef]
36. Telesca, L.; Lovallo, M.; Ramirez-Rojas, A.; Flores-Marquez, L. Relationship between the Frequency magnitude distri-bution and the visibility graph in the synthetic seismicity generated by a simple stick-slip system with asperities. *PLoS ONE* **2014**, *9*, e106233. [CrossRef]
37. Perez-Oregon, J.; Lovallo, M.; Telesca, L. Visibility graph analysis of synthetic earthquakes generated by the Olami–Feder–Christensen spring-block model. *Chaos Interdiscip. J. Nonlinear Sci.* **2020**, *30*, 093111. [CrossRef]
38. Pérez-Campos, X.; Kim, Y.; Husker, A.; Davis, P.M.; Clayton, R.W.; Iglesias, A.; Pacheco, J.F.; Singh, S.K.; Manea, V.C.; Gurnis, M. Horizontal subduction and truncation of the Cocos Plate beneath central Mexico. *Geophys. Res. Lett.* **2008**, *35*, L18303. [CrossRef]
39. Carciumaru, D.; Ortega, R.; Castellanos, J.C.; Huesca-Pérez, E. Crustal Characteristics in the Subduction Zone of Mexico: Implication of the Tectonostratigraphic Terranes on Slab Tearing. *Seism. Res. Lett.* **2020**, *91*, 1781–1793. [CrossRef]
40. Manea, V.; Manea, M.; Ferrari, L.; Orozco-Esquivel, T.; Valenzuela, R.; Husker, A.; Kostoglodov, V. A review of the geodynamic evolution of flat slab subduction in Mexico, Peru, and Chile. *Tectonophysics* **2017**, *695* (Suppl. C), 27–52. [CrossRef]
41. Bandy, W.L.; Hilde, T.W.C.; Yan, C.Y. The Rivera–Cocos plate boundary: Implications for Rivera Cocos relative motion and plate fragmentation. In *Cenozoic Tectonics and Volcanism of Mexico*; Delgado-Granados, H., Aguirre-Diaz, G., Stock, J.M., Eds.; Special Paper Geological Society of America: Boulder, CO, USA, 2000; Volume 334, pp. 1–28.
42. Dougherty, S.L.; Clayton, R.W.; Helmberger, D.V. Seismic structure in central Mexico: Implications for fragmentation of the subducted Cocos plate. *J. Geophys. Res. Solid Earth* **2012**, *117*, B09316. [CrossRef]
43. Stubailo, I.; Beghein, C.; Davis, P.M. Structure and anisotropy of the Mexico subduction zone based on Rayleigh-wave analysis and implications for the geometry of the Trans-Mexican Volcanic Belt. *J. Geophys. Res. Atmos.* **2012**, *117*, B05303. [CrossRef]
44. Dougherty, S.L.; Clayton, R.W. Seismicity and structure in central Mexico: Evidence for a possible slab tear in the South Cocos plate. *J. Geophys. Res. Solid Earth* **2014**, *119*, 3424–3447. [CrossRef]
45. Castellanos, J.C.; Clayton, R.W.; Pérez-Campos, X. Imaging the Eastern Trans-Mexican Volcanic Belt With Ambient Seismic Noise: Evidence for a Slab Tear. *J. Geophys. Res. Solid Earth* **2018**, *123*, 7741–7759. [CrossRef]

46. Letort, J.; Retailleau, L.; Boue, P.; Radiguet, M.; Gardonio, B.; Cotton, F.; Campillo, M. Lateral variations of the Guerrero–Oaxaca subduction zone (Mexico) derived from weak seismicity (Mb3.5+) detected on a single array at teleseismic distance. *Geophys. J. Int.* **2018**, *213*, 1002–1012. [CrossRef]
47. Ramírez-Herrera, M.T.; Kostoglodov, V.; Urrutia-Fucugauchi, J. Overview of Recent Coastal Tectonic Deformation in the Mexican Subduction Zone. *Pure Appl. Geophys.* **2011**, *168*, 1415–1433. [CrossRef]
48. Singh, S.K.; Reinoso, E.; Arroyo, D.; Ordaz, M.; Cruz-Atienza, V.; Pérez-Campos, X.; Iglesias, A.; Hjörleifsdóttir, V. Deadly Intraslab Mexico Earthquake of 19 September 2017 (Mw 7.1): Ground Motion and Damage Pattern in Mexico City. *Seism. Res. Lett.* **2018**, *89*, 2193–2203. [CrossRef]
49. Golt and Flores, 1997, Real-Time Earthquake Early Warning and Public Policy: A Report on Mexico City's Sistema de Alerta Sismica. *Seismol. Res. Lett.* **1997**, *68*, 727–733. [CrossRef]
50. Rosas, V. Estudio de microzonificación sísmica para la ciudad de Uruapan, Michoacán, aplicando el método SPAC. Bachelor's Thesis, Universidad Michoacana de San Nicolás de Hidalgo, Fac. de Ingeniería Civil, Michoacán, Mexico, 2002.
51. Mendoza-Ponce, A.; Figueroa-Soto, A.; Soria-Caballero, D.; Garduño-Monroy, V.H. Active faults sources for the Pátzcuaro–Acambay fault system (Mexico): Fractal analysis of slip rates and magnitudes Mw estimated from fault length. *Nat. Hazards Earth Syst. Sci.* **2018**, *18*, 3121–3135. [CrossRef]
52. Ramírez-Rojas, A.; Flores-Márquez, E.L. Nonlinear Statistical Features of the Seismicity in the Subduction Zone of Tehuantepec Isthmus, Southern México. *Entropy* **2022**, *24*, 480. [CrossRef] [PubMed]
53. Müller, R.D.; Seton, M.; Zahirovic, S.; Williams, S.E.; Matthews, K.J.; Wright, N.M.; Shephard, G.E.; Maloney, K.T.; Barnett-Moore, N.; Hosseinpour, M.; et al. Ocean Basin Evolution and Global-Scale Plate Reorganization Events Since Pangea Breakup. *Annu. Rev. Earth Planet. Sci.* **2016**, *44*, 107–138. [CrossRef]
54. Mandujano-Velazquez, J.J.; Keppie, J.D. Middle Miocene Chiapas fold and thrust belt of Mexico: A result of collision of the Tehuantepec Transform/Ridge with the Middle America Trench. *Geol. Soc. Lond. Spéc. Publ.* **2009**, *327*, 55–69. [CrossRef]
55. Keppie, J.D.; Morán-Zenteno, D.J. Tectonic Implications of Alternative Cenozoic Reconstructions for Southern Mexico and the Chortis Block. *Int. Geol. Rev.* **2005**, *47*, 473–491. [CrossRef]
56. Gutenberg, B.; Richter, C.F. *Seismicity of the Earth and Associated Phenomena*, 2nd ed.; Princeton University Press: Princeton, NJ, USA, 1954.
57. Perez-Oregon, J.; Muñoz-Diosdado, A.; Rudolf-Navarro, A.H.; Guzmán-Sáenz, A.; Angulo-Brown, F. On the possible correlation between the Gutenberg-Richter parameters of the frequency-magnitude relationship. *J. Seism.* **2018**, *22*, 1025–1035. [CrossRef]
58. Flores-Márquez, E.L.; Ramírez-Rojas, A.; Perez-Oregon, J.; Sarlis, N.V.; Skordas, E.S.; Varotsos, P.A. Natural Time Analysis of Seismicity within the Mexican Flat Slab before the M7.1 Earthquake on 19 September 2017. *Entropy* **2020**, *22*, 730. [CrossRef] [PubMed]


**Disclaimer/Publisher's Note:** The statements, opinions and data contained in all publications are solely those of the individual author(s) and contributor(s) and not of MDPI and/or the editor(s). MDPI and/or the editor(s) disclaim responsibility for any injury to people or property resulting from any ideas, methods, instructions or products referred to in the content.

*entropy*

*Article*

# Frequency Seismic Response for EEWS Testing on Uniaxial Shaking Table

**Codrin Donciu, Elena Serea * and Marinel Costel Temneanu**

Faculty of Electrical Engineering, "Gheorghe Asachi" Technical University of Iași, 700050 Iași, Romania; cdonciu@tuiasi.ro (C.D.)
* Correspondence: edanila@tuiasi.ro; Tel.: +40-740481758

**Abstract:** Earthquake early warning systems are used as important tools in earthquake risk management, providing timely information to residents and both public and private emergency managers. By doing this, the potential impact of large magnitude seismic events is significantly reduced. These systems use seismic sensors in order to acquire real-time data for the weaker but fast moving P wave (usually the first 3–5 s of the earthquake) and specific algorithms to predict the magnitude and the arrival time of the slower but more destructive surface waves. Most of these projection algorithms make use only of the vertical component of the acceleration and need extensive training in earthquake simulators in order to enhance their performance. Therefore, a low-inertial-mass uniaxial shaking table is proposed and analyzed in terms of frequency response in this paper, providing an effective cost/control ratio and high daily duty cycle. Furthermore, with the large variety of prediction algorithms, which use different frequency ranges, a new concept of selective frequency band error is also introduced and discussed in this paper as being a necessary tool for the final assessment of magnitude estimation algorithm error.

**Keywords:** earthquake; seismic response; frequency; shaking table

**Citation:** Donciu, C.; Serea, E.; Temneanu, M.C. Frequency Seismic Response for EEWS Testing on Uniaxial Shaking Table. *Entropy* **2023**, *25*, 655. https://doi.org/10.3390/e25040655

Academic Editor: Georgios Michas

Received: 17 March 2023
Revised: 12 April 2023
Accepted: 12 April 2023
Published: 14 April 2023

**Copyright:** © 2023 by the authors. Licensee MDPI, Basel, Switzerland. This article is an open access article distributed under the terms and conditions of the Creative Commons Attribution (CC BY) license (https://creativecommons.org/licenses/by/4.0/).

## 1. Introduction

The lithospheric plates of Earth are continuously set in motion by their thermal energy through the convection currents inside the mantle, generated by the difference between the high temperatures of the nucleus and the low temperatures of the mantle. When the plates slide past one another, this energy is internally stored and when a fault interaction or rupture zone occurs, the energy is suddenly released as radiated energy, fracture energy and thermal energy [1]. The share of the potential energy that radiates propagates through the crust outward in all directions in the form of seismic (elastic) waves. Only rapidly slipping (or seismic) events generate elastic waves, while the energy released during slow-slip events (aseismic events) would mostly dissipate [2]. The radiated energy is ordinarily calculated by measuring the energy flux at the point where the wave field of an earthquake is recorded, and from that inferring the total energy flux in all directions from the epicenter [3]. For spectral models and ground motion prediction, it is assumed that earthquakes radiate about half of the available strain energy into the surrounding medium [4]. Radiation efficiency, as the ratio between seismic radiated energy and the available strain energy, is correlated with fault geometry and symmetry and with rupture velocity.

Based on the energy released, an earthquake is quantitatively mainly inferred after moment magnitude $M_W$ (logarithmic scale). One magnitude corresponds to several distributed intensities, depending on the distance from the hypocenter: as distance decreases, the intensity increases [5]. Being assessed as a scattered seismic effect, earthquake intensity is a qualitative measure, which together with the direction of the fault rupture propagation and with the surface geology gives the weight of the potential damage [6]. Co-seismic, post-seismic and inter-seismic impact regarding crustal deformations, structural damage [7] and humming to the endangerment of any integrity is contingent on the approach to disaster preparedness. For the most exposed areas with the highest density and vulnerability to earthquakes, it is necessary to ensure a good resilience of the assets [8], namely non-structural or structural systems, of which the most exposed are the buildings. Each presents a degree of seismic fragility that can be effectively predicted using an analytical–mechanical-based procedure, revealing the seismic behavior of building portfolios on a typological base [9]. The seismic response can be accurately predicted for reinforced concrete buildings as well, with different degrees of diaphragm flexibility, and for non-structural components [10], facilitating the creation of a system model that is as realistic as possible in order to include the seismic effects in a local resilience plan. This new concept of resilience comprises the capabilities of a complex system, composed of interacting physical and social components, to withstand external stress and return to a state of equilibrium or bounce forward to improved new states of equilibrium [11]. These capabilities are bordered by the disaster response and community preparedness in case of an earthquake. Whereas some capabilities must be mobilized subsequently to support disaster responses, others need to serve a particular response situation continuously and accurately. A mechanism that upholds this type of capability is the earthquake early warning system (EEWS), an integrated architecture of hazard monitoring, forecasting and prediction, disaster risk assessment and communication [12]. EEWSs are critical not only in proximate areas but also far from the epicenter because earthquake magnitude and location cannot be precisely detected in the first instance [13]. Typically, earthquake parameters are estimated based on simultaneous real-time waveform records from three to six stations [14]. The detection is based on the primary P wave propagation and the EEWS trigger is settled after computing the secondary S wave travel time, proportional to the distance from the epicenter. The warning window can range from a few seconds to tens of seconds (Table 1), depending on the size of the earthquake and on the number and type of sensors in the EEWS architecture [15]. S wave arrival time is defined as the duration in which the S wave travels from the first to the farthest observation sensor from the epicenter. The arrival values are computed using point source algorithms (based on ground motion prediction-specific equations), which retrieve information on the P wave from at least three stations in the vicinity of the epicenter. The prediction algorithms neglect the tectonic stratigraphy or the depth of the earthquake and serve as an approximate guide because the shaking intensity at any specific location can vary compared to the average shaking at that distance according to ground motion prediction equations; thus, the strongest shaking can actually be felt later than the S wave arrival time [16].

**Table 1.** Example of EEWS appraisal for seismic parameters (after [17]).

| Earthquake Moment Magnitude ($M_w$) | Approximate Fault Length (km) | Maximum Epicentral Distance Where Earthquake Is Expected (km) | S Wave Arrival at Maximum Distance Where Earthquake Is Expected (s) | Potential Damage |
|---|---|---|---|---|
| 5 | 1 | 10 | 4 | insignificant |
| 5.9 | 6 | 40 | 10 | minor |
| 6 | 50 | 200 | 40 | moderate |
| 6.9 (crustal) | 400 | 700 | 200 | moderate to major |
| 7 (subduction) | 1000 | 1000 | 300 | major |

Seismic sensors are the most critical elements of seismographs or EEWS, as they must measure dynamic ground motion higher than the ground noise distinctly from human-induced artificial noise. Their frequency band must cover 0.01 to 100 Hz and ground motions from 1 nm to 10 mm [18]. From classical seismic receivers placed at or close to a designated area [19] to downhole receivers immersed at typical depths [20] or to latest optical fiber network enhancing [21], seismic sensors detect certain ground vibrations with the highest precision and process the information in real-time to trigger an alert. Lately, seismic recording instruments have gained particular importance in assisting the monitoring processes of other natural phenomena, such as rainfall episodes or thunderstorms [22], or for target detection and activity recognition [23]. Seismic nodal sensors differ in terms of performance, ease of deployment, size, power consumption, data format and storage and battery lives [24], characteristics that are reflected in their cost and further in the spatial distribution and density of the earthquake monitoring networks.

Since both the sensor's technology (range, resolution, noise density and frequency response) and prediction algorithms (including data preprocessing) are continuously evolving, researchers need an enormous volume of training data in order to match the seismic waveform parameters to the real-time (sensor-specific) acquired data and projection algorithm's error.

In order to fulfill the increasing demand of easy-to-use but also precise earthquake simulators, a low-inertial-mass (tailored to the purpose of carrying only the seismic sensor) uniaxial shaking table (because most algorithms use only the vertical component of the acceleration) is proposed in this paper.

By maintaining the hardware arrangement to a minimum, the pre-experimental phase is lowered and therefore the proposed solution provides an effective cost/control ratio and high daily duty cycle.

Its performances, in terms of frequency response, are compared with those reported in the literature. Furthermore, a new concept of selective frequency band error is also introduced and discussed in this paper, as being a necessary tool for the final assessment of magnitude estimation error.

## 2. Laboratory Small-Scale Shaking Table Design and Control

An earthquake laboratory reproduction is better realized with six-degrees-of-freedom vibrating tables, because structures are generally excited to three orthogonal components of ground motion: two orthogonal horizontal components in the principal directions of structures (chosen for convenience of understanding and analysis) and one vertical component of the earthquake [25]. As EEWS use only the vertical component for estimating earthquake parameters (the maximum acceleration is obtained from the vertical component of the waveform [26]), the designed shaking table for EEWS sensor testing is unidirectional (single axis and single degree of freedom).

The performance evaluation of unidirectional shaking tables has been issued in much research, addressing the interaction effects with different loading configurations [27], the accuracy of a small-scale low-cost electrodynamic shaker [28] and the bearing capacity issues when reproducing sinusoidal ground motion with frequencies up to 10 Hz [29], or revealing synchronization and tracking control advances [30]. The shaking table test outlasts the seismic test closest to the real seismic response, exhaustively appraising the impact of damping natural vibration frequency in all types of investigations, from laboratory validations to practical studies for settling the dynamic behavior of a dry-joint masonry arch [31] or to evaluate the effectiveness of coupling between a 2-degree-of-freedom shear-type frame system and a rigid block [32]. Intended for simulating all types of vibrations generated by sine waves, chirp signals or scaled earthquakes, the frequency range of small electrodynamic tables is 0~10 kHz, and for wide size tables for large process models it is 0~2 kHz. Sine wave and random wave tests can be realized; the acceleration waveform distortion is small, but executing large output and large displacement tests remains difficult [33]. To determine the essential characteristics of the seismic analysis for coupled systems or nonstructural components, a small-scale vibration table with a control algorithm (PID, adaptive, neural network or fuzzy control) provides an alternative cost-effective research method [34]. Reducing the vulnerability of non-structural components (such as EEWS) is crucial for achieving full earthquake resilience and for avoiding the loss of functionality and downtime. Although EEWSs are very efficient in reducing damage and have a reliable response thanks to quality control tests, they often show a brittle behavior that may strongly reduce global robustness in the case of extreme and rare seismic actions.

This highlights the need for additional studies with the purpose of the exact correlation of safety coefficients to required reliability levels [8].

The shaking table model from Figure 1, designed to test EEWS sensors, consists of three functional parts, with the following subcomponents:

- Mechanical part: (1) a standard steel plate, (2) a slide cart (linear rail system) with smooth surfaces to minimize the frictional resistance and auxiliary fixing elements.
- Electrodynamic part: (3) an actuator able to drive the mechanical component (a DC micromotor and an optical encoder), supplied through a power amplifier (4).
- Controlling part: computer with LabVIEW interface for generating input signals and (5) a data acquisition board through which command and data acquisition are executed.

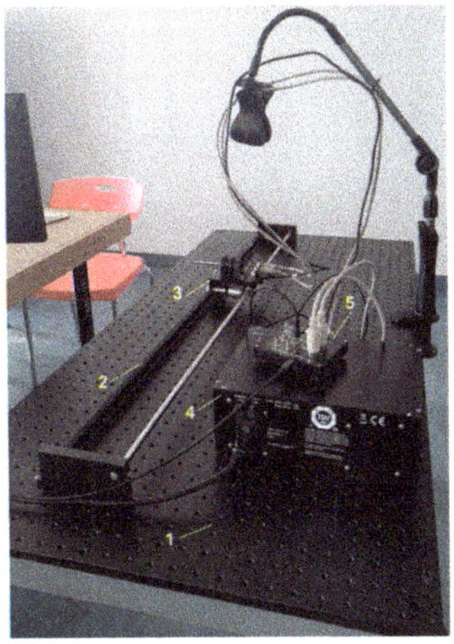


**Figure 1.** Small-scale unidirectional vibration table prototype.

Usually, to control table motions in order to reproduce seismic vibration, a PID algorithm for the shake table response is used [35]. Under the PID scheme the controller responds to the difference between the table command and feedback displacements (the error). This value is continously calculated and the displacement is corrected by the controller based on predefined ratios of proportional, integral, and derivative of the error [36]. The implemented control loop (Figure 2) operates in control displacement mode and to optimize the signal reproduction by controlling feedback parameters, which are continuously adjusted to increase the precision in acceleration.

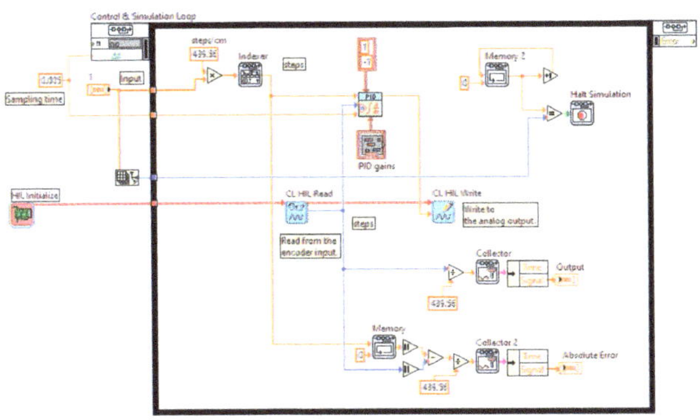


**Figure 2.** Block diagram of a unidirectional shaking table control with displacement feedback.

## 3. Test Results

To establish stability contingency, frequency domain analysis was performed for 12 earthquake sequences, as listed in Table 2.

**Table 2.** Seismic parameters used as the reference for testing.

| No. | Name | Year | Station Name | Magnitude (Mw) | Rjb (km) | Rrup (km) | Sampling Period (s) |
|---|---|---|---|---|---|---|---|
| 1 | Chi-Chi_Taiwan-02 | 1999 | CHY065 | 5.9 | 125.26 | 125.89 | 0.005 |
| 2 | Chi-Chi_Taiwan-02 | 1999 | CHY067 | 5.9 | 126.39 | 126.56 | 0.004 |
| 3 | Chi-Chi_Taiwan-02 | 1999 | CHY071 | 5.9 | 122.02 | 122.19 | 0.005 |
| 4 | Parkfield-02_CA | 2004 | Hollister—Airport Bldg #3 | 6 | 121.51 | 121.54 | 0.005 |
| 5 | Parkfield-02_CA | 2004 | Salinas—County Hospital Gnds | 6 | 120.74 | 120.79 | 0.005 |
| 6 | Chi-Chi_Taiwan-03 | 1999 | ILA006 | 6.2 | 129.11 | 129.4 | 0.004 |
| 7 | Chi-Chi_Taiwan-03 | 1999 | ILA007 | 6.2 | 127.25 | 127.54 | 0.004 |
| 8 | San Fernando | 1971 | Isabella Dam (Aux Abut) | 6.61 | 130 | 130.98 | 0.005 |
| 9 | San Fernando | 1971 | Bakersfield—Harvey Aud | 6.61 | 111.88 | 113.02 | 0.005 |
| 10 | El Alamo | 1956 | El Centro Array #9 | 6.8 | 121 | 121.7 | 0.005 |
| 11 | Hector Mine | 1999 | Bombay Beach Fire Station | 7.13 | 120.69 | 120.69 | 0.005 |
| 12 | Lander | 1992 | Covina—W Badillo | 7.28 | 128.06 | 128.06 | 0.005 |

These reference sequences, obtained from the PEER Ground Motion Database and available at https://ngawest2.berkeley.edu/, accessed on 17 October 2022, were selected to fulfill the following conditions:

- High enough magnitude because most of the EEWSs trigger the alarm at magnitudes higher than six;
- Different enough spectral content because the simulator's performances are evaluated in the frequency domain.

The amplitude of a seismic wave recorded in a site depends on two parameters: the magnitude of the earthquake measured in the epicenter and Rjb—the surface distance from the epicenter to the site. In order to compare the seismic waves, they were chosen from the database records with constant Rjb (as can be seen in Table 2, between 111 and 130 km) and variable magnitude in the coverage area of the EEWS (>5.9 Mw). Moreover, to verify the reproducibility of the data, a series of earthquakes with similar Mw and Rjb, but recorded at different stations, were selected: for Chi-Chi Taiwan-02, the data from three sites (stations) were processed, and for Parkfield-02 CA, Chi-Chi Taiwan-03 and San Fernando, the data from two different sites (stations) were processed. For the purposes of being informative, and for observing the depth at which the earthquake occurred, the parameter Rrup is given together with Rjb.

The experimental information flow from Figure 3 introduces the uniaxial shaking table as a displacement tracking system feed with the time domain displacement waveforms, as selected from the PEER database. Every seismic event from the sequence of 12 is treated individually, as the sapling parameters differ (from 0.004 s to 0.005 s). Both reference and output (slide cart) displacement are analyzed in the frequency domain using fast Fourier transform to obtain the frequency spectrum.

The performance in the frequency domain of the shaking table is further evaluated from two perspectives:

1. Locally, for each individual frequency magnitude, in terms of magnitude absolute error.
2. Globally, for the entire spectrum, in terms of root mean square error and normalized root mean square error, as to be compared with other reported results.

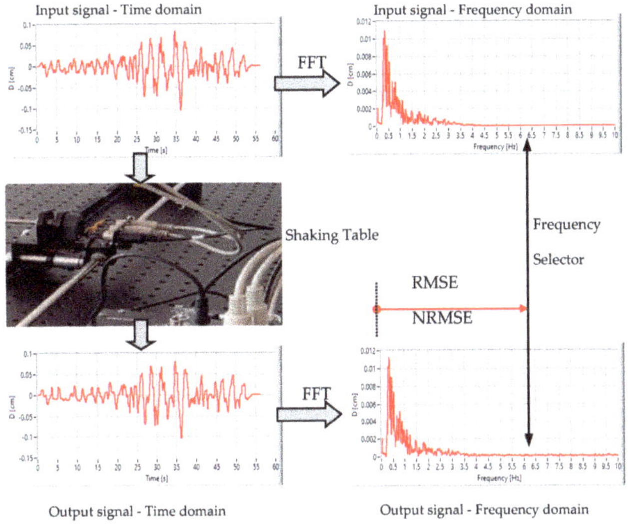


**Figure 3.** Experimental information flow: left panels: reference and output displacement waveforms and right panels: frequency spectrum with the error calculation point indicator.

The analysis of the results starts with the evaluation of the absolute error, a type of primary information about the system's performances.

In Figure 4, spectrograms of the reference signal (left column) and the absolute error for each spectral frequency (the difference between the amplitude of the same $i$ spectral component in the input $I_i$ and output $O_i$ spectrograms) are depicted. The evaluation range is from 0 to 10 Hz (the highest reachable frequency that corresponds to the natural frequency), as this is what typical seismographs routinely record [37].

From the graphs displayed in Figure 4, it can be seen that the absolute errors of the proposed shaking table model, calculated as the difference between the reference and the output frequency spectrums, are very small, with a band amplitude of 0.001 cm for small events and 0.0032 for large events.

However, the individual absolute errors do not give a relevant overview of the system's performances, with the universally recognized indicator for analyzing clustered data being the root mean square error (*RMSE*). Therefore, in Section 4, the results are discussed using this approach for comparing the model fitting for 12 seismic response variables and for assessing residual variance.

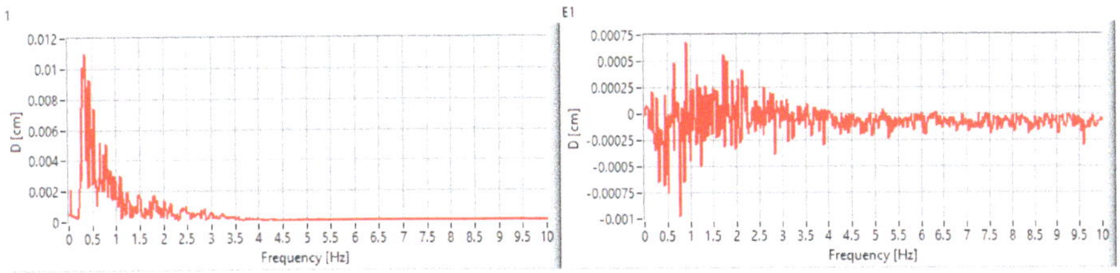


**Figure 4.** *Cont.*

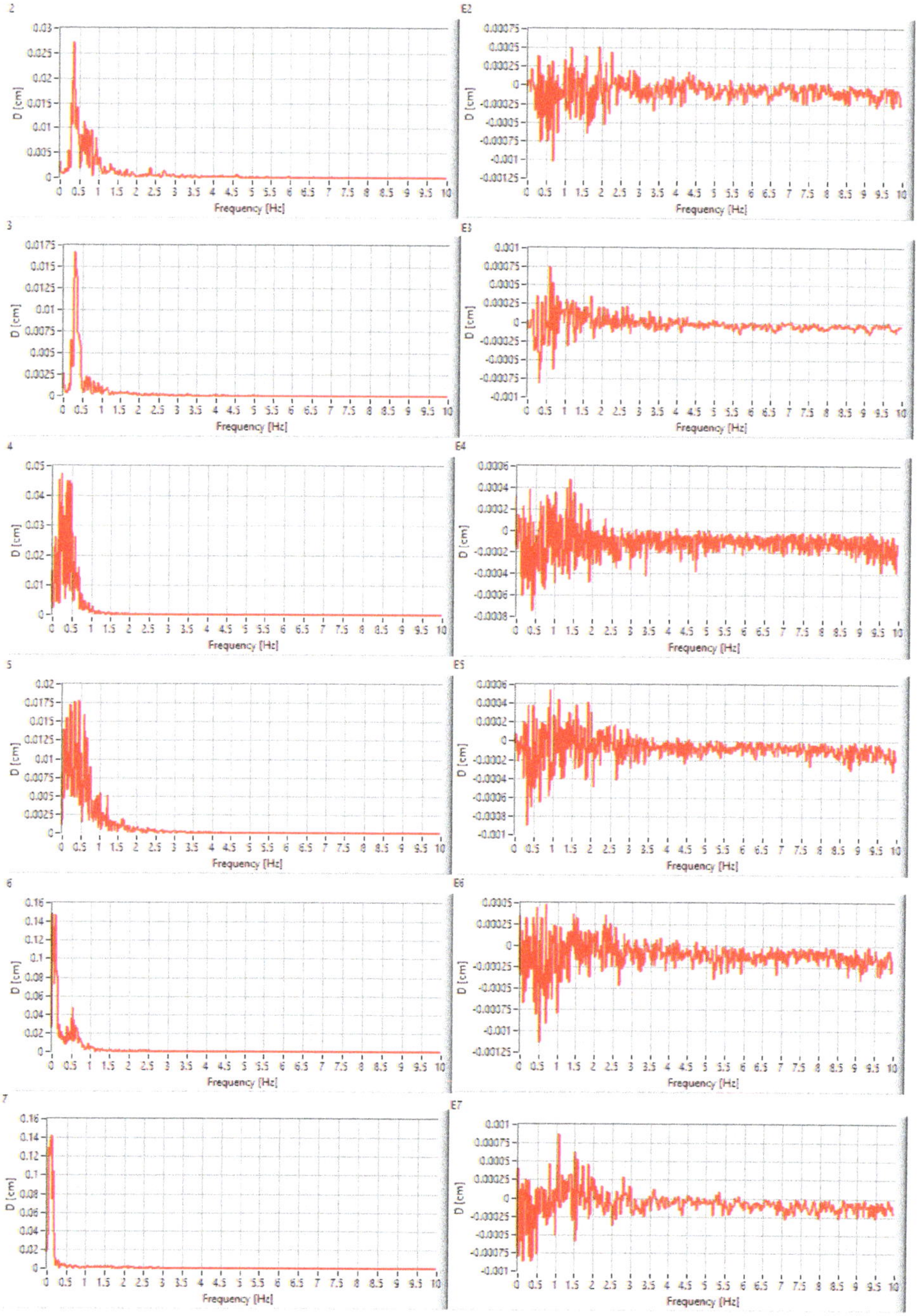


**Figure 4.** *Cont.*

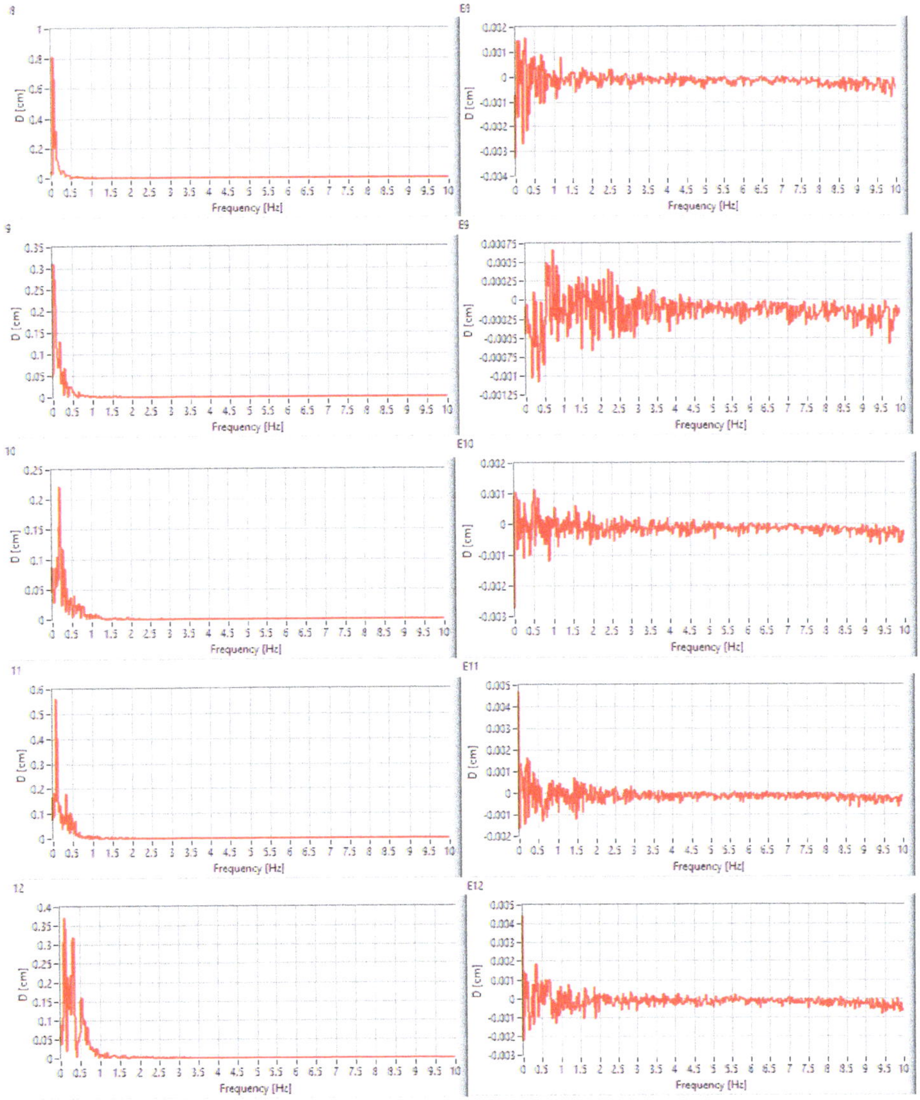

**Figure 4.** Shaking table frequency seismic response to unidirectional input ground motions ((**1–12**): the reference earthquake sequence; (**E1–E12**): the associated displacement absolute errors).

## 4. Discussion

An important aspect of the error metrics used for simulators' evaluation is their capability to discriminate among simulation results. Giving higher weighting to the unfavorable conditions, the root mean square error (*RMSE*) is better at revealing performance differences, being the widely used standard statistical metric that measures performance in natural phenomena studies [38]. For a specific spectrogram, with $n$ points in the frequency domain, *RMSE* is defined as

$$RMSE = \sqrt{\frac{\sum_{i=0}^{n}(I_i - O_i)^2}{n}}$$

with $I_i$ being the reference seismic frequency from the database (amplitude of spectral component $i$) and $O_i$ being the recorded seismic frequency from the shaking table (amplitude of spectral component $i$).

However, the use of this global performance indicator can lead to an erroneous assessment of the quality of the prediction algorithm.

As pointed out before, all prediction algorithms use extensive sets of experimental data, but these data are preprocessed prior to feeding the computational unit. The preprocessing includes one or more bandpass filtering stage, an essential but an unstandardized process since the high cutoff frequency varies from 3 to 10 Hz. It is important to use a performance indicator tailored to the specific bandpass of each algorithm if its quality is to be assessed.

Therefore, we will discuss the performance of the proposed shaking table in terms of selective frequency band error, with the process being depicted in Figure 5.

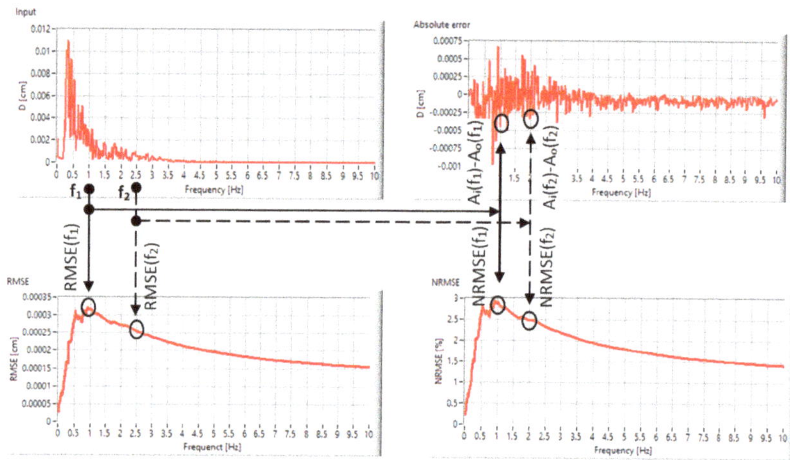


**Figure 5.** Error evaluation process using selective band frequency for two arbitrarily selected frequencies: *RMSE* (**left** panels) and *NRMSE* (**right** panels).

For different frequencies, different values of *RMSE* are computed using only the components lower than the frequency of interest:

$$RMSE_f = \sqrt{\frac{\sum_{i=0}^{\frac{f}{\Delta f}}(I_i - O_i)^2}{\frac{f}{\Delta f}}}[\mathrm{cm}]$$

where $f$ = the frequency of interest, $\Delta f$ = the resolution in frequency and $f/\Delta f$ = sample size, with the number of measuring points in the frequency spectrum of [0, f].

Normalization was further included to assess the simulator's performance, as it is sensitive to the amplitude and frequency errors and provides better identification of the output fitting correctness, similar to reference [39]. The normalized root mean square error was calculated using the following equation:

$$NRMSE = \frac{RMSE}{\max(I_i) - \min(I_i)}[\%]$$

*RMSE* and *NRMSE* variation are presented in Table 3, for four different band frequencies

**Table 3.** *RMSE* and *NRMSE* values for different band frequencies.

| Cutoff Frequency | *RMSE* (Cm) Min. | *RMSE* (cm) Max. | *NRMSE* (%) Min. | *NRMSE* (%) Max. |
|---|---|---|---|---|
| 3 Hz | 0.00021 | 0.00061 | 0.09 | 2.18 |
| 5 Hz | 0.00019 | 0.00047 | 0.058 | 1.75 |
| 7 Hz | 0.00017 | 0.00042 | 0.05 | 1.55 |
| 10 Hz | 0.00016 | 0.00038 | 0.044 | 1.48 |

The obtained results reveal two important aspects:

- Global *RMSEs*, computed for all spectral components to be compared with other reported results, have lower values than those reported in recent papers [39–42]. Even if simple control algorithms have been used, the minimization of the hardware arrangement, and the low weight of the moving parts (only 390 g for the cart and accelerometer), yields very good seismic-waveform-tracking characteristics. Additionally, the absence of some heavy actuators makes the pre-experimental phase very short, hence ensuring a high daily duty cycle.
- For lower cutoff frequencies, both *RMSE* and *NRMSE* increase their values. The assessment of the prediction algorithm accuracy by using the standard (full frequency domain) *RMSE* can lead to unrealistic expectations.

## 5. Conclusions

Research into earthquake simulation for particular vibration effects on structural and non-structural elements is mainly performed using a shaking table. It can realistically simulate the state of stress or deficiency of the tested system under a seismic action, but its dynamic behavior is poor because of the heavy moving parts and actuators. In order to perform seismic sensor testing, as required by EEWS algorithm development and optimization, a tailored uniaxial shaking table was developed and analyzed in terms of dynamic response.

Meeting a set of favorable characteristics, from minimal architecture with low-weight moving parts to a simple control algorithm, the proposed shaking table yield is effective in tracking fidelity and the cost/control ratio. Moreover, because of the absence of some heavy actuators, the pre-experimental phase is very short, ensuring a high daily duty cycle.

When analyzing the system's frequency domain performances, it is important to use an indicator tailored to the specific bandpass of each EEWS prediction algorithm. Therefore, a new frequency domain error evaluation tool is used and defined as being appropriate for contributing to a more accurate assessment of the quality of the prediction algorithms.

**Author Contributions:** Conceptualization, C.D. and M.C.T.; methodology, C.D. and M.C.T.; software, C.D.; validation, C.D., M.C.T. and E.S.; formal analysis, C.D.; investigation, C.D., M.C.T. and E.S.; resources, C.D.; data curation, C.D. and M.C.T.; writing—original draft preparation, C.D. and E.S.; writing—review and editing, E.S. and M.C.T.; visualization, C.D. and E.S.; supervision, C.D. and E.S.; project administration, M.C.T.; funding acquisition, M.C.T. All authors have read and agreed to the published version of the manuscript.

**Funding:** This research was funded by European Regional Development Fund, grant number 7227/19.11.2021 (SMIS code 137414)—"Seismic warning system with automatic unlocking of entrance doors with interphone". The APC was funded by "Gheorghe Asachi" Technical University of Iași, România.

**Data Availability Statement:** Not applicable.

**Conflicts of Interest:** The authors declare no conflict of interest.

## References


1. Kanamori, H.; Rivera, L. Energy partitioning during an earthquake. *Geophys. Monogr. Ser.* **2006**, *170*, 3–13.
2. Sheibani, F.; Hager, B.H. Seismic Efficiency vs. Fracability; Effects of Mechanical Properties on Radiated Elastic Waves with Application to Hydraulic-Fracturing-Induced Microseismicity. In Proceedings of the AAPG Annual Convention and Exhibition, Houston, TX, USA, 5 April 2017.
3. Beroza, G.C.; Kanamori, H. Earthquake seismology: An introduction and overview. In *Reference Module in Earth Systems and Environmental Sciences. Treatise on Geophysics*, 2nd ed.; Schubert, A., Ed.; Elsevier: Amsterdam, The Netherlands, 2015; Volume 4, pp. 1–50.
4. Ji, C.; Archuleta, R.J.; Wang, Y. Variability of Spectral Estimates of Stress Drop Reconciled by Radiated Energy. *Bull. Seismol. Soc. Am.* **2022**, *112*, 1871–1885. [CrossRef]
5. Das, R.; Joshi, S.; Meneses, C.; Biswas, T. Advanced Unified Earthquake Catalog for North East India. *Appl. Sci.* **2023**, *13*, 2812.
6. Ammon, C.J.; Velasco, A.A.; Lay, T.; Wallace, T.C. Rupture directivity. In *Foundations of Modern Global Seismology*, 2nd ed.; Academic Press, Ed.; Elsevier: London, UK, 2020; pp. 496–497.
7. Zhu, M.; Chen, F.; Zhou, W.; Lin, H.; Parcharidis, I.; Luo, J. Two-Dimensional InSAR Monitoring of the Co-and Post-Seismic Ground Deformation of the 2021 Mw 5.9 Arkalochori (Greece) Earthquake and Its Impact on the Deformations of the Heraklion City Wall Relic. *Remote Sens.* **2022**, *14*, 5212. [CrossRef]
8. Freddi, F.; Galasso, C.; Cremen, G.; Dall'Asta, A.; Di Sarno, L.; Giaralis, A.; Woo, G. Innovations in earthquake risk reduction for resilience: Recent advances and challenges. *Int. J. Disaster Risk Reduct.* **2021**, *60*, 102267. [CrossRef]
9. Ruggieri, S.; Calò, M.; Cardellicchio, A.; Uva, G. Analytical-mechanical based framework for seismic overall fragility analysis of existing RC buildings in town compartments. *Bull. Earthq. Eng.* **2022**, *20*, 8179–8216. [CrossRef]
10. Ruggieri, S.; Vukobratović, V. Acceleration demands in single-storey RC buildings with flexible diaphragms. *Eng. Struct.* **2023**, *275*, 115276. [CrossRef]
11. Koren, D.; Rus, K. The Potential of Open Space for Enhancing Urban Seismic Resilience: A literature Review. *Sustainability* **2019**, *11*, 5942. [CrossRef]
12. Esposito, M.; Palma, L.; Belli, A.; Sabbatini, L.; Pierleoni, P. Recent Advances in Internet of Things Solutions for Early Warning Systems: A Review. *Sensors* **2022**, *22*, 2124. [CrossRef]
13. Khan, I.; Kwon, Y.W. P-Detector: Real-Time P-Wave Detection in a Seismic Waveform Recorded on a Low-Cost MEMS Accelerometer Using Deep Learning. *IEEE Geosci. Remote Sens. Lett.* **2022**, *19*, 3006305. [CrossRef]
14. Kuyuk, H.S.; Susumu, O. Real-time classification of earthquake using deep learning. *Procedia Comput. Sci.* **2018**, *140*, 298–305. [CrossRef]
15. Prasanna, R.; Chandrakumar, C.; Nandana, R.; Holden, C.; Punchihewa, A.; Becker, J.S.; Jeong, S.; Liyanage, N.; Ravishan, D.; Sampath, R.; et al. "Saving Precious Seconds"—A Novel Approach to Implementing a Low-Cost Earthquake Early Warning System with Node-Level Detection and Alert Generation. *Informatics* **2022**, *9*, 25. [CrossRef]
16. Allen, R.M.; Melgar, D. Earthquake Early Warning: Advances, Scientific Challenges, and Societal Needs. *Annu. Rev. Earth Planet Sci.* **2019**, *47*, 361–388. [CrossRef]
17. Beltramone, L.; Gomes, R.C. Earthquake Early Warning Systems as an Asset Risk Management Tool. *CivilEng* **2021**, *2*, 120–133. [CrossRef]
18. Havskov, J.; Alguacil, G. Seismic sensors. In *Instrumentation in Earthquake Seismology. Modern Approaches in Geophysics*; Havskov, J., Alguacil, G., Eds.; Springer: Dordrecht, The Netherlands, 2004; Volume 22, pp. 11–76.
19. Choudhary, P.; Goel, N.; Saini, M. A Survey on Seismic Sensor based Target Detection, Localization, Identification, and Activity Recognition. *ACM Comput. Surv.* **2023**, *55*, 223. [CrossRef]
20. Lellouch, A.; Biondi, B.L. Seismic Applications of Downhole DAS. *Sensors* **2021**, *21*, 2897. [CrossRef] [PubMed]
21. Bogris, A.; Nikas, T.; Simos, C.; Simos, I.; Lentas, K.; Melis, N.S.; Chochliouros, I. Sensitive seismic sensors based on microwave frequency fiber interferometry in commercially deployed cables. *Sci. Rep.* **2022**, *12*, 14000. [CrossRef]
22. Diaz, J.; Ruiz, M.; Udina, M. Monitoring storm evolution using a high-density seismic network. *Sci. Rep.* **2023**, *13*, 1853. [CrossRef]
23. Saadia, B.; Fotopoulos, G. Characterizing Ambient Seismic Noise in an Urban Park Environment. *Sensors* **2023**, *23*, 2446. [CrossRef]
24. Anthony, R.E.; Ringler, A.T.; Wilson, D.C.; Wolin, E. Do Low-Cost Seismographs Perform Well Enough for Your Network? An Overview of Laboratory Tests and Field Observations of the OSOP Raspberry Shake 4D. *Seismol. Res. Lett.* **2018**, *90*, 219–228. [CrossRef]
25. Hussain, M.A.; Dutta, S.C.; Das, S. Effect of bidirectional ground shaking on structures in the elastic and post-elastic range: Adequacy of design provisions. *J. Build. Eng.* **2022**, *45*, 103656. [CrossRef]
26. Mărmureanu, A. Rapid magnitude determination for Vrancea early warning system. *Rom. J. Phys.* **2009**, *54*, 965–971.
27. Rinawi, A.M. Shaking Table-Structure Interaction. Ph.D. Thesis, University of California, Berkeley, CA, USA, 1991.
28. Baran, T.; Tanrikulu, A.K.; Dundar, C.; Tanrikulu, A.H. Construction and performance test of a low-cost shake table. *Exp. Tech.* **2011**, *35*, 8–16. [CrossRef]
29. Aldefae, A.H.; Shamkhi, M.S.; Khalaf, T. Design and manufacturing of geotechnical laboratory tools used in physical modeling. *Cogent Eng.* **2019**, *6*, 1637622. [CrossRef]
30. Ozcelik, O.; Conte, J.P.; Luco, J.E. Comprehensive mechanics-based virtual model of NHERI@ UCSD shake table—Uniaxial configuration and bare table condition. *Earthq. Eng. Struct. Dyn.* **2021**, *50*, 3288–3310. [CrossRef]

31. Gaetani, A.; Lourenço, P.B.; Monti, G.; Moroni, M. Shaking table tests and numerical analyses on a scaled dry-joint arch undergoing windowed sine pulses. *Bull. Earthq. Eng.* **2017**, *15*, 4939–4961. [CrossRef]
32. Pagliaro, S.; Aloisio, A.; Alaggio, R.; Di Egidio, A. Rigid block coupled with a 2 dof system: Numerical and experimental investigation. *Coupled Syst. Mech.* **2020**, *9*, 539–562.
33. Gao, C.; Yang, Y.; Wang, J. Development and application of a shaking table system. *Arab. J. Geosci.* **2022**, *15*, 1334. [CrossRef]
34. El Hoseny, M.; Ma, J.; Josephine, M. Effect of Embedded Basement Stories on Seismic Response of Low-Rise Building Frames Considering SSI via Small Shaking Table Tests. *Sustainability* **2022**, *14*, 1275. [CrossRef]
35. Damerji, H.; Yadav, S.; Sieffert, Y.; Debove, L.; Vieux-Champagne, F.; Malecot, Y. Design of a shake table with moderate cost. *Exp. Tech.* **2022**, *46*, 365–383. [CrossRef]
36. O'hagan, J.; Ma, T.M.Q. Experimental assessment of PID control for a uniaxial shake table. In Proceedings of the 15th World Conference on Earthquake Engineering, Lisbon, Portugal, 24–28 September 2012.
37. U.S. Geological Survey. Earthquake Hazards Program. Available online: https://earthquake.usgs.gov/monitoring/spectrograms/examples.php (accessed on 8 March 2023).
38. Chai, T.; Draxler, R.R. Root mean square error (RMSE) or mean absolute error (MAE)?–Arguments against avoiding RMSE in the literature. *Geosci. Model Dev.* **2014**, *7*, 1247–1250. [CrossRef]
39. Reyes, S.I.; Almazán, J.L.; Vassiliou, M.F.; Tapia, N.F.; Colombo, J.I.; de la Llera, J.C. Full-scale shaking table test and numerical modeling of a 3000-liter legged storage tank isolated with a vertical rocking isolation system. *Earthq. Eng. Struct. Dyn.* **2022**, *51*, 1563–1585. [CrossRef]
40. Chen, P.C.; Lai, C.T.; Tsai, K.C. A control framework for uniaxial shaking tables considering tracking performance and system robustness. *Struct. Health Monit.* **2017**, *24*, e2015. [CrossRef]
41. Yao, H.; Tan, P.; Yang, T.Y.; Zhou, F. Acceleration-based sliding mode hierarchical control algorithm for shake table tests. *Earthq. Eng. Struct. Dyn.* **2021**, *50*, 3670–3691. [CrossRef]
42. Larbi, S.H.; Bourahla, N.; Benchoubane, H.; Choutri, K.; Badaoui, M. Earthquake ground motion matching on a small electric shaking table using a combined NN-pdff controller. *Shock Vib.* **2020**, *2020*, 7260590. [CrossRef]

**Disclaimer/Publisher's Note:** The statements, opinions and data contained in all publications are solely those of the individual author(s) and contributor(s) and not of MDPI and/or the editor(s). MDPI and/or the editor(s) disclaim responsibility for any injury to people or property resulting from any ideas, methods, instructions or products referred to in the content.

*entropy* MDPI

*Article*

# Connection between Variations of the Probability Distribution of the Recurrence Time and Phases of the Seismic Activity

**Elisa Varini** [1,2,*,†] **and Renata Rotondi** [1,†]

[1] National Research Council of Italy, Institute for Applied Mathematics and Information Technologies, Via Corti 12, 20133 Milan, Italy; renata.rotondi@mi.imati.cnr.it
[2] ICSC National Research Centre for High Performance Computing, Big Data and Quantum Computing, Via Magnanelli 2, 40033 Casalecchio di Reno, Italy
* Correspondence: elisa.varini@mi.imati.cnr.it
† These authors contributed equally to this work.

**Abstract:** The probability distribution of the interevent time between two successive earthquakes has been the subject of numerous studies for its key role in seismic hazard assessment. In recent decades, many distributions have been considered, and there has been a long debate about the possible universality of the shape of this distribution when the interevent times are properly rescaled. In this work, we aim to discover if there is a link between the different phases of a seismic cycle and the variations in the distribution that best fits the interevent times. To do this, we consider the seismic activity related to the $M_w$ 6.1 L'Aquila earthquake that occurred on 6 April 2009 in central Italy by analyzing the sequence of events recorded from April 2005 to July 2009, and then the seismic activity linked to the sequence of the Amatrice-Norcia earthquakes of $M_w$ 6 and 6.5, respectively, and recorded in the period from January 2009 to June 2018. We take into account some of the most studied distributions in the literature: $q$-exponential, $q$-generalized gamma, gamma and exponential distributions and, according to the Bayesian paradigm, we compare the value of their posterior marginal likelihood in shifting time windows with a fixed number of data. The results suggest that the distribution providing the best performance changes over time and its variations may be associated with different phases of the seismic crisis.

**Keywords:** interevent time; probability distributions; probabilistic forecasting; seismic cycle; statistical seismology; statistical methods; Bayesian inference

**Citation:** Varini, E.; Rotondi, R. Connection between Variations of the Probability Distribution of the Recurrence Time and Phases of the Seismic Activity. *Entropy* **2023**, *25*, 1441. https://doi.org/10.3390/e25101441

Academic Editor: Georgios Michas

Received: 7 September 2023
Revised: 7 October 2023
Accepted: 10 October 2023
Published: 12 October 2023

**Copyright:** © 2023 by the authors. Licensee MDPI, Basel, Switzerland. This article is an open access article distributed under the terms and conditions of the Creative Commons Attribution (CC BY) license (https://creativecommons.org/licenses/by/4.0/).

## 1. Introduction

The time between two successive earthquakes, referred to as recurrence or waiting or interevent time, is one of the most studied quantities describing the seismic activity; it plays an important role in seismic hazard assessment being the main component of some stochastic processes—such as renewal processes, Markov processes—which model the temporal evolution of seismic phenomena. Several probability distributions have been proposed in the literature to model the recurrence time $\tau$, including the gamma, Weibull, lognormal, and exponential distributions, but the most remarkable result was that the shape of this distribution appeared to be universal when the times were suitably scaled by some critical indices, such as the Gutenberg–Richter $b$ value, the $\alpha$ exponent of the Omori law and the fractal dimension $d_f$ of the set of earthquake epicenters [1]. In other words, the distribution $F(\tau)$ would be independent of the spatial scale and of the magnitude threshold of the observations, which expresses a hierarchical organization in time, space, and magnitude. Further studies found that scaling by the average rate $\lambda$ of seismic activity—number of earthquakes per unit time—was sufficient to get approximately the same distribution in many different seismic regions [2]. A similar behavior was also obtained by Corral by fitting the density function given by

$$f(\tau) = \frac{C\delta}{a\Gamma(\gamma/\delta)}\left(\frac{\tau}{a}\right)^{\gamma-1} e^{-(\tau/a)^{\delta}} \qquad \gamma,\, \delta,\, a > 0 \tag{1}$$

to some regional data sets [3], where C is the normalizing constant, $a$ is a scale parameter, and $\gamma$ and $\delta$ control the shape for small and large $\tau$, respectively. This scaling function shows a power-law behavior for short times, and an exponential decrease for long times.

By simulations of the ETAS model with varying rate $\mu$ of independent events, which can be considered a proxy for regional size, Touati et al. [4] showed that the interevent time distribution is generally bimodal, and is best described as a mixture of two distributions: a gamma distribution for short waiting times between correlated events which belong to the same aftershock sequence, and an exponential distribution for longer waiting times between uncorrelated events. Completely general forms for the interevent time distribution can be obtained by resorting to a Bayesian nonparametric estimation method and considering the unknown distribution as a random measure [5].

Properties such as fractal structures and long-range correlations present in the earthquake activity have led to adopt theoretical tools of non-extensive statistical physics in the analysis of the statistical properties of some quantities describing the seismic activity in the space–time–magnitude domain. This approach is based on a generalization of the classic Boltzmann–Gibbs entropy proposed by Tsallis in 1988 [6]; by maximizing the non-extensive Tsallis entropy, a probability distribution, denoted as $q$-exponential distribution, was obtained and then successfully applied to investigate the distribution of various seismic quantities, such as magnitude ([7] and references therein), fault length [8], spatial distribution of epicenters [9], and interevent time [10]. To improve the fit of the gamma distribution by exploiting the results obtained through the $q$-exponential distribution, Michas et al. [11] used the $q$-generalized gamma distribution, borrowed from Queirós [12], which behaves as a power-law function for both short and large interevent times so as to provide a best fit when the seismicity is correlated at all timescales.

Generally, the aforementioned studies aim to obtain the best probability distribution for sets of interevent times which cover a large period, where the occurrences may involve a complex summation of triggered and/or independent events on different time scales. On the contrary, in this work, we wonder whether the probability distribution changes over time and whether these variations can be associated with different phases of the seismic process. To do this, we consider data from time windows with the same number of observations which shift at each new event; we compare the performance of some distributions chosen among the most studied ones by evaluating their posterior marginal likelihood in the Bayesian framework. We apply this procedure to two data sets related to the severe seismic sequences that hit central Italy in the last decades. The best probability distribution, i.e., the one that significantly outperforms the others, varies over time; in particular, we highlight that these changes characterize specific periods in the temporal evolution of seismic activity and therefore could be used for forecasting purposes.

## 2. Probability Distributions

Let us consider the sequence of $(N+1)$ seismic events that occurred at times $\{t_0, t_1, \dots, t_N\}$, and let $\mathcal{T} = \{\tau_i\}_{i=1}^{N}$ be the set of the interevent times $\tau$ between successive events defined as $\tau_i = t_i - t_{i-1}$, $i = 1, 2, \dots, N$. We assume that all events have a magnitude larger than, or equal to, the threshold $M_c$, which guarantees a sufficient degree of completeness for the data set. We present the main properties of the most studied probability distributions of the intervent time random variable.

### 2.1. Exponential Distribution

The simplest probability distribution for the interevent time is the exponential distribution with density function given by

$$f_1(\tau) = \lambda\, e^{-\lambda\,\tau}. \tag{2}$$

It describes the time between events in a homogeneous Poisson point process, i.e., a process in which events occur continuously and independently at a constant average rate; hence, the exponential distribution indicates uncorrelated behavior. Its key property is that it is memoryless, that is, the probability that the waiting time $\tau$ for an event exceeds a value $(t+s)$, conditioned on the fact that the time $t$ has already passed, is equal to the original probability of exceeding $s$:

$$Prob(\tau > t+s \,|\, \tau > t) = Prob(\tau > s) = e^{-\lambda\, s}. \tag{3}$$

Consequently, the exponential distribution is the only continuous probability distribution that has a constant hazard rate, equal to $\lambda$. According to Bayesian inference, the conjugate prior for the exponential distribution is the gamma distribution; hence, we consider the parameter $\lambda$ to be a gamma-distributed random variable with hyperparameters $a_0$, $b_0$ so that its posterior distribution is still a gamma distribution with parameters $(a_0 + N)$ and $(b_0 + \sum_{i=1}^{N} \tau_i)$.

### 2.2. Gamma Distribution

The gamma probability density function is given by

$$f_2(\tau) = \frac{\beta^\alpha}{\Gamma(\alpha)}\, \tau^{\alpha-1}\, e^{-\beta\,\tau} \tag{4}$$

where $\alpha$ is the shape parameter and $\beta$ is the scale parameter, both positive. In particular, the gamma distribution models the sum of exponentially distributed random variables; that is, if we consider a sequence of events such that each interevent time follows the exponential distribution with parameter $\beta$, then the waiting time of the $n$-th event is a gamma-distributed random variable with integer shape $\alpha = n$. In general, the extensive use of the gamma distribution is due to its ability to model the intertime between triggered aftershocks through its short-term scale power-law factor and the long-term scale Poissonian background activity through the exponential factor. Regarding parameter inference, since there is only the conjugate prior distribution of the scale parameter, we prefer to estimate the posterior distribution of both parameters through the Metropolis–Hastings algorithm, a stochastic simulation method of the class of Markov chain Monte Carlo (MCMC) methods. This algorithm generates a Markov chain that converges to the target distribution—in our case, the posterior distribution—using a proposal density for generating new candidate values and a method for rejecting some of the proposed values [13]. In this way, we obtain not only the parameter estimates, typically as their posterior means (average of the sampled, possibly thinned, values), but also a measure of their uncertainty as expressed through the simulated posterior distribution of each parameter.

### 2.3. Q-Exponential Distribution

Nonlinear dynamical systems showing fractal structures and long-range correlations are successfully studied in the framework of non-extensive statistical physics. The presence of the same properties also in seismicity [11] suggests analyzing the temporal behavior of the seismic activity through the $q$-exponential distribution:

$$f_3(\tau) = \frac{1}{\gamma}\left(1 - \frac{(1-q)}{(2-q)}\,\frac{\tau}{\gamma}\right)^{1/(1-q)}, \quad 1 < q < 2 \tag{5}$$

obtained by maximizing the non-additive Tsallis entropy $S_q$:

$$S_q = k_B \frac{1 - \int p^q(x)\,dx}{q-1} \tag{6}$$

under suitable constraints [6]; $q$ is called the entropic index, and $\gamma$ is the generalized expectation value, that is, the mean with respect to the *escort* probability distribution [14]:

$$f_q(x) = \frac{f^q(x)}{\displaystyle\int_0^{+\infty} f^q(x)\,dx}. \tag{7}$$

Given two independent systems $A$ and $B$, the Boltzmann–Gibbs entropy $S_{BG}$ satisfies the additive property $S_{BG}(A+B) = S_{BG}(A) + S_{BG}(B)$, whereas for the Tsallis entropy, we can verify that

$$\frac{S_q(A+B)}{k_B} = \frac{S_q(A)}{k_B} + \frac{S_q(B)}{k_B} + (1-q)\frac{S_q(A)}{k_B}\frac{S_q(B)}{k_B} \tag{8}$$

i.e., $S_q$ is nonadditive; in particular, the cases $q < 1$, $q = 1$, and $q > 1$ correspond to superadditivity (or superextensivity), additivity (extensivity), and subadditivity (subextensivity), respectively, and when $q \to 1$, $S_q$ recovers $S_{BG}$. Extensivity and additivity are terms often used interchangeably even though they are not equivalent because in most cases encountered in physics, additivity does imply extensivity [15]. In the Supplementary Material, we show the link between estimates of the $q$-index and specific states of the system.

For large $\tau$, the $q$-exponential density function (5) goes to zero as a power $\tau^{-1/(q-1)}$, and it is also always bounded below by the exponential density function; hence, it is a fat-tailed distribution and, in particular, a heavy-tailed distribution, being

$$\lim_{\tau\to+\infty} e^{t\tau}\bar{F}_3(\tau) = \lim_{\tau\to+\infty} e^{t\tau}\left(1 + \frac{q-1}{2-q}\frac{\tau}{\gamma}\right)^{-(2-q)/(q-1)} = +\infty \quad \forall\tau > 0 \tag{9}$$

where $\bar{F}_3(\tau) = 1 - F_3(\tau)$.

Since conjugate priors of the $q$ and $\gamma$ parameters with respect to the $q$-exponential function are unavailable, we again resort to Markov chain Monte Carlo methods to sample from the posterior probability distributions of the parameters after reparameterizing them as follows: $\theta = (2-q)/(q-1)$ with $\theta \in (0, +\infty)$.

### 2.4. Q-Generalized Gamma Distribution

Substituting the exponential term in (4) with the $q$-exponential function defined as

$$\exp_q(x) = [1 + (1-q)\,x]^{1/(1-q)}$$

borrowed from the nonextensive statistical mechanics, one obtains the so-called $q$-generalized gamma density function, which is characterized by two power-law regimes indicating clustering effects at all time scales and both short- and long-term memory in the seismogenic process. This distribution was proposed for the first time in the financial framework by Queirós [12]; it is based on local fluctuations of the $\omega$ mean value of the gamma distributed $\tau$ variable under study. Hence, scaling the variable by its mean value, we have the conditional density function

$$f(\tau \mid \omega) = \frac{\varphi^{\varphi}}{\omega\,\Gamma(\varphi)}\left(\frac{\tau}{\omega}\right)^{\varphi-1} \exp\left[-\frac{\varphi}{\omega}\,\tau\right] \tag{10}$$

where $E(\tau) = \omega$. Let us assume that $\omega$ varies over time and follows the inverse gamma distribution:

$$g(\omega) = \frac{\left(\frac{\varphi}{\lambda}\right)^{\delta}}{\Gamma(\delta)}\,\omega^{-\delta-1}\exp\left[-\frac{\varphi}{\omega\,\lambda}\right]. \tag{11}$$

Integrating the joint density $f(\tau \mid \omega)\,g(\omega)$ with respect to $\omega$, we obtain the marginal density function:

$$h(\tau) = \frac{\lambda\,\Gamma(\varphi+\delta)}{\Gamma(\varphi)\,\Gamma(\delta)}(\lambda\tau)^{\varphi-1}\,(1+\lambda\,\tau)^{-\varphi-\delta} \tag{12}$$

and carrying out the changes of variables

$$\lambda = \frac{\rho-1}{\xi}, \quad \delta = \frac{1}{\rho-1} - \varphi, \quad \alpha = \varphi - 1 \tag{13}$$

we have the $q$-generalized gamma probability density function

$$f_4(\tau) = \frac{(\rho-1)^{\varphi}\,\Gamma\left(\frac{1}{\rho-1}\right)}{\xi\,\Gamma\left(\frac{1}{\rho-1}-\varphi\right)\Gamma(\varphi)}\left(\frac{\tau}{\xi}\right)^{\varphi-1}\left[1+\frac{\rho-1}{\xi}\,\tau\right]^{1/(1-\rho)}. \tag{14}$$

In the limit $\rho \to 1$, it recovers the ordinary gamma distribution. Finally, as in the case of the $q$-exponential distribution (5), we reparametrize as follows: $\eta = \frac{(2-\rho)}{(\rho-1)}$, to simplify parameter estimation through MCMC methods.

## 3. Inferential Issues

We briefly give the basic concepts on the Bayesian approach that we followed in estimation and comparison of the four models. Let us assume that the data $\boldsymbol{\tau} = (\tau_1, \tau_2, \ldots, \tau_N)$ are realizations of a random variable $\mathcal{T}$ whose density function belongs to the parametric family $\mathcal{F} = \{f(\tau;\psi) \,:\, \psi \in \Psi\}$. Contrary to the frequentist approach, the parameter $\psi$ is considered a random variable and its *prior* distribution $p_0(\psi)$ collects our initial beliefs about the phenomenon under study. Through the Bayes' theorem, the information provided by the data and expressed by the likelihood $\mathcal{L}(f(\boldsymbol{\tau} \mid \psi)) = \prod_{i=1}^{N} f(\tau_i;\psi)$ is combined with that in the prior distribution into the *posterior* distribution:

$$p(\psi \mid \boldsymbol{\tau}) = \frac{p_0(\psi)\,\mathcal{L}(f(\boldsymbol{\tau} \mid \psi))}{\int_{\Psi} p_0(\psi)\,\mathcal{L}(f(\boldsymbol{\tau} \mid \psi))\,d\psi}. \tag{15}$$

From this distribution, we obtain not only the parameter estimate, typically as the posterior mean $E_p(\psi)$ but also indications on its accuracy through statistical summaries such as the posterior variance or the quartiles. The computational difficulties due to the evaluation of the integral in (15), which is often multi-dimensional as in our case, can be solved by resorting to the application of Markov chain Monte Carlo (MCMC) methods, which produce a sequence of random samples from the posterior distribution (15) through which the distribution can be approximated [16]. In particular, we apply the Metropolis–Hastings (MH) algorithm, which consists of the following steps: (a) generate an initial value $\psi_0$ from its prior distribution $p_0(\psi)$ and set $i = 0$, (b) for each iteration $i$, generate a next candidate sample $\tilde{\psi}$ from an arbitrary probability density $q(\psi \mid \psi_i)$, referred to as *proposal* or *jumping* density, (c) calculate the acceptance probability

$$\alpha_i = min\left(1,\ \frac{\mathcal{L}(f(\boldsymbol{\tau} \mid \tilde{\psi}))\,p_0(\tilde{\psi})\,q(\psi_i \mid \tilde{\psi})}{\mathcal{L}(f(\boldsymbol{\tau} \mid \psi_i))\,p_0(\psi_i)\,q(\tilde{\psi} \mid \psi_i)}\right),$$

and accept $\tilde{\psi}$ as $\psi_{i+1}$ with probability $\alpha_i$, or set $\psi_{i+1} = \psi_i$ with probability $(1-\alpha_i)$.

The initial values of the chain may be highly dependent on the starting value; for this, we neglect the first $k$ samples and use the sequence $\{\psi_i\}_{i=k+1}^{M+k}$, for large enough values of

$k$ and $M$, to estimate the posterior distribution $p(\psi \mid \boldsymbol{\tau})$ and to approximate the posterior marginal log-likelihood:

$$\log \mathcal{L}(f(\boldsymbol{\tau})) = \int_{\Psi} \log \mathcal{L}(f(\boldsymbol{\tau} \mid \psi))\, p\,(\psi \mid \boldsymbol{\tau})\, d\,\psi \approx 1/M \sum_{j=k+1}^{M+k} \log \mathcal{L}(f(\boldsymbol{\tau} \mid \psi_j)) \tag{16}$$

which enables us to verify how well the fitted model $f(\cdot)$ is able to describe the observed data. In general, given two statistical models $f_j(\cdot|\phi)$ and $f_k(\cdot|\eta)$, we can compare them using the difference of their posterior marginal log-likelihoods $\Delta_{jk} = \log\, \mathcal{L}\,(f_j(\boldsymbol{\tau})) - \log\, \mathcal{L}\,(f_k(\boldsymbol{\tau}))$, and then, similar to the Bayes factor, we establish the degree of evidence in favor of the first model according to the value $\mathcal{K}$ of the Jeffreys scale [17,18].

Considering our four probability models in detail, we note that only the $\lambda$ parameter of the exponential distribution has the gamma distribution as a conjugate prior. For each of the parameters of the other models, we choose a lognormal distribution both as a prior and as a proposal distribution in the Metropolis–Hastings algorithm. Let us consider, for instance, the parameter $\theta$ of the $q$-exponential distribution; we express our initial beliefs about it by specifying its prior distribution Lognormal$(mean_\theta, var_\theta)$, where $mean$ and $var$ indicate the mean and variance of the random variable $\theta$ and not the mean and variance of its logarithm, as in the common representation of the lognormal distribution. Moreover, regarding the generation of the Markov chain converging to the posterior distribution of $\theta$, at the $(i+1)$-th iteration, we adopt a Lognormal $(\theta_i, (\theta_i/\kappa_1)^2)$ as a proposal distribution for the next candidate value, where $\theta_i$ is the current value of the Markov chain and the value of $\kappa_1$ is calibrated through pilot runs so that the acceptance rate of the MH algorithm is approximately from 25% to 40%. The same applies for the other parameters.

## 4. Case Studies

To evaluate the performance of the four probability distributions of the interevent time presented in Section 2, we examine two sequences of earthquakes related to two strong seismic crises that occurred in central Italy in the last decades: the L'Aquila ($M_w$ 6.1) earthquake in 2009 and the Amatrice ($M_w$ 6)–Norcia ($M_w$ 6.5) earthquakes in 2016. These earthquakes have been associated with two composite seismogenic sources of the Database of Individual Seismogenic Sources (DISS, version 3.2.1) [19] that can have the potential for earthquakes up to $M_{max}$ 6.5. Both data sequences analyzed in this study are taken from the Italian National Institute of Geophysics and Volcanology (*Istituto Nazionale di Geofisica e Vulcanologia*; INGV) web services: Italian Seismological Instrumental and Parametric Database (ISIDe) working group (2016), version 1.0, accessible at http://cnt.rm.ingv.it/en/iside accessed on 6 September 2023 [20], in which the size of the events is expressed in different magnitude units, as local magnitude $M_L$, duration magnitude $M_D$, and moment magnitude $M_w$. We therefore applied the orthogonal regression relationships: $M_w = 1.066\, M_L - 0.164$ and $M_w = 1.718\, M_D - 1.897$ [21] to convert $M_L$ and $M_D$ to $M_w$ in order to construct the homogeneous data sets. The spatial extension of the areas under examination is established by taking into account the empirical relationship between the magnitude and rupture length $RL$—$\log_{10}(RL) = -3.22 + 0.69\, M_{max}$—by Wells and Coppersmith [22].

For each of the two sequences, we calculate the time between pairs of successive earthquakes in order to obtain a set of $N$ observed interevent times; then, we consider time windows that consist of $n = 100$ observations and that shift at each new event through the seismic sequence under examination. In this way, we obtain $(N - n + 1)$ data sets on which to evaluate the fitting of the probability distributions listed in Section 2 and to investigate whether the best distribution is unique or varies over time with the change of the seismic phases.

### 4.1. L'Aquila Sequence

On April 6, 2009 (01:13:40 UTC), a $M_w$ 6.1 shock was recorded with an epicenter at latitude 42.342 and longitude 13.380 near L'Aquila, the capital of the Abruzzo region in Central Italy. Consistent with the above criteria, we choose the rectangular area centered on the epicenter, of latitude size (41.8, 43.0) degrees and longitude size (12.8, 13.8) degrees as the study area, and we analyze the temporal period from 7 April 2005 to the end of July 2009, taking $m_0 = 2$ as the magnitude threshold to ensure the completeness of the data set, except, at most, a few hours after the main shock when temporary partial incompleteness can be observed due to the well-known difficulty in recording all the events at the beginning of the aftershock sequence, and especially those of low magnitude [23]. The main shock was preceded by a $M_w$ 4 foreshock on 30 March 2009 [24], and was followed by an aftershock sequence, which lasted more than a year, of which the strongest was of $M_w$ 5.4 occurred on 7 April 2009 [25].

Overall, we have 2725 events, that is, $N = 2724$ interevent times through which we construct $m = 2625$ temporal windows, each of 100 consecutive observations, shifting at each new event. To observe how the temporal distribution changes, we fit the four probability models (Section 2) to the data set associated with each time window, and then compare the pairwise differences of their posterior marginal log-likelihoods with the value $\mathcal{K} = 2.3026$ on the Jeffreys scale, indicating strong evidence in favor of the first model. In the case of the exponential probability density, $f_1(\tau) = \lambda \exp\{-\lambda \tau\}$, we adopt the conjugate Gamma(2,1) distribution as a prior distribution of the $\lambda$ parameter so that the expected seismic rate is approximately 2 (time in days). For the other three probability models under examination, Table 1 reports the parameters of the prior distributions and the $\kappa$ coefficients used in the proposal distributions of the MH algorithm to obtain suitable acceptance rates.

**Table 1.** Parameters of the prior distributions and $\kappa$ coefficients used in the proposal distributions in the MH algorithm.

| | | L'Aquila Case | | | Amatrice-Norcia Case | | |
|---|---|---|---|---|---|---|---|
| **Model** | **Parameters** | **Mean$_0$** | **Var$_0$** | $\kappa$ | **Mean$_0$** | **Var$_0$** | $\kappa$ |
| Gamma | $\alpha$ | 0.04 | 0.01 | 3.0 | 0.8 | 0.15 | 3.0 |
| | $\beta$ | 0.1 | 0.01 | 2.0 | 10.0 | 50.0 | 1.5 |
| $Q$-exponential | $\theta$ | 3.0 | 9.0 | 0.8 | 7.0 | 9.0 | 2.5 |
| | $\gamma$ | 3.0 | 9.0 | 2.1 | 0.3 | 4.0 | 1.3 |
| $Q$-generalized gamma | $\xi$ | 5.5 | 12.25 | 2.2 | 3.5 | 2.0 | 1.3 |
| | $\eta$ | 6.5 | 6.25 | 2 | 9.0 | 2.5 | 1.6 |
| | $\varphi$ | 0.7 | 0.04 | 4.0 | 0.7 | 0.02 | 3.5 |

Figure 1 shows the estimated density functions and the histogram of the interevent times belonging to the time window in which each density function, represented by a solid line, provides, respectively, the best fit to the data. In particular, (a) the window in the left top panel refers to the events that occurred from 1 April 2009 up to two hours after the main shock, 90% of which are aftershocks; the $q$-exponential density function has both its largest posterior marginal likelihood and the maximum difference from the second best density function ($q$-generalized gamma). (b) Right top panel: the window contains the events from 30 January 2009 to 1 h after the main shock; the $q$-generalized gamma density function is the best model but it is not worth more than a mere mention for the evidence shown by its fit to the data with respect to the second-best density (gamma). (c) Left bottom panel: the window is all made up of aftershocks that occurred in the first three hours after the main shock; the gamma density function is the best model, and it has a decisive strength of the evidence against the second best model ($q$-exponential). (d) Right bottom panel: the window covers the period from 24 April to 30 April 2009; there is nothing more than evidence barely worthy of note in favor of the exponential density against the $q$-exponential and gamma density functions. We note that the heavy-tailed $q$-exponential distribution best describes the data set with the mass most concentrated on the left and skewed to the right, while the gamma density fits well the unimodal histogram associated with the aftershock set immediately following the main shock. The relative lack of very short interevent times is probably related to the temporary incompleteness of the catalog in that period.

Figure 2 shows the largest value among the posterior marginal likelihoods of the four probability models at each time window; different colors highlight how the probability distribution that fits best the observations varies over time. To understand the motivations of these changes, we analyze the characteristics of each data set by showing some of their statistical summaries, such as first and third quartiles, median, mean and skewness. We remind that skewness is a measure of the asymmetry of the probability distribution. When it is positive, as in our cases, the right tail is longer, and the mean is greater than the median; more precisely, the greater the skewness, the more the distribution is left-leaning, that is, the more the observed intertimes are concentrated around short values; viceversa smaller skewness corresponds to more homogeneously distributed data where the median is closer to (to the left of) the mean.

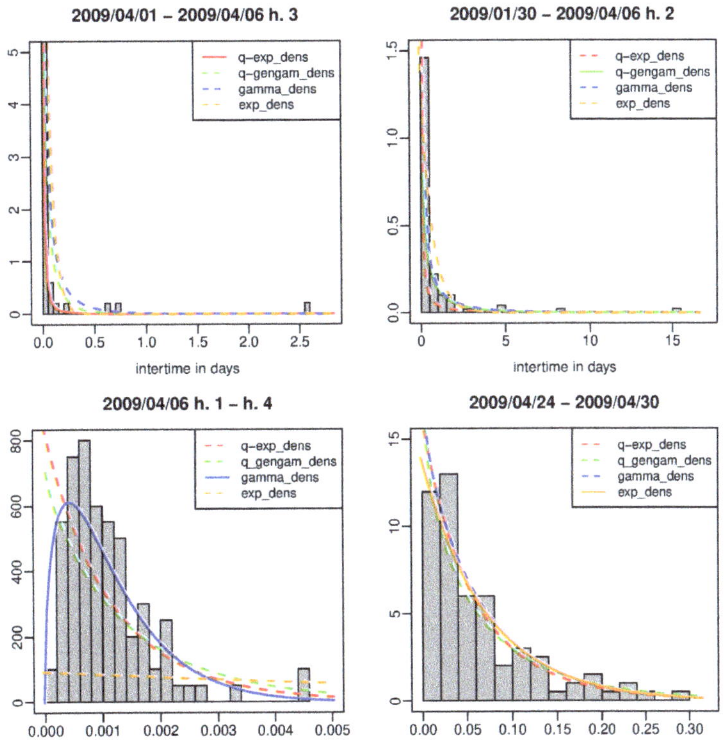


**Figure 1.** L'Aquila case—time window in which the $q$-exponential (**left top panel**), the $q$-generalized gamma (**right top panel**), the gamma (**left bottom panel**), and the exponential (**right bottom panel**) distribution provides the best fit to the data in terms of posterior marginal likelihood.

The time windows are divided as follows: in 1745 (66%), the $q$-exponential distribution represents the best model; in 68 (2.6%), the $q$-generalized gamma distribution; in 760 (29%), the gamma distribution; and in 52 (2%), the exponential distribution. We point out that the strength of the evidence in favor of the best distribution over the second-best model is strong in only about half of the time windows, particularly for the $q$-exponential and the gamma distribution in 828 and 334 windows, respectively, concentrated in the hours after the main shock, whereas for the $q$-generalized gamma and the exponential distribution, in no window. Moreover, we note that the difference between all four models is not particularly significant in 143 ($\sim$5%, approximately from the 2300-th to 2450-th window) time windows mainly concentrated between May and June 2009; we address this issue in more detail in the Supplementary Material. For brevity, we will say hereinafter that the models are *interchangeable* when the difference between their posterior marginal log-likelihoods does not exceed the threshold $\mathcal{K} = 2.3026$.

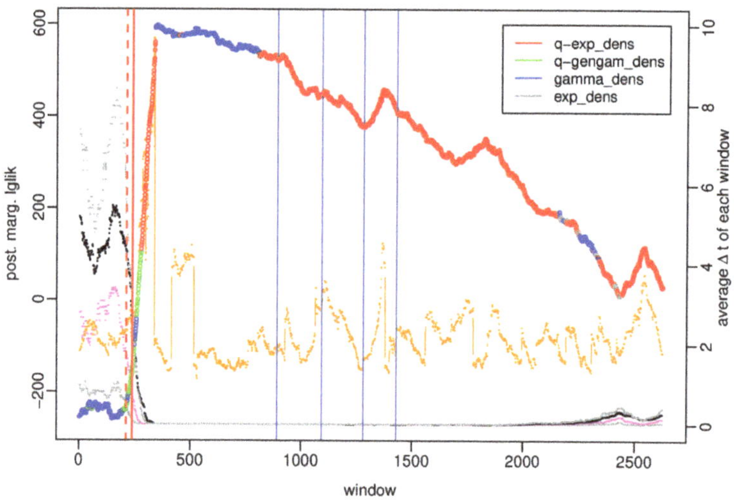


**Figure 2.** L'Aquila case—value of the posterior marginal log-likelihood of the probability distribution which provides the best fit to the data in each time window of the period (7 April 2005–31 July 2009) versus the number of the time window. Statistical summaries of the data set in each window: first and third quartile (gray dotted lines), median (violet dotted line), mean (black dotted line), skewness (orange points and line). Vertical bars indicate the occurrence time of the 30 March 2009 $M_w$ 4 earthquake (red dashed line), L'Aquila $M_w$ 6.1 earthquake (red solid line), and events of $5.0 \leq M_w < 5.5$ (blue line) respectively.

We examine the characteristic features of each probability model; the $q$-exponential distribution shows very strong evidence with respect to the other distributions (the $q$-generalized gamma is the second best model) in the time windows over the main shock, that is, which include some pre-main shock event and some aftershock and during the aftershock sequence, particularly since the end of June. The gamma distribution exceeds the other distributions on the day of the main shock—6 April 2009—and is the second-best model early in the aftershock sequence, often interchangeable with the best $q$-exponential model. Substantially, the exponential distribution outperforms the other distributions with slight evidence only in a few time windows in May–June 2009.

To better understand what happens before the main shock, in Figure 3, we zoom in on the first 350 time windows covering the period from 7 April 2005 to 6 April 2009 h. 4. The $q$-generalized gamma distribution is the best model in the period from 12 March 2009 to 6 April 2009, one hour after the main shock; that is, the period that we can denote as the preparatory phase in the seismic crisis because it includes the foreshock occurred on 30 March 2009 and is denoted by a red dotted line in Figure 3. In this period, the $q$-generalized gamma distribution is interchangeable with the gamma distribution but exceeds the other distributions with strong evidence. The role exchanges in the preceding period of background activity between April 2005 and March 2009, in which the gamma distribution is the best model, is interchangeable with the $q$-generalized gamma distribution and exceeds the other distributions with strong evidence (see the Supplementary Material).

As regards the statistical summaries, we note that the preparatory phase is characterized by a constant decrease in the mean and in the median of the data sets approximately from the beginning of 2009, while overall, the skewness increases up to 2 April 2009 and then decreases. This means that the interevent times get shorter between January and March 2009; obviously, the minimum values of the mean and median are observed during the aftershock sequence.

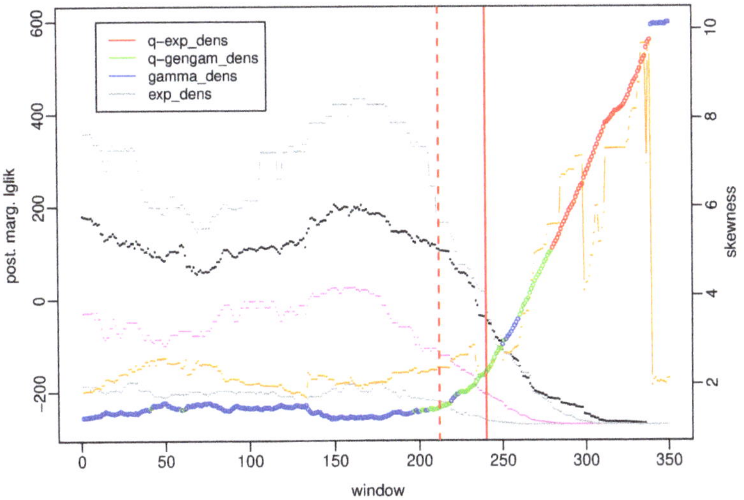


**Figure 3.** L'Aquila case—value of the posterior marginal likelihood of the probability distribution which provides the best fit to the data in the first 350 time windows covering the period (7 April 2005–6 April 2009 h. 4) versus the number of the time window. Statistical summaries of the data set in each window: first and third quartile (gray dotted lines), median (violet dotted line), mean (black dotted line), and skewness (orange points and line). Vertical red bars indicate the occurrence time of the 30 March 2009 $M_w$ 4 earthquake (red dashed line) and L'Aquila $M_w$ 6.1 earthquake (red solid line), respectively.

### *4.2. Amatrice-Norcia Sequence*

In 2016–2017, the junction area of the three regions Lazio, Marche, and Umbria in Central Italy was hit by a complex sequence of destructive seismic events; on 24 August 2016 (01:36:32 UTC, latitude 42.698, longitude 13.234), an earthquake of $M_w$ 6 shook the city of Amatrice and caused about 300 fatalities. This shock, initially considered the main shock, later proved to be the foreshock of the $M_w$ 6.5 strongest shock that struck the city of Norcia on 30 October 2016 (06:40:17 UTC, latitude 42.830, longitude 13.109). The aftershock sequence lasted roughly up to July 2017 [25] and recorded four $M_w$ 5+ earthquakes on 18 January 2017. We consider 5062 events ($N = 5061$ interevent times), which fall in the rectangular area of latitude size (42.3, 43.2) degrees and longitude size (12.7, 13.5) degrees and span the temporal period from January 2014 to June 2018, taking $m_0 = 2.5$ as the magnitude threshold in order to guarantee the completeness of the data set apart from the first hours following the main shock on 30 October 2016.

We investigate the behavior of the four probability distributions, given in Section 2, in $m = 4962$ data sets, each of which obtained, by shifting at each new event, a time window constituted of 100 consecutive waiting times. First, we evaluate the posterior marginal log-likelihood of each distribution in every time window by applying the MH algorithm to estimate the posterior distribution of the model parameters; Table 1 shows the parameters of the prior distributions and the $\kappa$ coefficients used in the proposal distributions of the MH algorithm to obtain suitable acceptance rates. Comparing the pairwise differences between the four posterior marginal log-likelihoods with the value $\mathcal{K} = 2.3026$ of the Jeffreys scale, we obtain the probability distribution with the best performance in each time window and the strength of its evidence with respect to the other distributions.

Figure 4 shows the estimated density functions and the histogram of the interevent times belonging to the time window in which each density function, represented by a solid line, provides, respectively, the best fit to the data, that is, it has its largest posterior marginal likelihood and the maximum difference from the likelihood of the second-best density function. In particular, (a) the left top panel refers to the time window which includes the first 98 aftershocks covering the two hours following the Amatrice shock, which was preceded by two waiting times of approximately 28 and 63 days; the data set is therefore very right skewed with a long right tail, and it is best described by the $q$-exponential density function, which outperforms the other distributions with decisive evidence. (b) Right top panel: the $q$-generalized gamma distribution is the best model in the time window from 26 June 2009 to 15 January 2010, which includes the final part of the L'Aquila aftershock sequence. (c) Left bottom panel: the time window includes the aftershocks that occurred between h. 8 and h. 10 of the occurrence day of the Norcia main shock (30 October 2016, 06:40:17 UTC); the gamma density function shows a decisive strength of evidence against the other probability distributions by adapting very well to the unimodal histogram of the interevent times, which are probably missing the shorter ones due to the temporary incompleteness of the catalog after the strongest event. (d) Right bottom panel: the exponential distribution is interchangeable with the gamma and $q$-exponential distribution in the time window from 25 December 2016 to 18 January 2017, when the first of the four $M_w$ 5+ events occurred at Capitignano.

Figure 5 shows the value of the largest posterior marginal log-likelihood at each time window, and the different colors indicate which probability model this value corresponds to; the $x$ axis represents the window number in the left panel and the time in the right panel. The first and third quartiles, median, mean and skewness of each data set are shown to highlight how the distribution of the observations changes over time. The time windows are divided as follows: in 1825 (36.8%), the $q$-exponential distribution represents the best model; in 360 (7.3%), the $q$-generalized gamma distribution; in 2748 (55.4%), the gamma distribution; and in 29 (0.6%), the exponential distribution. The strength of the evidence in favor of the best distribution with respect to the second-best model is strong or decisive in only about a third of the time windows, in which the outperforming distribution is especially $q$-exponential or gamma; in particular, as for the $q$-exponential, the $q$-generalized gamma and the gamma distribution in 613, 12 and 1025 windows, respectively, are essentially concentrated in the hours after the strongest shocks, whereas for the exponential distribution, in no window. The Supplementary Material provides a detailed visualization of the time windows in which the strength of the evidence in favor of a probability model is particularly significant and those in which it is not.

Let us take a more thorough look at the behavior of the various distributions. The $q$-exponential distribution is significantly the best model essentially in the first hours following the strongest earthquakes: from h. 5 to h. 8 of the day of occurrence of the L'Aquila earthquake (6 April 2009), from h. 2 to h. 3 of the day of occurrence of the Amatrice earthquake (24 August 2016), in the time windows including the first aftershocks of the Norcia earthquake (30 October 2016), and in the days of occurrence of the $M_w$ 5+ shocks on 18 and 19 January 2017; in the other time windows in which it has the largest log-likelihood, always during the aftershock sequences, it is interchangeable with the gamma distribution (see the Supplementary Material). The $q$-generalized gamma distribution characterizes the periods from January 2010 to November 2012 and from September 2017 to the end of our study; the difference from the other distributions has strong evidence only in January 2010, while in the other time windows, the $q$-generalized gamma density is interchangeable with the gamma density (see the Supplementary Material).

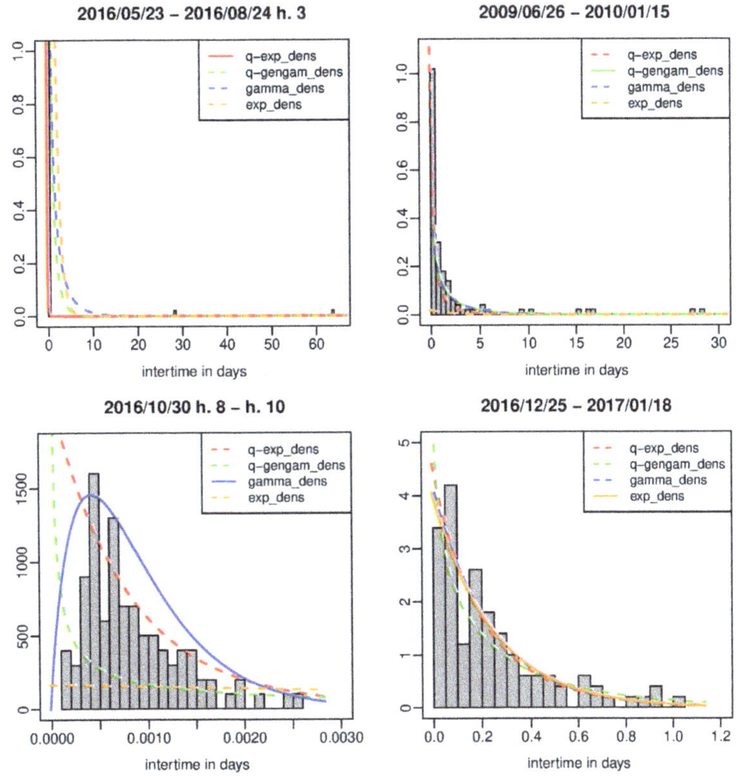


**Figure 4.** Amatrice-Norcia case—time window in which the $q$-exponential (**left top panel**), the $q$-generalized gamma (**right top panel**), the gamma (**left bottom panel**), and the exponential (**right bottom panel**) distribution provide the best fit to the data in terms of posterior marginal likelihood.

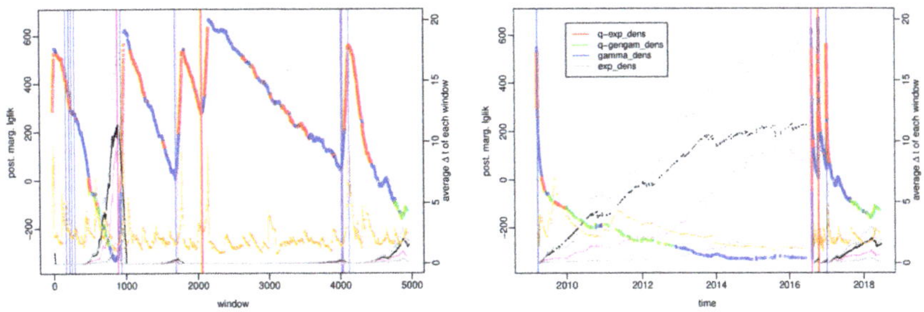


**Figure 5.** Amatrice-Norcia case—value of the posterior marginal log-likelihood of the probability distribution which provides the best fit to the data in each time window of the period (1 January 2009–31 June 2018) vs. (**left**) the number of the time window and (**right**) the time. Statistical summaries of the data set in each window: first and third quartiles (gray dotted lines), mean (black dotted line), median (violet dotted line), skewness (orange points and line). Vertical bars indicate the occurrence time of Amatrice 24 August 2016 $M_w$ 6.0 earthquake (magenta line), Norcia 30 October 2016 $M_w$ 6.5 earthquake (red line), events of $5.0 \leq M_w < 5.5$ (blue line) and of $5.5 \leq M_w < 6.0$ (violet line) respectively.

The gamma distribution has the largest value of the posterior marginal log-likelihood in most time windows—in almost half of them with strong or decisive evidence—and precisely in those covering both part of the aftershock sequences (in particular, the windows, including the first forty aftershocks following the Amatrice shock and the first hours after the main shock in Norcia) and the quiescence period from December 2012 to August 2016; in some of the first ones, the gamma model is interchangeable with the $q$-exponential model, while in the second ones, it is interchangeable with the $q$-generalized gamma model (see the Supplementary Material).

In Figure 6, we distinguish the results obtained before the start of the Amatrice-Norcia seismic crisis (left panels) from those produced by the sequence of earthquakes following the Amatrice shock (right panels); the value of the largest posterior marginal log-likelihood is plotted versus the time window number in the top panels and versus time in the bottom panels. We note that the average interevent time and the likelihood are inversely correlated, i.e., the more the observed recurrence times are concentrated in short times, the greater the likelihood, or, in other words, the likelihood decreases as the seismic activity decreases. Furthermore, the gamma distribution is the best model in correspondence with the minimum values of likelihood: local minima before the shocks of greater magnitude and the absolute minimum reached on 11 February 2015. The years leading up to the onset of the Amatrice-Norcia seismic crisis are characterized by decreasing values of the log-likelihood and by the gamma distribution as the best model but with barely noteworthy evidence with respect to the $q$-generalized gamma distribution, which is the second-best model; vice versa, the $q$-generalized gamma distribution is the best model, and it is interchangeable with the gamma distribution in the years between the end of the L'Aquila aftershock sequence and the beginning of 2013. Since 2009, the average interevent time has a constantly increasing trend, which becomes almost flat starting from 2015 at the same time that the value of the median approaches that of the average.

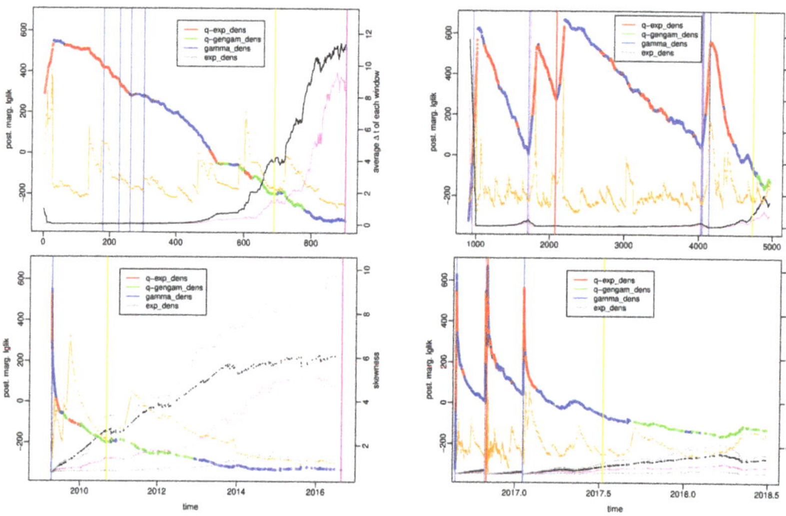


**Figure 6.** Amatrice-Norcia case—value of the posterior marginal log-likelihood of the probability distribution which provides the best fit to the data in the first 1000 time windows covering the period (1 January 2009–24 August 2016 h. 3) (**left panels**) and in the remaining time windows up to 31 June 2018 (**right panels**), versus the number of the time window (**top panels**) and versus the time (**bottom panels**). Statistical summaries of the data set in each window: mean (black dotted line), median (violet dotted line), skewness (orange points and line). Vertical bars indicate the occurrence time of Amatrice 24 August 2016 $M_w$ 6.0 earthquake (magenta line), Norcia 30 October 2016 $M_w$ 6.5 earthquake (red line), events of $5.0 \leq M_w < 5.5$ (blue line) and of $5.5 \leq M_w < 6.0$ (violet line), respectively.

## 5. Discussion

We investigated the probability distributions of the time between two successive earthquakes with the aim of finding out if there are links between the seismic phases and variations of the probabilistic model that best fits the data in those phases. To this end, we examined two seismic crises that hit central Italy and are related to the L'Aquila earthquake in 2009 and the Amatrice-Norcia shocks in 2016. Their retrospective analysis showed that the first crisis had a foreshock of $M_w$ 4 on 30 March 2009, while the second had one so strong—$M_w$ 6—that it was initially mistaken for the main shock. Overall, the Amatrice-Norcia sequence turned out to be more complex with greater energy release. As for the relationships between seismic phases and variations of the best probability distribution of the recurrence time, the two events share some features:

- Most of the probability densities estimated in the various time windows have a decreasing shape as an inverse power law.
- The time windows with many very short interevent times—like those in the aftershock sequences—are associated with great likelihood, while the data sets which are less concentrated around short times, typical of the quiescence period, have smaller likelihood.
- The $q$-exponential distribution outperforms the other distributions in the initial part of the aftershock sequence, and it becomes interchangeable with the gamma distribution.
- The $q$-generalized gamma distribution is associated with time intervals following aftershock sequences, such as, for example, the years from 2010 to 2012 following the aftershock sequence of L'Aquila earthquake and from September 2017 after the $M_w$ 5+ Capitignano shocks; in these periods, it slightly exceeds the gamma distribution. Only in the case of the L'Aquila earthquake, the $q$-generalized gamma distribution characterizes the initial phase of the activation between the fore- and the main shock.
- The gamma distribution is interchangeable with the $q$-generalized gamma distribution in the periods of low seismic activity and with the $q$-exponential distribution in part of the aftershock sequences.

Similar results have also been found in other studies, in which the phases of a seismic cycle are related to changes in the probability distribution of the magnitude [26] and of the spatial location of the epicenters [9].

Another remark can be made on the value of the log-likelihood in the case of the Amatrice-Norcia crisis. As we expect, it decreases during the aftershock sequences; however, we note that it becomes even negative just over two months after the Capitignano $M_w$ 5+ earthquakes, while it remains high when the same time has elapsed after the Amatrice and Norcia shocks. This could suggest that we should expect further energy releases after the Amatrice and Norcia events.

As regards the effects on the results of the possible partial incompleteness of the catalog in the few hours following a strong shock, we recall that a sensitivity analysis for the $q$-exponential distribution of the magnitude was carried out in [26] and highlighted a substantial similarity in the behavior of the entropy and of the $q$-index curves obtained from the different completeness magnitudes.

## 6. Conclusions

If, upon repeating the same analyses on different cases and in different seismotectonic contexts, the connection between changes in the probability distributions and seismic phases were confirmed, it could be hypothesized that these observations have the value of precursors of the activation level in a seismic region. The existence of the seismic precursors is still a widely open problem; but, already in 1999, Evison claimed that "despite the difficulties confronting experimentation in this field, there is growing empirical evidence that precursors exist." [27]. Just as it is true that the most informative way to convey an earthquake prediction, including the uncertainties, is by means of probability distributions in the space–time–magnitude domain [28], it could be the same probability distributions that play the role of precursors.

We conclude by adopting Evison's statement [27]: "Prediction is ubiquitous in science as a test of understanding: to the extent that a phenomenon is understood, it can be predicted, and vice-versa. If earthquakes were unpredictable, seismogenesis would be a closed book, research would be futile, and the earthquake would remain, in the words of Alexander McKay [29] 'a visitation and a mistery'".

**Supplementary Materials:** The following supporting information can be downloaded at: https://www.mdpi.com/article/10.3390/e25101441/s1.

**Author Contributions:** Conceptualization, E.V. and R.R.; methodology, E.V. and R.R.; software, R.R.; validation, E.V.; formal analysis, R.R.; investigation, E.V. and R.R.; resources, E.V.; data curation, R.R.; writing—original draft preparation, R.R.; writing—review and editing, E.V.; visualization, R.R.; supervision, E.V.; funding acquisition, E.V. All authors have read and agreed to the published version of the manuscript.

**Funding:** This work was supported by the ICSC *National Research Centre for High Performance Computing, Big Data and Quantum Computing* (CN00000013, CUP B93C22000620006) within the European Union-NextGenerationEU program.

**Institutional Review Board Statement:** Not applicable.

**Data Availability Statement:** The data sets analyzed for this study can be found in the Italian Seismological Instrumental and Parametric Database (ISIDe) [20] at http://iside.rm.ingv.it (last accessed on 31 October 2021).

**Acknowledgments:** The authors are thankful to the editor and two anonymous reviewers for their very useful comments and suggestions, as well as to F. Ruggeri for his help in the proofreading.

**Conflicts of Interest:** The authors declare no conflict of interest. The views and opinions expressed are those of the authors alone and do not necessarily reflect those of the European Union or the European Commission. Neither the European Union nor the European Commission can be held responsible for them.

## References

1. Bak, P.; Christensen, K.; Danon, L.; Scanton, T. Unified scaling law for earthquakes. *Phys. Rev. Lett.* **2002**, *88*, 178501. [CrossRef] [PubMed]
2. Saichev, A.; Sornette, D. "Universal" distribution of interearthquake times explained. *Phys. Rev. Lett.* **2006**, *97*, 078501. [CrossRef] [PubMed]
3. Corral, A. Universal local versus unified global scaling laws in the statistics of seismicity. *Phys. A* **2004**, *340*, 590–597. [CrossRef]
4. Touati, S.; Naylor, M.; Main, I.G. Origin and nonuniversality of the earthquake interevent time distribution. *Phys. Rev. Lett.* **2009**, *102*, 168501. [CrossRef] [PubMed]
5. Rotondi, R. Bayesian nonparametric inference for earthquake recurrence time distributions in different tectonic regimes. *J. Geophys. Res.* **2010**, *115*, B01302. [CrossRef]
6. Tsallis, C. Possible generalization of Boltzmann-Gibbs statistics. *J. Stat. Phys.* **1988**, *52*, 479–487. [CrossRef]
7. Vallianatos, F.; Michas, G.; Papadakis, G. Nonextensive Statistical Seismology: An overview. In *Complexity of Seismic Time Series; Measurement and Application*; Chelidze, T., Telesca, L., Vallianatos, F., Eds.; Elsevier: Amsterdam, The Netherlands, 2018; pp. 25–59.
8. Vallianatos, F.; Sammonds, P. A non-extensive statistics of the fault-population at the Valles Marineris extensional province, Mars. *Tectonophysics* **2011**, *509*, 50–54. [CrossRef]
9. Rotondi, R.; Varini, E. Temporal Variations of the Probability Distribution of Voronoi Cells Generated by Earthquake Epicenters. *Front. Earth Sci.* **2022**, *10*, 928348. [CrossRef]
10. Abe, S.; Suzuki, N. Scale-free statistics of time interval between successive earthquakes. *Phys. A* **2005**, *350*, 588–596. [CrossRef]
11. Michas, G.; Vallianatos, F.; Sammonds, P. Non-extensivity and long-range correlations in the earthquake activity at the West Corinth rift (Greece). *Nonlinear Processes Geophys.* **2013**, *20*, 713–724. [CrossRef]
12. Queirós, S.M.D. On the emergence of a generalised gamma distribution. Application to traded volume in financial markets. *Europhys. Lett.* **2005**, *71*, 339–345. [CrossRef]
13. Gilks, W.R.; Richardson, S.; Spiegelhalter, D.J. *Markov Chain Monte Carlo in Practice*, 1st ed.; Chapman & Hall: London, UK, 1996.
14. Tsallis, C. *Introduction to Nonextensive Statistical Mechanics. Approaching a Complex World*; Springer: New York, NY, USA, 2009. [CrossRef]
15. Touchette, H. When is a quantity additive, and when is it extensive? *Phys. A* **2002**, *305*, 84–88. [CrossRef]
16. Roberts, C.P.; Casella, G. *Monte Carlo Statistical Methods*; Springer: New York, NY, USA, 2004; ISBN 978-0-387-21239-5.
17. Kass, R.E.; Raftery, A.E. Bayes factor. *J. Am. Stat. Assoc.* **1995**, *430*, 773–795. [CrossRef]
18. Gelman, A.; Carlin, J.B.; Stern, H.S.; Rubin, D.B. *Bayesian Data Analysis*; Chapman & Hall: London, UK, 2004.

19. DISS Working Group, Database of Individual Seismogenic Sources (DISS). *Version 3.2.1: A Compilation of Potential Sources for Earthquakes Larger than M 5.5 in Italy and Surrounding Areas*; Istituto di Geofisica e Vulcanologia: Roma, Italy, 2018. [CrossRef]
20. ISIDe Working Group. *Italian Seismological Instrumental and Parametric Database (ISIDe)*; Istituto Nazionale di Geofisica e Vulcanologia (INGV): Roma, Italy, 2007. [CrossRef]
21. Gasperini, P.; Lolli, B.; Vannucci, G. Empirical calibration of local magnitude data sets versus moment magnitude in Italy. *Bull. Seism. Soc. Am.* **2013**, *103*, 2227–2246. [CrossRef]
22. Wells, D.L.; Coppersmith, K.J. New empirical relationships among magnitude, rupture length, rupture width, rupture area, and surface displacement. *Bull. Seism. Soc. Am.* **1992**, *84*, 974–1002.
23. Hainzl, S. Rate-dependent incompleteness of earthquake catalogs. *Seismol. Res. Lett.* **2016**, *87*, 337–344. [CrossRef]
24. Papadopoulos, G.A.; Charalampakis, M.; Fokaefs, A.; Minadakis, G. Strong foreshock signal preceding the L'Aquila (Italy) earthquake ($M_w$ 6.3) of 6 April 2009. *Nat. Hazards Earth Syst. Sci.* **2010**, *10*, 19–24. [CrossRef]
25. Sebastiani, G.; Govoni, A.; Pizzino, L. Aftershocks patterns in recent central Apennines sequences. *J. Geophys. Res.* **2019**, mboxemph124, 3881–3897. [CrossRef]
26. Rotondi, R.; Bressan, G.; Varini, E. Analysis of temporal variarions of seismicity through nonextensive statistical physics. *Geophys. J. Int.* **2022**, *230*, 1318–1337. [CrossRef]
27. Evison, F. On the existence of earthquakes precursors. *Ann. Geofis.* **1999**, *42*, 763–770. [CrossRef]
28. UNESCO/IASPEI Working Group. Code of practice for earthquake prediction. *IUGG Chron.* **1984**, *165*, 115–124.
29. McKay, A. *Report on the Recent Seismic Disturbances within Cheviot County, in the Northern Canterbury, and the Amuri District of Nelson, New Zealand*; Government Printer: Wellington, New Zealand, 1984.


**Disclaimer/Publisher's Note:** The statements, opinions and data contained in all publications are solely those of the individual author(s) and contributor(s) and not of MDPI and/or the editor(s). MDPI and/or the editor(s) disclaim responsibility for any injury to people or property resulting from any ideas, methods, instructions or products referred to in the content.

entropy

MDPI

*Article*

# Universal Non-Extensive Statistical Physics Temporal Pattern of Major Subduction Zone Aftershock Sequences

**Eleni-Apostolia Anyfadi [1,2], Sophia-Ekaterini Avgerinou [1,2], Georgios Michas [1,2] and Filippos Vallianatos [1,2,*]**

[1] Section of Geophysics-Geothermics, Department of Geology and Geoenvironment, National and Kapodistrian University of Athens, 15784 Athens, Greece

[2] Institute of Physics of Earth's Interior and Geohazards, UNESCO Chair on Solid Earth Physics and Geohazards Risk Reduction, Hellenic Mediterranean University Research & Innovation Center, 73133 Chania, Greece

* Correspondence: fvallian@geol.uoa.gr

**Abstract:** Large subduction-zone earthquakes generate long-lasting and wide-spread aftershock sequences. The physical and statistical patterns of these aftershock sequences are of considerable importance for better understanding earthquake dynamics and for seismic hazard assessments and earthquake risk mitigation. In this work, we analyzed the statistical properties of 42 aftershock sequences in terms of their temporal evolution. These aftershock sequences followed recent large subduction-zone earthquakes of $M \geq 7.0$ with focal depths less than 70 km that have occurred worldwide since 1976. Their temporal properties were analyzed by investigating the probability distribution of the interevent times between successive aftershocks in terms of non-extensive statistical physics (NESP). We demonstrate the presence of a crossover behavior from power-law ($q \neq 1$) to exponential ($q = 1$) scaling for greater interevent times. The estimated entropic $q$-values characterizing the observed distributions range from 1.67 to 1.83. The $q$-exponential behavior, along with the crossover behavior observed for greater interevent times, are further discussed in terms of superstatistics and in view of a stochastic mechanism with memory effects, which could generate the observed scaling patterns of the interevent time evolution in earthquake aftershock sequences.

**Keywords:** subduction zones; megathrust earthquakes; aftershock sequences; interevent times; superstatistics; Tsallis entropy

**Citation:** Anyfadi, E.-A.; Avgerinou, S.-E.; Michas, G.; Vallianatos, F. Universal Non-Extensive Statistical Physics Temporal Pattern of Major Subduction Zone Aftershock Sequences. *Entropy* **2022**, *24*, 1850. https://doi.org/10.3390/e24121850

Academic Editor: Cataldo Godano

Received: 4 November 2022
Accepted: 16 December 2022
Published: 19 December 2022

**Copyright:** © 2022 by the authors. Licensee MDPI, Basel, Switzerland. This article is an open access article distributed under the terms and conditions of the Creative Commons Attribution (CC BY) license (https://creativecommons.org/licenses/by/4.0/).

## 1. Introduction

Megathrust faults, which bring together the surfaces of overthrusting and underthrusting plates in subduction zones, host the largest earthquakes on Earth. The downward motion of the subducting plate can generate exceptionally larger earthquakes with magnitudes even greater than $M_w$ 7, releasing tremendous amounts of seismic energy that can have devastating repercussions across wide geographic regions. These large earthquakes generate prolonged aftershock sequences, which may last days to months or even years [1–6]. Not only can the mainshock of a megathrust earthquake cause extensive damages to many structures in its vicinity and the loss of lives, but the aftershocks that follow may also have destructive consequences. As a result, emergency preparedness and planning must include the effects of aftershocks on susceptible sectors of society and infrastructure [1–4]. This is achieved by estimating the parameters of various well-established empirical laws for aftershock sequences following major mainshocks in a specific seismogenic region [6–8]. In a subduction zone, the intense friction between the descending and overriding plates may generate shallow, intermediate, or deep earthquakes, which may occur both within the descending and overriding plates as well as along the interface between the two plates [9,10]. The magnitudes of such earthquakes vary greatly, depending on the sorts of boundaries that cause them. Almost all major earthquakes occur along boundaries of convergence or transform. Megathrust earthquakes are the most devastating events, with magnitudes reaching or exceeding $M_W$ 9.0 in some cases. As a result, a deeper understanding is crucial for earthquake physics and earthquake early warning systems. Aftershock sequences following megathrust earthquakes can raise the level of seismic risk, even in distant areas far away from the mainshock's fault zone.

To this aim, the aftershock sequences of significant subduction-zone earthquakes with magnitudes greater than 7.0 that occurred between 1976 and 2020 were investigated, with a focus on their temporal distributions in view of the ideas of statistical physics. The statistical physics approach expresses the basic principles of the evolution of seismicity using the universal principle of entropy. In particular, we used the non-extensive statistical physics (NESP) framework, which provides a generalization of the ordinary Boltzmann–Gibbs (BG) statistical physics [11–17]. The main advantage of using NESP is that all length correlations and memory effects are considered in the temporal evolution of seismicity, including BG statistical physics as a particular case [18–22]. In this work, the temporal scaling characteristics of major subduction zone aftershock sequences in terms of NESP were studied by estimating the probability distribution of the interevent times between successive aftershocks and its non-additive entropic parameter ($q$) [12,18,23–29]. We show that the observed probability distributions present a universal behavior with a crossover from power-law ($q \neq 1$) to exponential ($q = 1$) scaling at longer interevent times.

## 2. Materials and Methods

Non-extensive statistical physics (NESP), based on the generalization of BG statistics, is an appropriate method to describe complex systems with (multi)fractality, long-range interactions, and long-term memory effects, resulting in power-law asymptotic behavior that is widely observed in nature [12,30–40]. Central to NESP is the expression of Tsallis entropy ($S_q$) [12,31–33], which is given as

$$S_q = k_B \frac{1 - \sum p^q(X)}{q - 1}, \tag{1}$$

in terms of a fundamental parameter's probability distribution ($p(X)$). In the present work, $X$ is the interevent time ($T$), i.e., the time interval between successive aftershock events [35,41], and $k_B$ is Boltzmann's constant. The $q$ parameter is the so-called entropic index, which signifies the degree of non-additivity in the system [12]. For the particular case of $q = 1$, $S_q = S_{BG}$ and the BG entropy is recovered. Despite the fact that $S_q$ and $S_{BG}$ have many similarities, such as non-negativity, expansibility, and concavity, there is a major difference between the two formalisms. The BG entropy is additive, meaning that the entropy of a coupled system is equal to the sum of the entropies of its constituent components, whereas the Tsallis entropy ($S_q$) (with $q \neq 1$) is non-additive [12,18,30,42,43]. Tsallis entropy satisfies the following condition for any two probabilistically independent systems, $A$ and $B$:

$$S_q(A + B) = S_q(A) + S_q(B) + \frac{(q-1)}{k_B} S_q(A) S_q(B), \tag{2}$$

In particular, $q < 1$ corresponds to superadditivity, and $q > 1$ corresponds to subadditivity, while the right-hand side of Equation (2) disappears when $q = 1$, so it corresponds to the additivity characteristics [12,24,44–56].

If $X$ is a physical parameter that characterizes the system, such as earthquake interevent times ($T$), the probability distribution ($p(T)$) is determined using the Lagrange multipliers method by maximizing the entropy under suitable constraints [12]. Using the previous approach, Equation (1) leads to the probability distribution function

$$p(T) = \frac{\left[1 - (1 - q)\left(\frac{T}{T_q}\right)\right]^{\frac{1}{1-q}}}{Z_q} = \frac{exp_q(-T/T_q)}{Z_q}, \tag{3}$$

where $Z_q$ is the $q$-partition function

$$Z_q = \int_0^\infty exp_q(-T/T_q)dT, \tag{4}$$

and $T_q$ is a generalized scaled interevent time. The nominator of Equation (3) is the $q$-exponential function, which is defined as [12]

$$exp_q(X) = [1 + (1-q)X]^{\frac{1}{1-q}}, \tag{5}$$

for $1 + (1-q)X \geq 0$, while in other cases $exp_q(X) = 0$.

The cumulative distribution function $P(>T) = N(>T)/N_o$, where $N(>T)$ is the number of interevent times with values greater than $T$ and $N_o$, their total number, is estimated as [24]

$$P(>T) = exp_Q(-T/T^*), \tag{6}$$

which has the mathematical form of a $q$-exponential function with $T^* = T_qQ$ and $q = 2 - (1/Q)$. The inverse function of the Q-exponential function is the so-called Q-logarithmic function, which is defined as [24]

$$ln_Q P(>T) = \frac{P(>T)^{1-Q} - 1}{1-Q}, \tag{7}$$

From Equation (7), it follows that $ln_Q P(>T) = -\frac{T}{T^*}$, indicating a straight line with slope $-1/T^*$.

### 3. Data Selection and Analysis

Herein, the statistical properties of the temporal evolution of aftershock sequences that followed large subduction-zone earthquakes worldwide during the last few decades are presented. The analyzed earthquakes occurred in various subduction zones all around the edge of the Pacific Ocean, Canada, Alaska, Russia, Indonesia, and Japan (Figure 1). We used 42 aftershock sequences generated by mainshocks of $M_w$ 7.0 and greater with focal depths less than 70 km that occurred from January 1976 to July 2020 and were located at a maximum distance of 100 km from the main event. The aftershock catalogues were extracted from the United States Geological Survey (USGS) database [48]. To create the catalogues of the aftershock sequences, an elliptical region with a maximum distance of 100 km from the mainshock was selected for each main event, based on the distribution of aftershocks, to designate a probable aftershock zone. Then, all earthquake events that occurred within this region, for a maximum period of two years after the mainshock, were included in the catalogue. Once the catalogues were defined, the frequency–magnitude distribution was used to estimate the completeness magnitude ($M_c$) of each aftershock sequence [49,50]. The catalogues that were analyzed consisted of at least 100 events. Ultimately, the final list included 42 mainshocks and their aftershock sequences (see Table 1).

Figure 1 depicts the geographical locations of the 42 studied mainshocks that occurred in subduction zones around the world. The event indexes correspond to the number of each mainshock in Table 1, which are listed chronologically from the oldest to the most recent. The parameters of each mainshock and its aftershock sequence, along with the entropic parameter ($q$), were calculated for the interevent time distribution and are presented in Table 1, along with the parameter $T_c$, which marks the crossover points between the non-additive and additive behavior in each aftershock sequence.

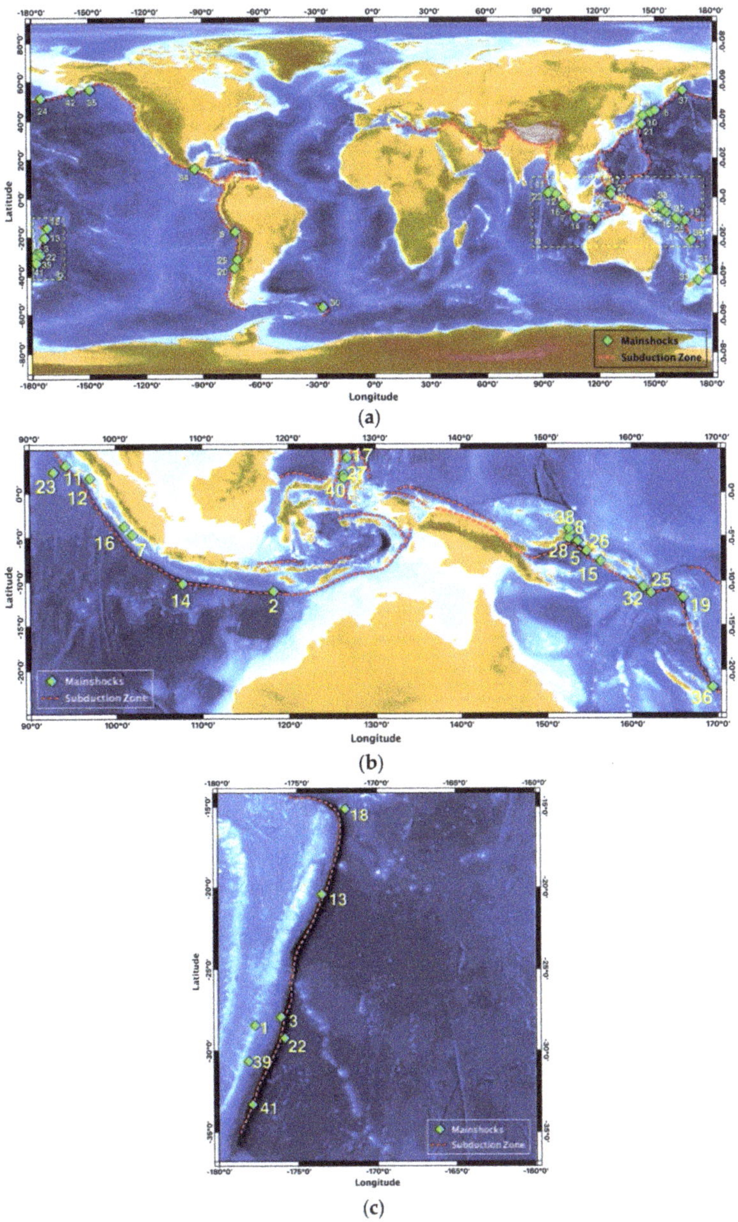


**Figure 1.** (**a**) Geographical distribution of the 42 subduction-zone earthquakes that were studied. The event indexes correspond to those listed in Table 1, while the dashed lines represent the subduction zones. The details of earthquakes located on the dashed squares are extracted in panels (**b**,**c**).

**Table 1.** Summary of the results for the 42 subduction zone aftershock sequences. $M_c$ is the completeness magnitude of each catalogue, $N$ is the number of aftershocks, $q$ the entropic index of the interevent time distribution, $T_q$ is the generalized scaled interevent time, and $T_c$ is the critical interevent time, where a crossover from NESP to BG statistical mechanics appeared (see the text for details).

| Index Number | Date | Epicenter (Lat, Lon) | Depth (km) | Duration (days) | Mainshock Magnitude ($M_w$) | $M_c$ | $N$ | $q_T$ | $T_q$ (s) | $T_c$ (s) |
|---|---|---|---|---|---|---|---|---|---|---|
| 1 | 14 January 1976 | −28.43, −177.66 | 33.00 | 346 | 8.0 | 5.4 | 101 | 1.78 | 4444 | $4 \times 10^5$ |
| 2 | 19 August 1977 | −11.14, 118.23 | 23.30 | 649 | 8.3 | 5.3 | 124 | 1.69 | 3750 | $5 \times 10^5$ |
| 3 | 20 October 1986 | −27.93, −176.07 | 50.40 | 732 | 7.7 | 5.3 | 103 | 1.74 | 1579 | $10 \times 10^5$ |
| 4 | 4 October 1994 | 43.60, 147.63 | 68.20 | 248 | 8.3 | 5.2 | 219 | 1.72 | 472 | $2 \times 10^5$ |
| 5 | 16 August 1995 | −5.51, 153.64 | 45.60 | 89 | 7.7 | 5.2 | 100 | 1.77 | 2791 | $4 \times 10^5$ |
| 6 | 3 December 1995 | 44.82, 150.17 | 25.90 | 333 | 7.9 | 5.2 | 138 | 1.83 | 200 | $2 \times 10^5$ |
| 7 | 4 June 2000 | −4.73, 101.94 | 43.90 | 719 | 7.9 | 5.2 | 162 | 1.82 | 1636 | $3 \times 10^5$ |
| 8 | 16 November 2000 | −4.56, 152.79 | 24.00 | 357 | 8.0 | 5.5 | 165 | 1.73 | 822 | $5 \times 10^5$ |
| 9 | 23 June 2001 | −17.28, −72.71 | 29.60 | 710 | 8.4 | 5.2 | 109 | 1.79 | 250 | $20 \times 10^5$ |
| 10 | 25 September 2003 | 42.21, 143.84 | 28.20 | 708 | 8.3 | 5.1 | 110 | 1.78 | 4000 | $6 \times 10^5$ |
| 11 | 26 December 2004 | 3.09, 94.26 | 28.60 | 692 | 9.0 | 5.1 | 356 | 1.69 | 3063 | $3 \times 10^5$ |
| 12 | 28 March 2005 | 1.67, 97.07 | 25.80 | 709 | 8.6 | 5.1 | 210 | 1.83 | 667 | $2 \times 10^5$ |
| 13 | 3 May 2006 | −20.39, −173.47 | 67.80 | 713 | 8.0 | 5.0 | 100 | 1.75 | 22,500 | $3 \times 10^5$ |
| 14 | 17 July 2006 | −10.28, 107.78 | 20.00 | 186 | 7.7 | 5.3 | 133 | 1.67 | 433 | $0.2 \times 10^5$ |
| 15 | 1 April 2007 | −7.79, 156.34 | 14.10 | 687 | 8.1 | 5.2 | 115 | 1.81 | 2692 | $8 \times 10^5$ |
| 16 | 12 September 2007 | −3.78, 100.99 | 24.40 | 722 | 8.5 | 5.1 | 174 | 1.83 | 833 | $1 \times 10^5$ |
| 17 | 11 February 2009 | 3.92, 126.81 | 23.90 | 699 | 7.2 | 5.2 | 126 | 1.81 | 346 | $20 \times 10^5$ |
| 18 | 29 September 2009 | −15.13, −171.97 | 12.00 | 728 | 8.1 | 5.2 | 228 | 1.83 | 1500 | $1 \times 10^5$ |
| 19 | 7 October 2009 | −11.86, 166.01 | 41.70 | 695 | 7.8 | 5.1 | 154 | 1.80 | 1600 | $4 \times 10^5$ |
| 20 | 27 February 2010 | −35.98, −73.15 | 23.20 | 720 | 8.8 | 5.0 | 190 | 1.76 | 6098 | $3 \times 10^5$ |
| 21 | 11 March 2011 | 37.52, 143.05 | 20.20 | 717 | 9.1 | 5.0 | 435 | 1.74 | 26,316 | $1 \times 10^5$ |
| 22 | 6 July 2011 | −29.22, −175.83 | 32.50 | 364 | 7.6 | 5.1 | 140 | 1.71 | 6286 | $4 \times 10^5$ |
| 23 | 11 April 2012 | 2.35, 92.82 | 45.60 | 359 | 8.6 | 5.2 | 129 | 1.80 | 286 | $8 \times 10^5$ |
| 24 | 30 August 2013 | 51.54, −175.23 | 29.00 | 724 | 7.0 | 4.8 | 133 | 1.80 | 224 | $10 \times 10^5$ |
| 25 | 12 April 2014 | −11.35, 162.24 | 27.30 | 411 | 7.6 | 4.9 | 177 | 1.75 | 173 | $60 \times 10^5$ |
| 26 | 19 April 2014 | −6.64, 154.67 | 43.40 | 722 | 7.5 | 4.7 | 107 | 1.83 | 776 | $2 \times 10^5$ |
| 27 | 15 November 2014 | 1.98, 126.37 | 45.00 | 664 | 7.1 | 4.6 | 119 | 1.81 | 472 | $3 \times 10^5$ |
| 28 | 29 March 2015 | −5.18, 152.59 | 37.60 | 719 | 7.5 | 4.7 | 245 | 1.71 | 2857 | $2 \times 10^5$ |
| 29 | 16 September 2015 | −31.57, −71.67 | 22.40 | 575 | 8.3 | 4.6 | 213 | 1.78 | 222 | $6 \times 10^5$ |
| 30 | 28 May 2016 | −56.24, −26.94 | 68.00 | 616 | 7.2 | 4.9 | 107 | 1.78 | 3333 | $3 \times 10^5$ |
| 31 | 1 September 2016 | −37.36, 179.15 | 19.00 | 580 | 7.0 | 4.7 | 166 | 1.74 | 263 | $3 \times 10^5$ |
| 32 | 13 November 2016 | −42.74, 173.05 | 15.10 | 648 | 7.8 | 4.9 | 144 | 1.72 | 556 | $50 \times 10^5$ |
| 33 | 8 December 2016 | −10.68, 161.33 | 40.00 | 479 | 7.8 | 5.0 | 100 | 1.72 | 278 | $10 \times 10^5$ |
| 34 | 8 September 2017 | 15.02, −93.90 | 47.40 | 737 | 8.2 | 4.8 | 252 | 1.74 | 921 | $3 \times 10^5$ |
| 35 | 23 January 2018 | 56.00, −149.17 | 14.10 | 437 | 7.9 | 4.5 | 113 | 1.81 | 367 | $2 \times 10^5$ |
| 36 | 5 December 2018 | −21.95, 169.43 | 10.00 | 670 | 7.5 | 4.8 | 166 | 1.78 | 217 | $1 \times 10^5$ |
| 37 | 20 December 2018 | 55.10, 164.70 | 16.60 | 662 | 7.3 | 4.6 | 131 | 1.78 | 435 | $8 \times 10^5$ |
| 38 | 14 May 2019 | −4.05, 152.60 | 10.00 | 573 | 7.6 | 4.9 | 103 | 1.86 | 411 | $4 \times 10^5$ |
| 39 | 15 June 2019 | −30.64, −178.10 | 46.00 | 530 | 7.3 | 4.8 | 217 | 1.75 | 2250 | $3 \times 10^5$ |
| 40 | 14 November 2019 | 1.62, 126.42 | 33.00 | 364 | 7.1 | 4.8 | 191 | 1.82 | 143 | $2 \times 10^5$ |
| 41 | 18 June 2020 | −33.29, −177.86 | 10.00 | 102 | 7.4 | 4.9 | 121 | 1.72 | 694 | $0.6 \times 10^5$ |
| 42 | 22 July 2020 | 55.07, −158.60 | 28.00 | 138 | 7.8 | 4.2 | 207 | 1.69 | 313 | $0.5 \times 10^5$ |

For each aftershock sequence, the cumulative interevent time distribution was estimated, and the corresponding fitting with the Q-exponential function, up to the value $T_c$, provided the Q and $q$ parameter values. We noted that, for large values of $T$, with

$T > T_c$, a deviation from the Q-exponential function was observed (e.g., Figure 2a). In addition, the Q-logarithmic function of $P(>T)$ as a function of the interevent times ($T$) was plotted using the Q value estimated in the previous analysis. Then, the range of interevent times where $\ln_Q P(>T)$ vs. $T$ was a straight line, given by Equation (8), was defined with its correlation coefficient. The deviation from the linearity at $T_c$ indicated the crossover between NESP and BG statistical physics. Furthermore, the evolution of the interevent times ($T$) as a function of time ($t$) since the main event further indicated that, in short time scales after the mainshock, the main driving mechanism is governed by NESP, while, as the aftershock sequence evolves at $T > T_c$, the system is governed by BG statistical mechanics. The aforementioned earthquake statistical analysis and processing was carried out for all major earthquake aftershock sequences listed in Table 1, with results consistent with the previous analysis for all aftershock sequences, indicating a universal behavior in their temporal evolution.

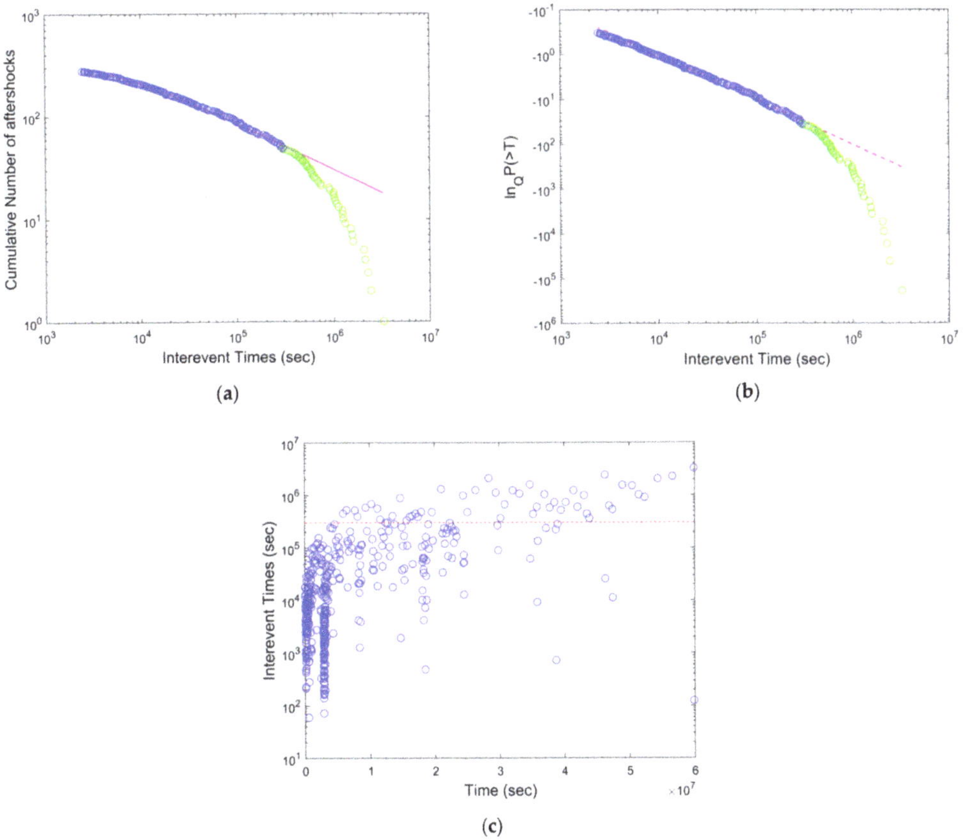


**Figure 2.** (**a**) The cumulative distribution function of the interevent times for the *2004 $M_w$ 9.0 Sumatra–Andaman Islands Earthquake*. The magenta line is the Q-exponential function fitting with $q = 1.69$. (**b**) The Q-logarithmic function of $P(>T)$ as a function of the interevent times ($T$), where the dashed line is the Q-exponential function fitting with $q = 1.69$, exhibiting a correlation coefficient of 0.9923. The deviation from linearity suggests a $T_c$ value close to $3 \times 10^5$ s. (**c**) The evolution of interevent times ($T$) as a function of the time ($t$) since the main event. The $T_c$ value is indicated by the red dashed line.

In Table 1, the results of the analysis of the 42 subduction zone aftershock sequences are summarized. For each aftershock sequence, the entropic index ($q$) of the interevent time distribution, along with the generalized scaled interevent time ($T_q$) and the critical interevent time ($T_c$), are presented. In the following section, we present some characteristic cases referring to the strongest events during the last few decades: the 2004 $M_W$ 9.0 Sumatra–Andaman Islands Earthquake and the 2011 $M_W$ 9.1 Great Tohoku (Japan) Earthquake. The corresponding results for the other aftershock sequences listed in Table 1 are provided in the Supplementary Materials.

### 3.1. The 2004 $M_w$ 9.0 Sumatra–Andaman Islands Earthquake

The $M_W$ 9.0 Sumatra earthquake, which occurred on 26 December 2004 at a focal depth of 30 km, was the fourth largest earthquake recorded since 1900. It originated from thrust faulting between the meeting point of the Indian plate and the Burma micro-plate. According to the USGS database, 356 aftershocks occurred with magnitudes greater than $M_W$ 5.1 in a period of two years after the mainshock. One of the greatest disasters recorded in human history was brought on by the tsunami generated by the mainshock. More than 283,000 people were killed in total, while the severeness and impact of the earthquake were demonstrated by the fact that the tsunami crossed the Pacific and Atlantic Oceans and was recorded in New Zealand as well as along the west and east coasts of South and North America. Tsunamis continued to occur in Mozambique, South Africa, Australia, and Antarctica. The mainshock even caused eruptions in a mud volcano at Baratang, Andaman Islands, on 28 December 2004 [51,52].

The cumulative distribution function of the interevent times for the 2004 $M_W$ 9.0 Sumatra–Andaman Islands Earthquake was fitted for $T < T_c$ with the Q-exponential function for $q = 1.69$ (Figure 2). The corresponding Q-logarithmic function of $P(>T)$, as a function of the interevent times for $q = 1.69$, was fitted by a straight line, as given by Equation (8), with a correlation coefficient of 0.9923. The deviation from linearity was observed at a $T_c$ value close to $3 \times 10^5$ s, indicating the crossover point between NESP and BG statistical physics. Furthermore, the evolution of the interevent times ($T$) as a function of the time ($t$) since the main event is presented in Figure 2c. The $T_c$ value (red dashed line in Figure 2c) indicates that the majority of interevent times in the early period following the mainshock had $T$ values less than $T_c$, suggesting that the Tsallis entropic mechanism was predominant in the immediate part of the aftershocks' evolution. As time evolved, some of the characteristics of the early aftershock sequence, such as that of the long-range memory related to NESP, were less predominant and the BG statistical physics recovered (i.e., $q = 1$).

### 3.2. The 2011 $M_w$ 9.1 Great Tohoku (Japan) Earthquake

The $M_W$ 9.1 Tohoku earthquake, which occurred on 11 March 2011, was generated by thrust faulting at the boundary of the subduction zone between the Pacific and North American plates. It was located close to Honshu, on Japan's northeast coast. This earthquake was preceded by many strong foreshocks that occurred over a period of two days prior to the mainshock, starting on March 9 with an $M_W$ 7.2 earthquake and continuing the same day with three more earthquakes greater than $M_W$ 6.0. Additionally, during the period of two years after the mainshock, 435 aftershocks of magnitudes equal or greater than $M_W$ 5.0 occurred. The mainshock generated a tsunami reaching heights up to 40.5 m, with a devastating impact on Japan's northern island coastal areas, before spreading all over the Pacific coasts of North and South America, from Alaska to Chile. The event's destructive tsunami waves are estimated to have contributed to 19,759 fatalities and 6242 injuries. The Fukushima nuclear power plant was damaged as a result of the tsunami, which led to significant radioactive pollution [52,53].

The cumulative distribution function of the interevent times, presented in Figure 3a, presented similar characteristics as the previous case of the $M_W$ 9.0 Sumatra earthquake. For $T < T_c$, it was fitted by a Q-exponential function with $q = 1.74$ (Figure 3a), while the corresponding Q-logarithmic function of $P(>T)$ presented a correlation coefficient of

0.9828 for $T$ values up to $T_c$ (Figure 3b). The deviation from linearity was observed at a $T_c$ value close to $1 \times 10^5$ s, indicating the crossover point between NESP and BG statistical physics (Figure 3c).

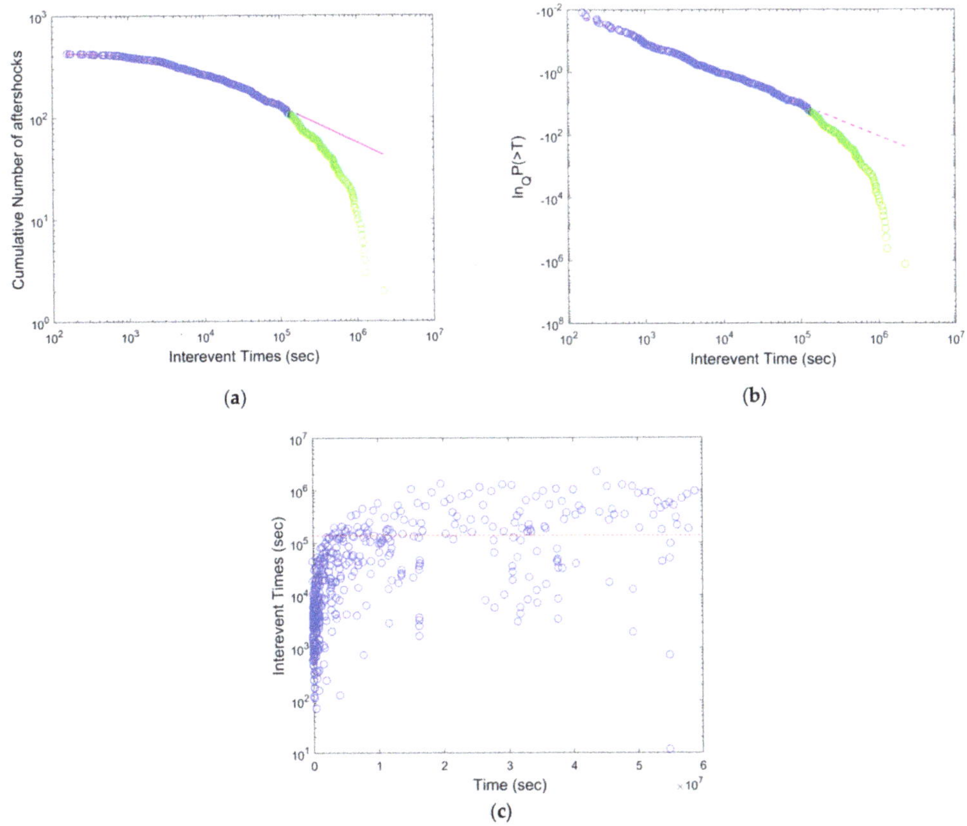


**Figure 3.** (**a**) The cumulative distribution function of the interevent times ($T$) for the *2011 Mw 9.1 Great Tohoku (Japan) Earthquake*. The magenta line is the Q-exponential function fitting with $q = 1.74$. (**b**) The Q-logarithmic function of $P(>T)$ as a function of the interevent times ($T$). The dashed line is the Q-exponential function fitting with $q = 1.74$, presenting the correlation coefficient of 0.9828. The deviation from linearity suggests a $T_c$ value close to $1 \times 10^5$ s. (**c**) The evolution of the interevent times ($T$) as a function of the time ($t$) since the main event. The $T_c$ value is indicated by the red dashed line.

Since the results have demonstrated that $P(>T) = \exp_Q(-T/T^*)$ for $T < T_c$, we introduced a new variable, $x = T/T_c$, for which $x < 1$ suggests the range where the Tsallis entropic mechanism is predominant, while $x > 1$ is related to the exponential roll-off in the tail of the distribution. In Figure 4, the cumulative interevent time distributions are presented for all aftershock sequences listed in Table 1. For all analyzed aftershock sequences, a deviation from the Q-exponential function existed for $x > 1$ (i.e., $T > T_c$). It is straightforward that $P(>x) = \exp_Q(-x/x^*)$, where $x^* = T^*/T_c$. An inspection of Figure 4 suggested that for $0.01 < x < 1$ a power-law scaling range was observed for all aftershock sequences, with an average slope of 0.33, which corresponds to a value of $q \approx 1.75$, in general agreement with the range of $q$ values reported in Table 1.

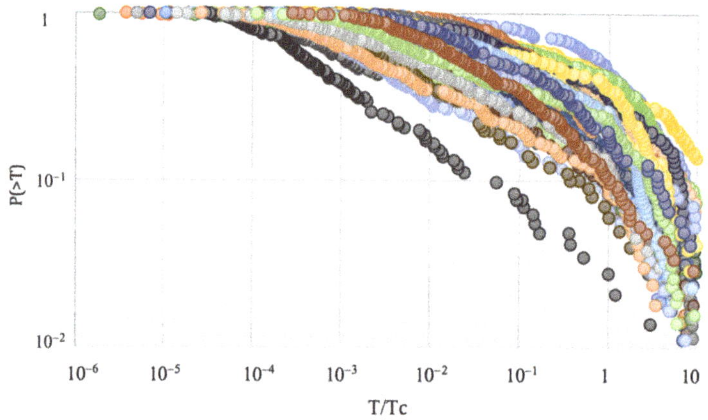


**Figure 4.** The cumulative interevent time distributions for all analyzed aftershock sequences as functions of $T/T_c$. For all sequences, a deviation from the Q-exponential function is presented for $T/T_c > 1$.

## 4. Discussion

In the present work, the temporal patterns of major subduction zone aftershock sequences that occurred from 1976 to 2020 were analyzed in terms of non-extensive statistical physics. We observed that in all cases a Q-exponential function described the cumulative distribution $P(>T)$ of the aftershock interevent times for short timescales, while for large values of $T$ ($T > T_c$), where $T_c$ was a critical crossover interevent time between the non-additive and additive behavior, a deviation from the Q-exponential function appeared. For each aftershock sequence, the entropic parameter ($q$) was estimated by fitting a Q-exponential function to the observed data up to a value near $T_c$. Thus, the applicability of non-extensive statistical physics to the cumulative distribution functions of interevent times and the presence of a crossover behavior from power-law ($q \neq 1$) to exponential ($q = 1$) scaling for greater interevent times was demonstrated. The latter implies a sub-additive process with $q$-values greater than one, supporting the concept of long-range memory in the temporal evolution of aftershocks for $T < T_c$. Furthermore, most of the estimated non-extensive $q$-values that characterized the observed distributions were within the range of 1.67–1.83.

The observed deviation from the Q-exponential function for longer interevent times can be described as the superposition of two aftershock mechanisms. The first mechanism, described by Tsallis entropy, was dominant for interevent times with $T < T_c$, whereas the second, characterized by an exponential function, became evident for $T > T_c$. To incorporate a crossover from anomalous ($q \neq 1$) to normal BG ($q = 1$) statistical physics, we introduced the generalization presented in [54,55] where

$$\frac{dp(T)}{dT} = -\beta_1 p - (\beta_q - \beta_1)p^q, \tag{8}$$

whose solution is

$$p(T) = C\left[1 - \frac{\beta_q}{\beta_1} + \frac{\beta_q}{\beta_1}e^{(q-1)\beta_1 T}\right]^{1/1-q}, \tag{9}$$

where $C$ is a normalization factor and $p(T)$ decreases monotonically with increasing $T$ for positive $\beta_q$ and $\beta_1$. As a result, in the case where $(q-1)\beta_1 \ll 1$, Equation (9) is approximated with a $q$-exponential, $p(T) \approx Cexp_q(-T/T_q)$, where $T_q = 1/\beta_q$, whereas for $(q-1)\beta_1 \gg 1$, the asymptotic behavior of the probability distribution $p(T) \propto \left(\frac{\beta_1}{\beta_q}\right)^{1/(q-1)} e^{-\beta_1 T}$ is an exponential function, where $T_c = 1/(q-1)\beta_1$ is the crossover point between the non-additive and additive behavior [11,32].

The $q$-exponential scaling behavior of interevent times for $T < T_c$ can be originated from a simple mechanism, namely a gamma-distributed allocated parameter ($\beta$) of the local Poisson process, and may be used to explain the interevent time distribution in aftershock sequences. The $T_c$ value indicated that in the early aftershock period the majority of interevent times had $T$ values lower than $T_c$ and their distributions were described by NESP, while properties such as long-range memory, associated with NESP, became less prominent as the system relaxed and the BG statistical physics recovered [56–61].

The $q$-exponential behavior of the interevent times can further be viewed in terms of superstatistics, which are based on a superposition of ordinary local equilibrium statistical mechanics with a suitable intensive parameter ($\beta$) that varies as a gamma distribution on a reasonably wide temporal scale and is supplementary to NESP [19,23,61–65].

Then, a superstatistical approach for the interevent times of the earthquake aftershock sequences can be used, where the local Poisson process $p(T|\beta) = \beta e^{-\beta T}$ with $\beta$ as an intensive fluctuating parameter has a particular value denoted by the equation $p(T|\beta)$. On a long time scale, this parameter is distributed with the probability density ($f(\beta)$) [19,62–66]. Then, the probability distribution ($p(T)$) is given as:

$$p(T) = \int_0^{\infty} f(\beta)\beta e^{-\beta T} d\beta, \tag{10}$$

In the case where the probability density of β is given by a gamma distribution:

$$f(\beta) = \frac{1}{\Gamma(n/2)} \left(\frac{n}{2\beta_0}\right)^{n/2} \beta^{\frac{n}{2}-1} exp\left(-\frac{n\beta}{2\beta_0}\right), \tag{11}$$

the integral (10) can easily be evaluated [67] and $p(T) \approx C(1 + B(q-1)T)^{1/(1-q)}$ is obtained, which is exactly the result estimated in the frame of NESP, with $q = 1 + (\frac{2}{n+2})$ and $B = 2\beta_0/(2-q)$ [23]. Since the $q$ value was in the range of 1.67–1.81, it suggested that the system was derived by a low number of degrees of freedom, possibly close to one.

This implied that a stochastic mechanism with memory effects can be the driving mechanism in the temporal evolution of an aftershock sequence. In agreement with [68] (see also [59]), we may consider the following stochastic differential equation for the evolution of seismicity:

$$dT = -\gamma(T - \langle T \rangle)dt + \varphi\sqrt{T}\, W_t, \tag{12}$$

where the temporal occurrence of earthquakes is represented by the interevent time series ($T$) after some time ($t$). The latter stochastic equation manifests two parts controlling the evolution of seismicity. The first deterministic part aims to keep the seismic rate ($R$) stable to the typical value of $R = 1/\langle T \rangle$, according to a restoring constant ($\gamma$) that represents the rate of relaxation to the mean waiting time ($\langle T \rangle$). The second stochastic part represents memory effects in the evolution of seismicity. The stochastic term $W_t$ is the standard Wiener process following a Gaussian distribution with a zero mean and unitary variance that mimics the microscopic effects in the evolution of interevent times in the aftershock sequence. Due to its random sign, $W_t$ leads to an increase ($W_t > 0$) or decrease ($W_t < 0$) in $T$. We note that large values of $T$ provoke large amplitudes in the stochastic term, leading to an increase or decrease in $T$, depending on the sign of $W_t$. The term $\varphi$ adds some noise to the process and can be expressed as a function of the mean interevent time ($\langle T \rangle$) and the restoring constant ($\gamma$) as $\varphi = \sqrt{2\gamma\langle T \rangle}$.

The stochastic differential equation given in Equation (12) is a classic example of multiplicative noise, further known in statistics as the Feller process [67,69].

To determine the evolution of the interevent time series ($T$) after some time ($t$), given by the probability distribution $f(T,t)$, we can write the corresponding Fokker–Planck equation for Equation (12) [70,71]:

$$\frac{\partial f(T,t)}{\partial t} = \frac{\partial}{\partial T}[\gamma(T - \langle T \rangle)f(T,t)] + \frac{\partial^2}{\partial T^2}[T\langle T \rangle\gamma f(T,t)], \tag{13}$$

The stationary solution of the latter Fokker–Planck equation, Equation (13), is the distribution [70]:

$$p(T/\langle T\rangle) = f(T) = \frac{1}{\langle T\rangle} e^{-\frac{\gamma}{\langle T\rangle}T}, \tag{14}$$

In this case, Equation (14) provides the conditional probability of *T*, given <*T*>.

Furthermore, we can consider local fluctuations in the seismic rate ($R = 1/<T>$), which are associated with non-stationarities in the evolution of the earthquake activity over time scales much larger than $\gamma^{-1}$, which is necessary for Equation (12) to reach stationarity. In this case, local fluctuations in the mean interevent time (<*T*>) appear, and we may assume that these fluctuations follow the stationary gamma distribution:

$$f(\langle T\rangle) = \frac{\left(\frac{1}{\lambda}\right)^{\delta}}{\Gamma[\delta]} \langle T\rangle^{-(1+\delta)} e^{-\frac{1}{\lambda\langle T\rangle}}, \tag{15}$$

The marginal probability of *T*, independent of <*T*>, is then given by [68]:

$$p(T) = \int_0^{\infty} p(T/\langle T\rangle) f(\langle T\rangle) d\langle T\rangle, \tag{16}$$

Performing the integration, we obtain the solution for varying <*T*>:

$$p(T) = \frac{\lambda\Gamma[1+\delta]}{\Gamma[\delta]} (1+\lambda T)^{-(1+\delta)}, \tag{17}$$

By further carrying out the changes in the variables:

$$\lambda = \frac{q-1}{T_o} \text{ and } \delta = \frac{1}{q-1} - 1 = \frac{2-q}{q-1} \tag{18}$$

and considering the *q*-exponential function given in Equation (3), Equation (12) can be written as [26,68]:

$$p(T) = \frac{(q-1)\Gamma\left(\frac{1}{q-1}\right)}{T_o\Gamma\left[\frac{1}{q-1} - 1\right]} exp_q\left(-\frac{T}{T_o}\right) \tag{19}$$

The last equation, Equation (19), is the *q*-generalized gamma function [68], and the last term on the right-hand side has the exact form of the *q*-exponential function given in Equation (3). Equation (19) was derived by the stochastic model, Equation (12), for a varying mean interevent time (<*T*>), i.e., non-stationary earthquake activity.

## 5. Conclusions

In order to statistically analyze the temporal patterns of aftershock sequences in major subduction zones, we examined the interevent time distribution for each sequence. The NESP approach to the interevent time distribution indicated a system in anomalous equilibrium, with a crossover behavior from anomalous ($q > 1$) to normal ($q = 1$) statistical physics for greater interevent times. The range of the non-extensive parameter ($q_T$) for all analyzed sequences was between 1.67 and 1.83. The models used in the analysis fit the observed distributions rather well, indicating the usefulness of NESP in investigating such phenomena. Finally, the superstatistical approach led us to the conclusion that significant non-additive characteristics and a high-level organizational structure describe the earthquake aftershock sequences that occurred in subduction zones [55].

**Supplementary Materials:** The following supporting information can be downloaded at: https://www.mdpi.com/article/10.3390/e24121850/s1.

**Author Contributions:** Conceptualization, F.V.; methodology, F.V., E.-A.A., S.-E.A. and G.M.; software, G.M. and E.-A.A.; validation, E.-A.A. and S.-E.A.; formal analysis, F.V., E.-A.A., S.-E.A. and G.M.; investigation, F.V., E.-A.A., S.-E.A. and G.M.; resources, F.V., E.-A.A., S.-E.A. and G.M.; data curation, F.V., E.-A.A., S.-E.A. and G.M.; writing—original draft preparation, FV, E.-A.A., S.-E.A. and G.M.; writing—review and editing, F.V., E.-A.A. and G.M.; visualization, F.V. and E.-A.A.; supervision, F.V. All authors have read and agreed to the published version of the manuscript.

**Funding:** This research received no external funding.

**Data Availability Statement:** Not applicable.

**Conflicts of Interest:** The authors declare no conflict of interest.

## References

1. Shcherbakov, R.; Goda, K.; Ivanian, A.; Atkinson, G.M. Aftershock Statistics of Major Subduction Earthquakes. *Bull. Seism. Soc. Am.* **2013**, *103*, 3222–3230. [CrossRef]
2. Becker, T.W.; Faccenna, C. A Review of the Role of Subduction Dunamics for Regional and Global Plate Motions. In *Subduction Zone Geodynamics*, 1st ed.; Lallemand, S., Funiciello, F., Eds.; Springer: Berlin/Heidelberg, Germany, 2009; pp. 3–5. [CrossRef]
3. Stern, R.J. Subduction Zones. *Rev. Geophys.* **2002**, *40*, 1012. [CrossRef]
4. Lobkovsky, L.I.; Kerchman, V.I.; Baranov, B.V.; Pristavakina, E.I. Analysis of Seismotectonic Processes in Subduction Zones from the Standpoint of a Keyboard Model of Great Earthquakes. *Tectonophysics* **1991**, *199*, 211–212. [CrossRef]
5. Heuret, A.; Lallemand, S.; Funiciello, F.; Piromallo, C.; Faccena, C. Physical characteristics of subduction interface type seismogenic zones revisited. *Geochem. Geophys. Geosyst.* **2011**, *12*, Q01004. [CrossRef]
6. Bilek, S.L.; Lay, T. Subduction Zone Megathrust Earthquakes. *Geosph.* **2018**, *14*, 1468–1470. [CrossRef]
7. Shcherbakov, R.; Yakovlev, G.; Turcotte, D.L.; Rundle, J.B. Model for the Distribution of Aftershock Interoccurrence Times. *Phys. Rev. Lett.* **2005**, *95*, 218501. [CrossRef]
8. Shcherbakov, R.; Turcotte, D.L.; Rundle, J.B. Aftershock Statistics. *Pure Appl. Geophys.* **2005**, *162*, 1053–1054. [CrossRef]
9. Ruff, L.; Kanamori, H. Seismicity and the Subduction Process. *Phys. Earth Plan. Int.* **1980**, *23*, 240–250. [CrossRef]
10. Zhang, L. Spatiotemporal Seismic Hazard and Risk Assessment of Global M9.0 Megathrust Earthquake Sequences. Ph.D. Thesis, University of Bristol, Bristol, UK, 24 March 2020.
11. Sugiyama, M. Introduction to the topical issue: Nonadditive entropy and non-extensive statistical mechanics. *Contin. Mech. Thermodyn.* **2004**, *16*, 221–222. [CrossRef]
12. Tsallis, C. *Introduction to Nonextensive Statistical Mechanics*; Springer: Berlin/Heidelberg, Germany, 2009; pp. 41–218. [CrossRef]
13. Cohen, E.G.D. Boltzmann and Einstein: Statistics and dynamics—An unsolved problem. *Pramana—J. Phys.* **2005**, *64*, 635–643. [CrossRef]
14. Mathai, A.M.; Haubold, H.J. Pathway model, superstatistics, Tsallis statistics, and a generalized measure of entropy. *Phys. A Stat. Mech. Appl.* **2007**, *375*, 110–122. [CrossRef]
15. Baldovin, F.; Moyano, L.; Tsallis, C. Boltzmann-Gibbs thermal equilibrium distribution for classical systems and Newton law: A computational discussion. *Eur. Phys. J. B* **2006**, *52*, 113–117. [CrossRef]
16. Schwämmle, V.; Tsallis, C. Two-parameter generalization of the logarithm and exponential functions and Boltzmann-Gibbs-Shannon entropy. *J. Math. Phys.* **2007**, *48*, 113301. [CrossRef]
17. Fukuda, I.; Nakamura, H. Deterministic generation of the Boltzmann-Gibbs distribution and the free energy calculation from the Tsallis distribution. *Chem. Phys. Lett.* **2003**, *382*, 110–122. [CrossRef]
18. Tsallis, C. Nonextensive Statistical Mechanics: Construction and Physical Interpretation. In *Nonextensive Entropy: Interdisciplinary Applications*; Murray, G.M., Tsallis, C., Eds.; Oxford Academic: Oxford, NY, USA, 2004; pp. 1–53. [CrossRef]
19. Beck, C.; Cohen, E.G.D. Superstatistics. *Phys. A* **2003**, *322*, 267–269. [CrossRef]
20. Lyra, M.L.; Tsallis, C. Nonextensivity and Multifractality in Low-Dimensional Dissipative Systems. *Phys. Rev. Lett.* **1998**, *80*, 53–56. [CrossRef]
21. Silva, R.; França, G.S.; Vilar, C.S.; Alcaniz, J.S. Nonextensive models for earthquakes. *Phys. Rev. E* **2006**, *73*, 026102. [CrossRef]
22. Sarlis, N.V.; Skordas, E.S.; Varotsos, P.A. Nonextensivity and natural time: The case of seismicity. *Phys. Rev. E* **2010**, *82*, 0211. [CrossRef]
23. Vallianatos, F.; Pavlou, K. Scaling properties of the Mw 7.0 Samos (Greece), 2020 aftershock sequence. *Acta Geophys.* **2021**, *69*, 5–7. [CrossRef]
24. Vallianatos, F.; Papadakis, G.; Michas, G. Generalized statistical mechanics approaches to earthquake and tectonics. *Proc. R. Soc. A* **2016**, *472*, 20160497. [CrossRef]
25. Vallianatos, F. A description of seismicity based on non-extensive statistical Physics. In *Earthquakes and Their Impact on Society*; D'Amico, S., Ed.; Springer Nature Hazzrds: Berlin/Heidelberg Germany, 2016; pp. 1–42. [CrossRef]
26. Michas, G.; Pavlou, K.; Avgerinou, S.-E.; Anyfadi, E.-A.; Vallianatos, F. Aftershock patterns of the 2021 Mw 6.3 Northern Thessaly (Greece) earthquake. *J. Seismol.* **2022**, *26*, 201–225. [CrossRef]

27. Ramírez-Rojas, A.; Sigalotti, L.G.; Márquez, E.L.F.; Rendón, O. Non-extensive statistcs in time series: Tsallis theory. In *Time Series Analysis in Seismology: Practical Applications*; Ramírez-Rojas, A., Sigalotti, L.G., Márquez, E.L.F., Rendón, O., Eds.; Elsevier: Amsterdam, The Netherlands, 2019; pp. 147–164. [CrossRef]
28. Vallianatos, F.; Karakostas, V.; Papadimitriou, E. A Non-Extensive Statistical Physics View in the Spatiotemporal Properties of the 2003 (Mw6.2) Lefkada, Ionian Island Greece, Aftershock Sequence. *Pure Appl. Geophys.* **2013**, *171*, 1443–1449. [CrossRef]
29. Papadakis, G.; Vallianatos, F.; Sammonds, P. A Nonextensive Statistical Physics Analysis of the 1995 Kobe, Japan Earthquake. *Pure Appl. Geophys.* **2014**, *172*, 1923–1927. [CrossRef]
30. Vallianatos, F.; Michas, G.; Papadakis, G.; Sammonds, P. A Non-Extensive Statistical Physics View to the Spatiotemporal Properties of the June 1995, Aigion Earthquake (M6.2) Aftershock Sequence (West Corinth Rift, Greece). *Acta Geophys.* **2012**, *60*, 759–765. [CrossRef]
31. Tsallis, C. Possible generalization of Boltzmann-Gibbs statistics. *J. Stat. Phys.* **1988**, *52*, 479–487. [CrossRef]
32. Tsallis, C. Nonadditive entropy and non-extensive statistical mechanics—An overview after 20 years. *Brazil. J. Phys.* **2009**, *39*, 337–339. [CrossRef]
33. Queirós, S.; Moyano, L.; de Souza, J. A nonextensive approach to the dynamics of financial observables. *Eur. Phys. J. B* **2007**, *55*, 161–167. [CrossRef]
34. Tzanis, A.; Efstathiou, A.; Vallianatos, F. Are Seismogenetic Systems Random or Organized? A Treatise of Their Statistical Nature Based on the Seismicity of the North-Northeast Pacific Rim. In *Complexity of Seismic Time Series: Measurement and Application*, 1st ed.; Chelidze, T., Vallianatos, F., Telesca, L., Eds.; Elsevier: Amsterdam, The Netherlands, 2018; pp. 370–377.
35. Michas, G.; Vallianatos, F.; Sammonds, P. Non-extensivity and long-range correlations in the earthquake activity at the West Corinth rift (Greece). *Nonlin. Process. Geophys.* **2013**, *20*, 713–724. [CrossRef]
36. Vallianatos, F.; Koutalonis, I.; Chatzopoulos, G. Evidence of Tsallis entropy signature on medicane induced ambient seismic signals. *Phys. A* **2019**, *520*, 36–42. [CrossRef]
37. Vallianatos, F.; Michas, G.; Benson, P.; Sammonds, P. Natural time analysis of critical phenomena: The case of acoustic emissions in triaxially deformed Etna basalt. *Phys. A Stat. Mech. Appl.* **2013**, *392*, 5172–5178. [CrossRef]
38. Chochlaki, K.; Michas, G.; Vallianatos, F. Complexity of the Yellowstone Park Volcanic Field Seismicity in Terms of Tsallis Entropy. *Entropy* **2018**, *20*, 721. [CrossRef] [PubMed]
39. Michas, G.; Sammonds, P.; Vallianatos, F. Dynamic Multifractality in Earthquake Time Series: Insights from the Corinth Rift, Greece. *Pure Appl. Geophys.* **2015**, *172*, 1909–1921. [CrossRef]
40. Caruso, F.; Pluchino, A.; Latora, V.; Vinciguerra, S.; Rapisarda, A. Analysis of self-organized criticality in the Olami-Feder-Christensen model and in real earthquakes. *Phys. Rev. E* **2007**, *75*, 055101(R). [CrossRef] [PubMed]
41. Michas, G.; Papadakis, G.; Vallianatos, F. A Non-Extensive approach in investigating Greek seismicity. *Bull. Geol. Soc. Greece* **2016**, *XLVII*, 1178–1181. [CrossRef]
42. Papadakis, G.; Vallianatos, F.; Michas, G. The earthquake interevent time distribution along the Hellenic subduction Zone. *Bull. Geol. Soc. Greece* **2013**, *XLVII*, 1194–1200. [CrossRef]
43. Vallianatos, F.; Benson, P.; Meredith, P.; Sammonds, P. Experimental evidence of a non-extensive statistical physics behavior of fracture in triaxially deformed Etna basalt using acoustic emissions. *EPL* **2012**, *97*, 58002. [CrossRef]
44. Efstathiou, A.; Tzanis, A.; Vallianatos, F. On the nature and dynamics of the seismogenetic systems of North California, USA: An analysis based on Non-Extensive Statistical Physics. *Phys. Earth Planet. Inter.* **2017**, *270*, 46–72. [CrossRef]
45. Vallianatos, F.; Michas, G.; Papadakis, G.; Tzanis, A. Evidence of non-extensivity in the seismicity observed during the 2011–2012 unrest at the Santorini volcanic complex, Greece. *Nat. Hazards Earth Syst. Sci.* **2013**, *13*, 177–185. [CrossRef]
46. Vallianatos, F.; Sammonds, P. Evidence of non-extensive statistical physics of the lithospheric instability approaching the 2004 Sumatran-Andaman and 2011 Honshu mega-earthquakes. *Tectonophysics* **2013**, *590*, 52–58. [CrossRef]
47. Papadakis, G.; Vallianatos, F.; Sammonds, P. Evidence of Nonextensive Statistical Physics behavior of the Hellenic Subduction Zone seismicity. *Tectonophysics* **2013**, *608*, 1037–1048. [CrossRef]
48. U.S. Geological Survey. Available online: https://www.usgs.gov/natural-hazards/earthquake-hazards/earthquakes (accessed on 30 September 2020).
49. Nava, F.A.; Márquez-Ramírez, V.H.; Zúñiga, F.R.; Ávila-Barrientos, L.; Quinteros, C.B. Gutenberg-Richter b-values maximum likelihood estimation and sample size. *J. Seism.* **2016**, *21*, 127–128. [CrossRef]
50. Gutenberg, B.; Richter, C.F. Magnitude and energy of earthquakes. *Ann. Geophys.* **1956**, *9*, 1–14. [CrossRef]
51. Lay, T.; Kanamori, H.; Ammon, C.; Nettles, M.; Ward, S.; Aster, R.; Beck, S.; Bilek, S.; Brudzinski, M.; Butler, R.; et al. The Great Sumatra-Adaman Earthquake of 26 December 2004. *Science* **2005**, *308*, 1127–1133. [CrossRef] [PubMed]
52. Hayes, G.P.; Myers, E.K.; Dewey, J.W.; Briggs, R.W.; Earle, P.S.; Benz, H.M.; Smoczyk, G.M.; Flamme, H.E.; Barnhart, W.D.; Gold, R.D.; et al. *Tectonic Summaries of Magnitude 7 and Greater Earthquakes from 2000 to 2015*; U.S. Geological Survey: Reston, VA, USA, 2017. [CrossRef]
53. Shao, G.; Li, X.; Ji, C.; Maeda, T. Focal mechanism and slip history of the 20011 Mw 9.1 off the Pacific coast of Took Earthquake, constrained with te;eseismic body and surface waves. *Earth Planet Space* **2011**, *63*, 559–564. [CrossRef]
54. Vallianatos, F. A non-extensive statistical physics approach to the polarity reversals of the geomagnetic field. *Phys. A Stat. Mech. Appl.* **2011**, *390*, 1773–1778. [CrossRef]

55. Vallianatos, F.; Michas, G.; Papadakis, G. Non-Extensive Statistical Seismology: An overview. In *Complexity of Seismic Time Series: Measurement and Application*, 1st ed.; Chelidze, T., Vallianatos, F., Telesca, L., Eds.; Elsevier: Amsterdam, The Netherlands, 2018; pp. 25–53.
56. Vallianatos, F.; Sammonds, P. Is plate tectonics a case of non-extensive thermodynamics? *Phys. A Stat. Mech. Appl.* **2010**, *389*, 4989–4993. [CrossRef]
57. Vallianatos, F.; Telesca, L. Statistical mechanics in earth physics and natural hazards. *Acta Geophys.* **2012**, *60*, 499–501. [CrossRef]
58. Chochlaki, K.; Vallianatos, F.; Michas, G. Global regionalized seismicity in view of Non-Extensive Statistical Physics. *Phys. A Stat. Mech. Appl.* **2018**, *493*, 276–285. [CrossRef]
59. Michas, G.; Vallianatos, F. Scaling properties, multifractality and range of correlations in earthquake timeseries: Are earthquakes random? In *Statistical Methods and Modeling of Seismogenesis*; Limnios, N., Papadimitriou, E., Tsaklidis, G., Eds.; ISTE John Wiley: London, UK, 2021. [CrossRef]
60. Abe, S.; Suzuki, N. Scale-free statistics of time interval between successive earthquakes. *Phys. A Stat. Mech. Appl.* **2005**, *350*, 588–596. [CrossRef]
61. Sob'yanin, D.N. Generalization of the Beck-Cohen superstatistics. *Phys. Rev. E* **2011**, *84*, 051128. [CrossRef]
62. Beck, C. Superstatistics: Theory and applications. *Contin. Mech. Thermodyn.* **2004**, *16*, 294–298. [CrossRef]
63. Beck, C. Recent developments in superstatistics. *Braz. J. Phys.* **2009**, *39*, 357–362. [CrossRef]
64. Beck, C. Dynamical Foundations of Nonextensive Statistical Mechanics. *Phys. Rev. Lett.* **2001**, *87*, 180601. [CrossRef]
65. Beck, C. Superstatistical Brownian Motion. *Prog. Theor. Phys. Suppl.* **2006**, *162*, 29–31. [CrossRef]
66. Antonopoulos, C.G.; Michas, G.; Vallianatos, F.; Bountis, T.; Physica, A. Evidence of q-exponential statistics in Greek seismicity. *Phys. A Stat. Mech. Appl.* **2014**, *409*, 71–77. [CrossRef]
67. Feller, W. Two Singular Diffusion Problems. *Ann. Math.* **1951**, *54*, 173–182. [CrossRef]
68. Michas, G.; Vallianatos, F. Stochastic modeling of nonstationary earthquake time series with long-term clustering effects. *Phys. Rev. E* **2018**, *98*, 042107. [CrossRef]
69. Risken, H. *The Fokker-Planck Equation: Methods of Solution and Applications*, 2nd ed.; Springer: Berlin/Heidelberg, Germany, 1989. [CrossRef]
70. Mathai, A.M. A pathway to matrix-variate gamma and normal densities. *Lin. Al. Appl.* **2005**, *396*, 317–328. [CrossRef]
71. Gardiner, C.W. *Handbook of Stochastic Methods for Physics, Chemistry, and the Natural Sciences.*, 1st ed.; Springer: Berlin/Heidelberg, Germany, 1993.

entropy

MDPI

Article

# Forecasting Strong Subsequent Earthquakes in Greece with the Machine Learning Algorithm NESTORE

**Eleni-Apostolia Anyfadi [1,2], Stefania Gentili [3,*], Piero Brondi [3] and Filippos Vallianatos [1,2]**

[1] Section of Geophysics-Geothermics, Department of Geology and Geoenvironment, National and Kapodistrian University of Athens, 15784 Athens, Greece; elenanif1305@gmail.com (E.-A.A.); fvallian@geol.uoa.gr (F.V.)
[2] Institute of Physics of Earth's Interior and Geohazards, UNESCO Chair on Solid Earth Physics and Geohazards Risk Reduction, Hellenic Mediterranean University Research & Innovation Center, 73133 Chania, Greece
[3] National Institute of Oceanography and Applied Geophysics-OGS, 33100 Udine, Italy; pbrondi@ogs.it
* Correspondence: sgentili@ogs.it

**Abstract:** Aftershocks of earthquakes can destroy many urban infrastructures and exacerbate the damage already inflicted upon weak structures. Therefore, it is important to have a method to forecast the probability of occurrence of stronger earthquakes in order to mitigate their effects. In this work, we applied the NESTORE machine learning approach to Greek seismicity from 1995 to 2022 to forecast the probability of a strong aftershock. Depending on the magnitude difference between the mainshock and the strongest aftershock, NESTORE classifies clusters into two types, Type A and Type B. Type A clusters are the most dangerous clusters, characterized by a smaller difference. The algorithm requires region-dependent training as input and evaluates performance on an independent test set. In our tests, we obtained the best results 6 h after the mainshock, as we correctly forecasted 92% of clusters corresponding to 100% of Type A clusters and more than 90% of Type B clusters. These results were also obtained thanks to an accurate analysis of cluster detection in a large part of Greece. The successful overall results show that the algorithm can be applied in this area. The approach is particularly attractive for seismic risk mitigation due to the short time required for forecasting.

**Keywords:** NESTORE; machine learning algorithm; aftershocks; features; Greek seismicity; clusters; forecasting; training procedure

**Citation:** Anyfadi, E.-A.; Gentili, S.; Brondi, P.; Vallianatos, F. Forecasting Strong Subsequent Earthquakes in Greece with the Machine Learning Algorithm NESTORE. *Entropy* **2023**, *25*, 797. https://doi.org/10.3390/e25050797

Academic Editor: Cataldo Godano

Received: 5 April 2023
Revised: 5 May 2023
Accepted: 9 May 2023
Published: 13 May 2023

**Copyright:** © 2023 by the authors. Licensee MDPI, Basel, Switzerland. This article is an open access article distributed under the terms and conditions of the Creative Commons Attribution (CC BY) license (https://creativecommons.org/licenses/by/4.0/).

## 1. Introduction

It is widely known that large earthquakes are followed by other earthquakes, usually smaller and occurring in close proximity, days to years later, and that it takes some time for seismicity to return to normal levels [1–4]. However, it may happen that the following earthquake magnitude is comparable with the previous one. Moreover, aftershocks can affect numerous facilities in a city, and repeated earthquakes worsen the damage already inflicted upon vulnerable structures and infrastructure. Greece's location at the point of contact between the tectonic plates of Eurasia and Africa has resulted in several geodynamic processes and high seismicity, with multiple events of large magnitude recorded both in ancient and modern times. Greece ranks sixth in the world and first in the Mediterranean region for seismic energy emission [5,6]. The significant geotectonic phenomena, such as the continental convergence, where the oceanic crust of the North African plate is subducted under the European plate, are often associated in the literature with the high seismic activity in Greece. This migration was accompanied by significant crustal shortening and an uplift rate of a few millimeters per year along the Hellenic Arc because of the accretion of African plate sediments beneath the underlying Aegean plate. The rollback of the subducting African slab, resulting in high-rate extension in the back-arc region, is also a significant seismic source. Last but not least, seismic activity is caused by the North Aegean Trough

(NAT), the most notable tectonic feature of the North Aegean Sea and the Cephalonia Transform Fault Zone (CTFZ) [7].

The Mediterranean and Greek regions are particularly well-known both for intense seismicity and the large-scale earthquakes that have taken place both in recent years and in ancient times. A typical example is the devastating 365 earthquake of Crete, estimated to have had a moment magnitude of 8.5 or greater. The earthquake is believed to have caused the island of Crete to rise by nine meters, while a tsunami that followed the earthquake destroyed the southern and eastern Mediterranean coasts [8]. On 3 April 1881, the deadliest earthquake ($M_W$ 6.5) in Greece's seismic history devastated the SE Chios island. Numerous fatalities occurred, and the vast majority of facilities were totally destroyed [9,10]. The Great Kefalonia earthquake, which devastated the southern Ionian Islands in Greece in 1953 with a magnitude of Ms 7.2 and killed over 500 people, was another significant earthquake of the 20th century [11]. Among the earthquakes used for the analysis in this article, some are major and particularly important, such as the $M_W$ 6.4 Aigio (15 June 1995) [12], the $M_W$ 6.5 Andravida (8 June 2008) [13], the $M_W$ 6.9 Limnos (24 May 2014) [14], the $M_W$ 7.0 Samos (30 October 2020) [15], the $M_W$ 6.3 Elassona (3 March 2021) [16], and the $M_W$ 5.7 Arkalochori (27 September 2021) [17] earthquakes.

In such seismotectonic context, to help mitigate the seismic risk after a strong earthquake, it may be useful to develop and test an algorithm, based on the immediate mild aftershocks, for forecasting the occurrence of stronger subsequent earthquakes. For real-time or near-real-time applications carried out during a cluster occurrence, it is not known whether a first high-magnitude earthquake will be followed by one or more strong events. For this reason, we use the term "o-mainshock" (short for "operative-mainshock"), which refers to the first earthquake in the cluster that exceeds a certain magnitude threshold [18].

Many studies have focused on the value of $Dm$, which corresponds to the magnitude difference between the mainshock and the Strongest Subsequent Large Earthquake (SSLE) [19–24]. The magnitude of the SSLE increases as $Dm$ decreases, making the cluster more dangerous for the population. Using the assumption of the self-similarity theory of seismicity, which assumes similar behavior for shocks of different magnitudes, studies on this topic are based on $Dm$ rather than SSLE magnitude. This approach also has the advantage of using clusters characterized by mainshocks of lower magnitude, which are more frequent than others, and thus, by increasing the training and testing database, improve statistical reliability. Some studies investigating the relationship between $Dm$ and mainshock characteristics show that they vary considerably depending on the region [18].

In this paper, we propose a machine learning approach to the problem of $Dm$ forecasting during the occurrence of seismic clusters. NESTORE (NExt STrOng Related Earthquake) is a machine learning-based approach for SSLE forecasting that can be applied to clusters whose magnitude of completeness is at least equal to the mainshock magnitude minus 2 [18,25–27]. The clusters are divided into two groups based on the mainshock magnitude $M_m$ and the SSLE magnitude: Type A if $Dm \leq 1$ and Type B otherwise. The method is based on the analysis of the seismicity after the mainshock by extracting some features used for machine learning. The features describe the characteristics of the seismicity during the cluster in terms of radiated energy, number of events, and space and time distribution. NESTORE trains a one-node decision tree for each feature separately and evaluates thresholds so that clusters whose feature is above the threshold are classified as Type A and the others are classified as Type B. The probability of being a Type A cluster is independently estimated for each feature classifier from the percentage of Type A clusters below and above the threshold in the training set; these probabilities are combined for a final probability estimate using a Bayesian approach [25]. To simulate the increase in knowledge over time after the o-mainshock, the analysis was performed at different time intervals $T_i$, ranging from 6 h to 7 days after the mainshock. In this case, we applied the NESTORE algorithm to the Greek seismicity by using the NESTOREv1.0 software available on GitHub [26].

In the first tests, we performed the analysis for each region separately to account for the different seismotectonic regimes. However, the results were not statistically relevant due to the small number of clusters in each region. We performed an analysis to check if the clusters belong to the same population and we combined all the above regions into a single area (see Figure 2); we merged the original regions by adding the Peloponnese, Thessaly, Central Greece, and Crete, but omitting (1) the area of the Greek volcanic arc because of the possible different origins of the earthquakes, (2) the area of the subduction zone because of possible viscoelastic effects, and (3) the western Turkish coast and offshore regions because of the poor coverage by the national seismological network.

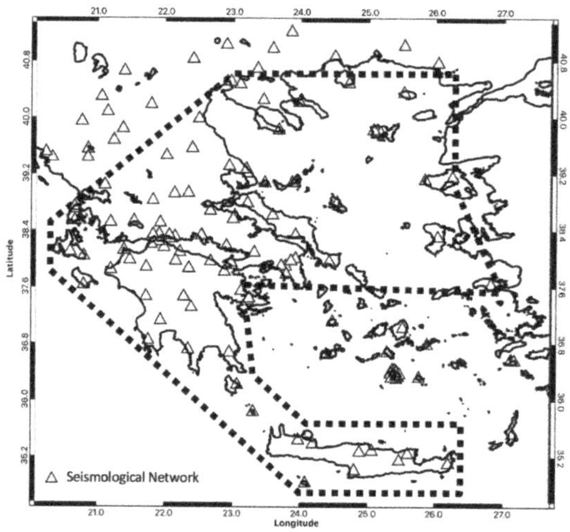


**Figure 2.** The region we examined in this paper is outlined by the black dashed polygon. The triangles show the locations of the installed Hellenic Unified Seismological Network (HUSN).

## 4. NESTORE Algorithm

The multiparameter machine learning approach called NESTORE examines the evolution of seismicity at various time intervals. Its main goal is the estimate of the probability that the analyzed clusters are of Type A. This machine learning approach is designed particularly for the analysis of seismicity problems and limited data, as there are typically tens of available clusters. In order to simulate the evolution of seismicity over time, the analysis was conducted on increasing time intervals $T_i$, beginning shortly after the mainshock. NESTOREv1.0 uses earthquakes with magnitude $M \geq M_m - 2$, and to avoid classifications of clusters in which the class has been already defined, it examines Type A clusters for time intervals shorter than the time difference between the mainshock and the first aftershock with magnitude $\geq M_m - 1$. For this reason, both the training and the test sets change depending on the considered time interval, because for longer time intervals, fewer Type A clusters are available.

A set of features (see Supplementary Materials for a detailed description) are extracted from spatio-temporal and energy distribution of seismicity, and for each feature independently, a simple threshold is used to distinguish between the classes. The analysis focuses on features based on the earthquakes following the o-mainshock attempting to capture high and irregular earthquake activity [55,56]. It is important to remark that the framework of the algorithm is independent of the specific features used, which can be adapted based on the study area's characteristics, including both aspects of seismicity and data availability. Before strong earthquakes, some variations and a change in earthquake flow, which becomes more intense and anomalous in space and time, have been reported and analyzed as a symptom of instability of a nonlinear system equivalent to seismic faults [57]; Vorobieva and Panza (1993) supposed that similar behavior can be detected if a strong subsequent earthquake is expected (Type A clusters) [19]. This is the assumption on which the features adopted by NESTOREv1.0 software package are based. From a physical standpoint, these variances are comprehensible because the lithosphere may respond to tectonic stress more strongly, and since those symptoms appear after a mainshock, they may be precursors to the occurrence of a second major event [58].

The NESTOREv1.0 software package is divided into four main modules. In our analysis, we used the cluster identification, training, and testing modules [26]. The cluster identification module identifies seismic clusters whose mainshock has a magnitude $M_m$ above a given threshold of magnitude $M_{th}$. The training module uses decision trees based on different features to find appropriate thresholds with the aim of discriminating clusters of known class (A or B) in a training database. The testing module is used to check the performances of the training; it uses the outputs of the training module to provide an estimate of the probability that a cluster is of Type A in a test database; then it compares the result with the already known actual class of the clusters. The last module, not used in this work, is the near-real-time classification module, which has been proposed recently [26] for new ongoing cluster classifications after validation of the method in an area. Such validation is the topic of this paper.

### 4.1. NESTORE Cluster Identification

Since cluster identification is a non-unique process, there are numerous methods in the literature that provide a range of results. It is easier to solve the task of declustering a catalogue by removing dependent earthquakes than to assign each dependent event to a particular cluster, since this may be controversial for clusters that are close in time and space. The events belonging to each cluster can be significantly affected by the choice of cluster identification procedure. Different methods have been used to identify clusters, depending on the research field. For example, a deep learnable scattering network had been used to cluster seismic events in continuous waveforms [59]. Another method related to distinguishing different structures of lightning phenomena in a multidimensional image developed an analysis pipeline using the t-distributed stochastic neighbor embedding (t-SNE) method and a DBSCAN algorithm for further cluster detection [60]. In space-time analysis of seismicity, many cluster identification algorithms are applied (for further details, see [61]). In particular, in our research, the NESTOREv1.0 cluster identification module implements a simple method of cluster identification, a window-based technique [61], where the cluster is defined as all events occurring within a time and space window around the mainshock, the size of which depends on the magnitude of the mainshock. Window techniques provide a quick and easy way to detect mainshocks and aftershocks, but it is necessary to define a region-dependent law for the maximum time interval $t(M_m)$ after the mainshock and the maximum distance $d(M_m)$ from the mainshock of the earthquakes inside the cluster [61].

### 4.2. NESTORE Training Procedure

NESTORE assesses a set of features individually before combining the best feature classification after training. In particular, each feature is assessed using a pattern recognition method that employs an independent decision tree [62,63] and the algorithm then determines a threshold $Th$ for each feature, $f$, so that if $f \geq Th$, the cluster is designated as A and otherwise as B. The features are calculated at time intervals $[s_1, s_2]$, where $s_1$ is the time after the mainshock used to guarantee that the completeness magnitude of $M_c \leq M_m - 2$ can be achieved [18] and $s_2$ corresponds to the ending time for the analysis. The features [27] used by NESTORE in this case are evaluated using events having magnitude $M \geq M_m - 2$ and correspond to nine seismicity parameters related to the number of events, their spatial distribution, magnitude, and energy trend over time in increasing time intervals following the occurrence of the mainshock [18,26]; see the Supplementary Materials for a detailed description.

The major goal of utilizing these features is to spot changes in the flow of earthquakes, such as irregularities in space, time, and magnitude that can be related to different seismic behaviors between Type A and B clusters. In addition, in order to achieve a balance between the need for as many clusters as possible for our study and the necessity for a strong enough statistic on the development of seismicity, we set up the analysis starting at the first 6 h (0.25 days) after the mainshock and at time intervals $T_i$ ending 0.25, 0.50, 0.75, 1, 2, 3, 4, 5, and 7 days after [18,25–27,64].

The training set of samples and the expected output class are inputs to the training procedure, which consists of the following sections: feature extraction, decision tree training, good interval identification, inheritance, and validation [18].

NESTORE algorithm is based on a supervised training approach. For each time interval following the mainshock, the algorithm extracts the desired features from the input training clusters (see Figure 3). To prevent a too complex structure of the classifier, which would lead to an overfitting of the data if the data are few, the training is performed using binary decision trees with a depth of 1, which splits classes based simply on a threshold. Using the information on the class of the training clusters, the threshold is chosen such that (most of) the clusters of Type A have features greater than or equal to the threshold, while (most) Type B clusters have values of the features under the threshold. If no tree can be found to solve the issue, the value NaN (Not a Number) is assigned as the threshold and the feature is ignored for that time interval. The performance could be poor even when the decision tree finds a threshold; in order to avoid this problem, the quality of threshold-based classifiers is estimated by performance evaluators.

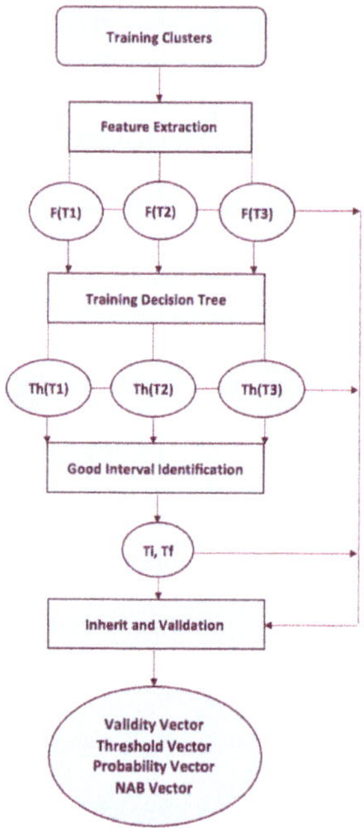


**Figure 3.** NESTORE training procedure template for a single feature.

Specifically, performance evaluators are Accuracy, Recall, Precision, and Informedness. The last evaluator is given between −1 and 1, where 1 is the best and −1 is the worst. The first previous three evaluators are specified between 0 and 1, where 1 is the best and 0 is the worst. As the observation time $T_i$ grows, the value of these performance indicators often increases until it reaches a peak, and then stabilizes or decreases with longer observation time. The algorithm chooses intervals $T_i$ in which three requirements are satisfied. The first states that Recall, Precision, and Accuracy should all be greater than 0.5. The second one specifies that Accuracy should be greater than or equal to the Accuracy we can obtain from a constant response corresponding to the most populous class (Class B). The last one states that Informedness should be greater than zero. Therefore, for each time interval, a set of reliable classifiers corresponding to a subset of the original features is selected.

When the greatest level of Informedness is reached for a feature at a particular $T_i$, the instances of that feature for each cluster of NESTORE are automatically set to the value they have for $T = T_i$. The time intervals smaller than or equal to $T_i$ that satisfy the previous three conditions are called good intervals. For longer time intervals, both the feature value and its threshold are set to the ones corresponding to the maximum value of Informedness. This procedure, called inheritance, is intended to use features with high performance in a given time period, even for longer periods when performance is becoming poorer. However, as $T_i$ increases, inherited features and thresholds may experience a fall in performance due to a selection effect on clusters, since for some features, Type A clusters with later SSLEs belong to a separate population. Over time, the percentage of these clusters in the dataset rises, resulting in a decline in feature performance. For this reason, the algorithm rechecks the performances. It determines if the percentage of Type A clusters properly categorized is higher than the percentage of Type B clusters mistakenly classified as Type A, for all inherited thresholds and features. If this does not happen, the interval $T_i$ is removed from the list of intervals associated with the feature [18].

### 4.3. NESTORE Testing Procedure

The NESTORE testing procedure (see Figure 4) uses the information from the training procedure to classify clusters of an independent test set and compares the obtained results with the actual value of the cluster typology. The classification is performed for all the time intervals and for all the classifiers considered reliable in those time intervals by the training procedure. For each time interval, combining the different classification results, a unique classification is produced, which can be binary ("class A" or "class B") or continuous ("class A probability"). A voting process is the simplest method for classifying combinations, where each classifier receives one vote (A or B). If the number of A votes exceeds the number of B votes by a certain amount, the classification is A; otherwise, it is B. The previous methods presuppose that all classifiers are equally reliable, but in reality, this is frequently not the case. Therefore, for each time interval and feature, NESTORE estimates the probability that a cluster belongs to Type A depending not only on whether it is above or below the corresponding threshold $Th$ for single features, but also on how reliable each feature classification is above or below $Th$. By combining many independent classifiers, NESTORE uses a Bayesian technique to determine the total probability [25].

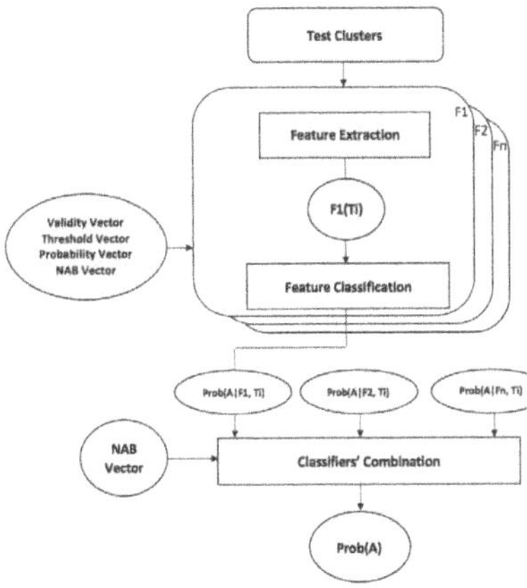


**Figure 4.** Classification procedure for a time interval $T_i$.

According to Bailer-Jones et al. (2011), utilizing independent information (feature) $D_1$, ..., $D_n$, the posterior probability that class is C is [65]:

$$P(C|D_1 \dots D_N) = \alpha \frac{\prod_{n=1}^{N} P(C|D_n)}{P(C)^{(N-1)}} \quad (1)$$

where $P(C)$ is the probability of having a class C and $P(C \mid D_n)$ is the posterior probability that the class C is at $D_n$. $N$ is the number of classifiers and $\alpha$ is the normalized factor such that

$$\sum_k P(C_k|D_1, \dots D_N) = 1 \quad (2)$$

where $C_k$ are the classes of the problem.

In our case, we have two classes, A and B and, assuming for each i-th time interval

$$P(A) = \frac{N(A)}{N(A)+N(B)} \quad \text{and} \quad P(B) = \frac{N(B)}{N(A)+N(B)} \quad (3)$$

where $N(A)$ and $N(B)$ are the number of $A$ and $B$ clusters in the dataset for the $i$-th interval, respectively. Equation (1) can be written as

$$P(A|D_1 \dots D_N) = \frac{[N(B)]^{N-1} \prod_{n=1}^{N} p_n}{[N(B)]^{N-1} \prod_{n=1}^{N} p_n + [N(A)]^{N-1} \prod_{n=1}^{N} (1-p_n)} \quad (4)$$

where $p_n = P(A|Dn)$ is the probability of having the cluster of Class A at a $D_n$ value for the feature $n$; $p_n$ is calculated from the training as the percentage of Type A clusters (divided by 100) that are above or below the output threshold and acts as a weight depending on different features' reliability. A benefit of this method is that it takes into consideration the number of Type A and B clusters in the dataset, which is crucial for imbalanced classes such as the one we have (e.g., the Type A clusters account for one-fifth of the total clusters in Greece). The testing supplies in output the Receiver Operating Characteristics (ROC) and the Precision–Recall graphs, which show the performances of the training on an independent test set (see the Supplementary Materials for a detailed description).

Binary classifiers distinguish between two classes, one positive (in our case, Class A) and one negative (in our case, Class B). To evaluate the effectiveness of single-features classifiers to determine if a strong aftershock would occur within a cluster, the output Prob(A) for each test set cluster is binarized, so that if Prob(A) $\geq 0.5$, the class is A; otherwise, it is B. Resulting classes are compared with the actual one for each cluster and the results are shown by using the ROC graph together with the Precision–Recall graph.

The ROC graph (see, e.g., 6th figure (a,c) in Section 5) shows the normalized percentage of positive instances correctly classified as positives (True Positive Rate or Recall) vs. the percentage of negatives incorrectly classified as positives (False Positive Rate). In the ROC graph, a discrete classifier generates some points whose coordinates graphically represent its performances [27]. The ideal classifier is represented by the point (0, 1) when all instances are correctly classified [27]. In fact, if a point in the space ROC is closer to the point (0, 1), it has a higher rate of positive and/or a lower rate of negative results, so it is preferable to other points. The diagonal line indicates random guessing, and any classifier that occurs in the lower right triangle performs worse than the random one and should be discarded.

The Precision–Recall graph (see, e.g., 6th figure (b,d) in Section 5) shows other useful information: the Precision, which corresponds to the percentage (normalized to 1) of clusters classified as A that are actually A. This information is important for evaluating performance on imbalanced datasets. While both Recall and the False Positive Rate are independent on the relative abundance of the classes, the abundance affects Precision and, therefore, the random guessing horizontal line in the Precision–Recall (PR) graph. As there are fewer A's over longer time periods, the random guessing line parallel to the x-axis has a decreasing y-intercept as $T_i$ increases. A classifier that lies below the random guessing line is characterized by unreliable results; the closer it is to the upper right corner, the more effective it is. The ideal classifier, shown by the upper right corner of the PR graph, correctly classifies all Type A clusters and misclassifies no Type B cluster as A. The best performance for the Precision–Recall graph corresponds to point (1,1) [27].

## 5. Results

In our case study, we applied NESTOREv1.0 to Greek seismicity using the AUTH earthquake catalogue, expressed in magnitude ML, for the period 1995–2022 with a maximum focal depth of 50 km; the analyzed region is shown in Figure 5.

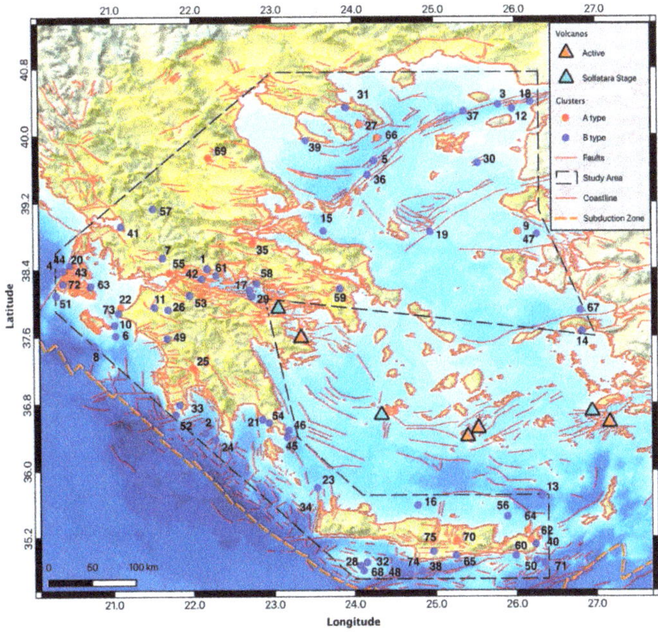


**Figure 5.** Analyzed region. The mainshocks of the clusters are shown by circles.

### 5.1. Cluster Identification and Completeness Magnitude Assessment in Greece

In window-based cluster identification applications, the first step is to evaluate how the temporal and spatial extent of the cluster depends on the magnitude of the mainshock. An incorrect assessment may lead to the loss of events belonging to some clusters, thus underestimating their impact on the analyzed area, or, conversely, in including background events or events belonging to other clusters, thus overestimating the impact of the clusters on the area. Since this pre-selection can influence the results of the following analysis, it is an important preliminary step of the cluster analysis. In order to understand which was the most appropriate law for window-based cluster identification in Greece, we compared several laws available in the literature that have been successfully applied to other parts of the world. In these laws, both the duration of the cluster and the radius of a circular area around the mainshock in which aftershocks occur are given as functions of the mainshock magnitude. We set the minimum magnitude of the mainshocks equal to 4, and we tested the equations for duration of Gardner and Knopoff (1974) [66], Lolli and Gasperini (2003) [67], Gentili and Bressan (2008) [68], and Uhrhammer (1986) [69]. For space windows, the equations of Kagan et al. (2002) [70], Uhrhammer (1986) [69], Gardner and Knopoff (2000) [71], and Gentili and Bressan (2008) [68] were tested, the last one with the addition of two kilometers to account for localization inaccuracies.

The choice of the best law for Greece was performed manually. First of all, we estimated manually the distance between the mainshock and the most distant aftershock for a large dataset of clusters and plotted this radius as a function of magnitude, comparing it with the curves representing the equations to be checked. The main idea was to select a curve that corresponds to the smallest radius that encompass most of the clusters, in order not to lose events belonging to the cluster but, on the other hand, to avoid the inclusion of independent events. Figure 6a shows such a plot on 177 clusters in the area.

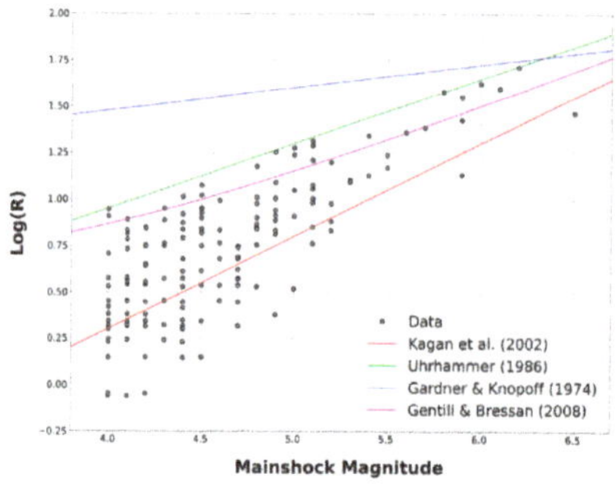


(**a**)

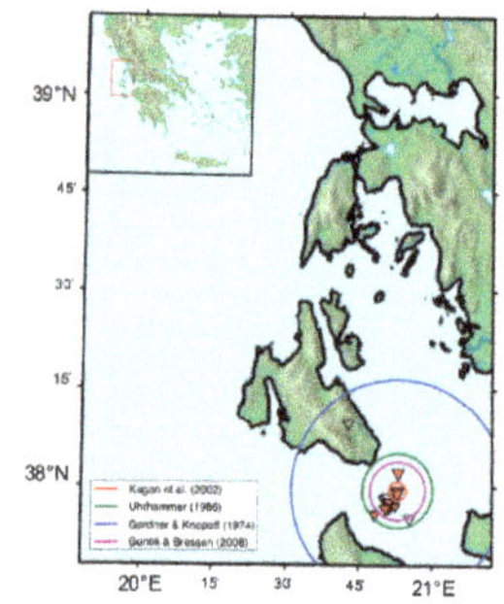


(**b**)

**Figure 6.** (**a**) Plot of the calculated radius vs. the magnitude of the cluster mainshock. The coloured lines indicate the radius estimation equations. (**b**) Determination of the best space-window law by map visualization [66–71].

In order to check the results on a larger dataset, we also manually inspected the maps of all the clusters obtained by imposing the larger radius equation (Gardner and Knopoff, 1974) [66] and comparing earthquakes' positions with the circles representing the checked equations (see Figure 6b). In both cases, the best choice was the equation proposed by Uhrhammer (1986) [69]. This equation provides a much smaller radius than the one proposed by Gardner and Knopoff (1974) [66], but also helps to avoid the inclusion of independent earthquakes in a cluster. In addition, it provides a larger radius compared to the equations proposed by Kagan et al. (2002) [70] and Gentili and Bressan (2008) [68], allowing more aftershocks to be included in the defined cluster. For time window, we plotted the magnitude vs. time for the obtained clusters (see Figure 7) and we compared it to the duration obtained by different equations. Again, the best choice was the equation proposed by Uhrhammer (1986) [69], which supplies a shorter $t(M_m)$ compared to the other methods (see Figure 7) and has the advantage of including highly dependent events in the cluster.

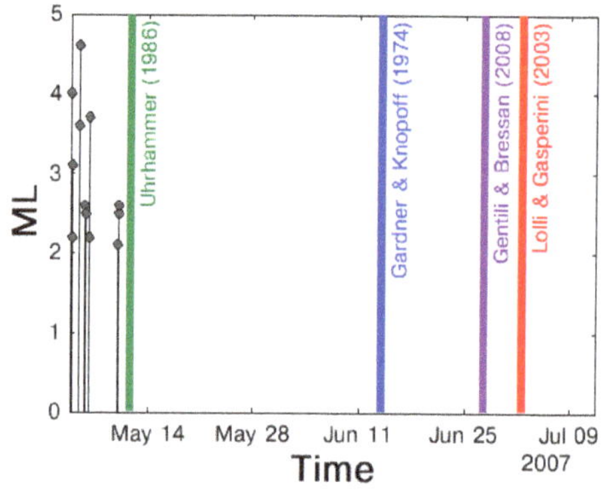


**Figure 7.** Determination of the best time-window law [67–69,71].

Equations (5) and (6) show the selected radius (in km) and duration (in days):

$$d = e^{-1.024+0.804Mm} \tag{5}$$

$$t = e^{-2.87+1.235Mm} \tag{6}$$

This procedure failed only in two earthquakes in the northern Gulf of Evia. The first occurred on 17 November 2014 and the second on 9 June 2015, with magnitudes ML of 5.3 and 5.1, respectively. As indicated by Ganas et al. (2016) [72], the above earthquakes belong to the same cluster, which is a Type A cluster according to the NESTORE classification. However, the applied method of cluster identification leads to an obvious classification failure as it splits the cluster into two parts. Since the NESTOREv1.0 module is independent of the others in the NESTOREv1.0 package, it can be substituted with a different cluster identification procedure. A more reliable cluster identification method will be used in the future for the analysis of the region. In this application, we removed the cluster from the analysis.

As previously stated, NESTOREv1.0 needs clusters with a completeness magnitude of $\leq M_m - 2$, where $M_m$ is the o-mainshock magnitude. When at least 80 earthquakes are available in a cluster, NESTOREv1.0 automatically evaluates the completeness magnitude for the cluster using the maximum curvature method (+0.2 to account for possible underestimates of the method); otherwise, it allows a default value. We considered a completeness magnitude value of 3.0 for clusters that occurred before 2009 and a magnitude value of 2.5 for those starting in 2009. This assumption is based on a general analysis of the completeness magnitude as a function of time for the analyzed area that we carried out using Zmap software [73] (see Figure 8).

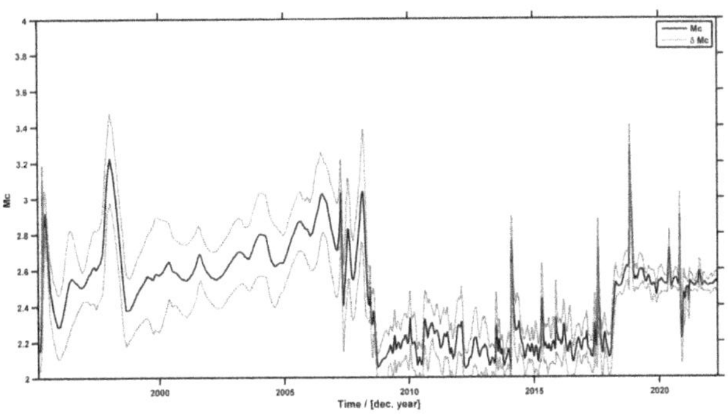


**Figure 8.** Plot of Mc vs. time for Greece using AUTH earthquake catalogue.

Besides the selection based on the completeness magnitude, another selection of Type A clusters was performed based on the time of the strongest aftershock: since the first analysis is performed 6 h after the mainshock, NESTOREv1.0 analyzed only the Type A clusters which did not have an aftershock with magnitude $\geq M_m - 1$ in the first 6 h.

At the end of the selection procedure, we detected 75 clusters satisfying NESTORE requirements, of which 12 are Type A and 63 are Type B. In Figure 5, we superimposed on the map of the studied area the locations of the o-mainshock of the clusters; we used red color for Type A clusters and blue for Type B clusters. The clusters are located both offshore and along the mainland.

Analyzing the characteristics of the clusters, we did not find any correlation between the type of cluster (A or B) and some parameters of the mainshock, such as the focal mechanism [37], location, depth, or magnitude.

### 5.2. *NESTOREv1.0 Application to the Current Dataset*

As described in Sections 2 and 3, Greece is extremely heterogenous from a seismo-tectonic point of view. For this reason, it is important to check that, given one type of cluster (Type A or Type B), all the clusters of that type have similar characteristics (i.e., they belong to the same population) according to the NESTORE model. If there is more than one population depending on the sub-region, the training of each sub-region must be performed separately. In order to check this, we trained NESTOREv1.0 with the whole dataset of 1995–2022 both as a training set and a test set (autotest). Figure 9 illustrates the probability P(A) of being a Type A cluster for different time intervals. The analysis was performed on increasing time intervals ending every 6 h in the first day and every day in the first week after the mainshock.

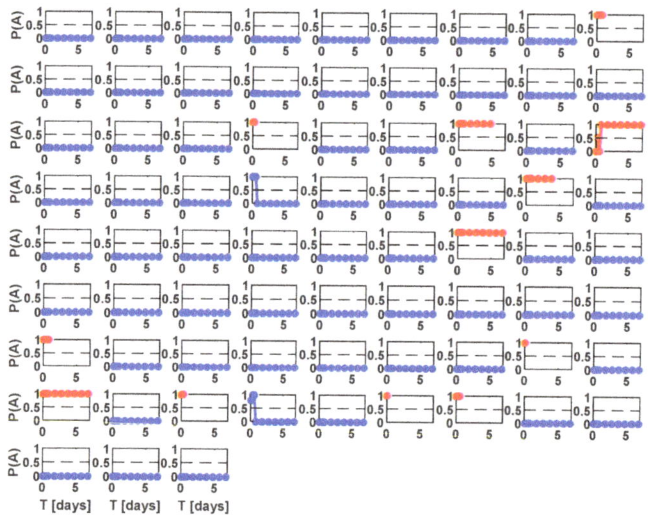

**Figure 9.** Estimated probability of being a Type A cluster vs. time for all the clusters in the dataset (autotest). Red points correspond to A type clusters, while blue points correspond to B type ones.

Each cell corresponds to a different cluster classification for different time intervals. Red circles correspond to Type A clusters and blue ones to Type B clusters.

The figure shows that for most time periods, the probability of being A is close to 1 and close to 0 for B. This is not an assessment of the performance of the method, since overfitting is an obvious risk when the training and test sets are coincident, but a preliminary check of the coherence of the dataset, showing that the two classes can be distinguished and that there are no obvious outliers. In detail, the good result in Figure 9 shows that the clusters of the same type in different parts of Greece belong to the same population from NESTOREv1.0's point of view, and the analysis can be performed on the whole area together.

To fully exploit the potential of the machine learning approach for Type A cluster forecasting, we created a test set separate from the training set that contains instances with known classes. In the testing procedure, the class of each cluster in the test set is evaluated using the information obtained from training. The forecasted cluster class is compared to the already known actual class to obtain an estimate of the training performance. The choice of the number of clusters to be selected for the training set and the test set could, in principle, affects the results. Especially when few data are available, it is important from one side to have enough data in the training set to have a good estimate of the parameters, but on the other hand to have enough data in the test set, such that the obtained performances are reliable. A rule of thumb often used in machine learning suggests that three-quarters of the total data should be used for training and the remaining one-quarter for testing [74]; however, the number of Type A clusters is only 12 in the dataset 6 h after the mainshock. This means there are only three clusters to check the results for a time interval of 6 h, and fewer for longer time periods, due to the decrease in the number of Type A clusters. Xu and Goodacre proposed a range between 50% and 70% [75], which is more suitable for our application. We used the years from 1995 to 2015 for the training and the following 7 years for testing (see Table 1). Table 1 shows in detail the number of clusters, particularly Type A clusters, in the training set and the test set.

**Table 1.** Training and testing dataset information.

| Training Period | Testing Period | No. of Clusters (Training Set) | No. of A Clusters (Training Set) | No. of Clusters (Test Set) | No. of A Clusters (Test Set) |
|---|---|---|---|---|---|
| 1995–2015 | 2016–2022 | 46 | 6 | 29 | 6 |

Figure 10 shows the performances of the method. The NESTOREv1.0 Bayesian classification performance for each time period $T_i$ is shown by magenta stars, and some examples of single-feature classifier performances are shown with different symbols.

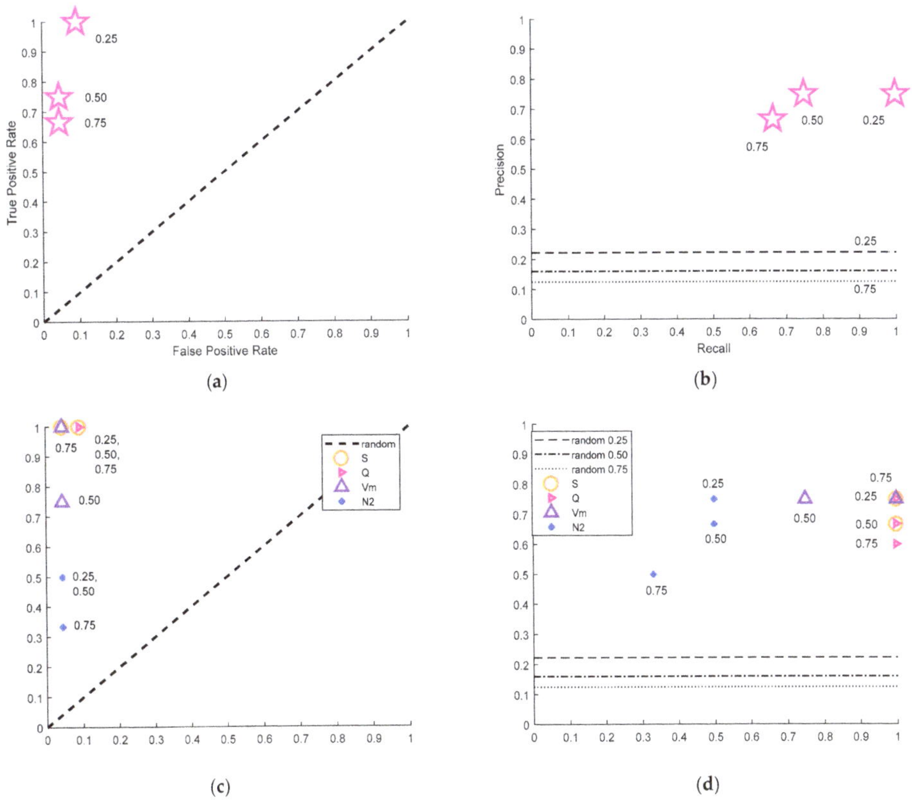


**Figure 10.** NESTOREv1.0 performances for different $T_i$ values which are listed close to the corresponding star. (**a**) ROC graph for NESTOREv1.0 Bayesian classification; (**b**) Precision–Recall (PR) graph for NESTOREv1.0 Bayesian classification; (**c**) ROC graph for selected features; (**d**) PR graph for selected features.

Figure 11 shows the probability vs. time of being a Type A cluster obtained for different time periods Ti for the clusters of the test set.

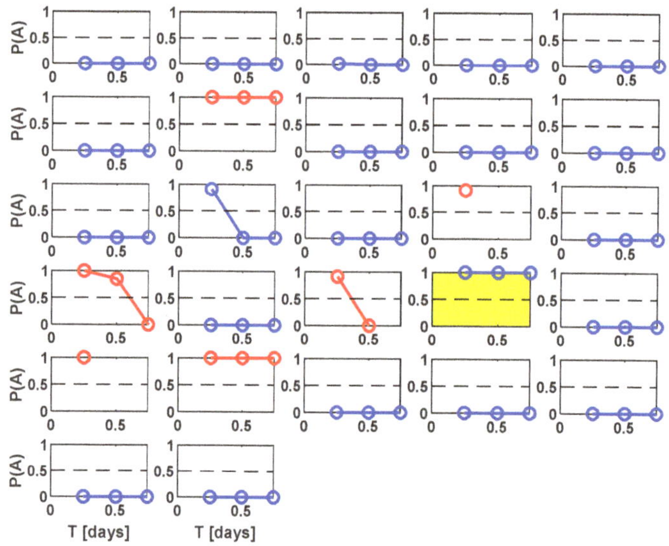


**Figure 11.** Estimated probability of being a Type A cluster vs. time for all the clusters in the time period 2016–2022 (training period 1995–2015).

Both the ROC and PR plots in Figure 10a,b show that NESTOREv1.0 Bayesian classification consistently lies within ranges corresponding to reliable classifiers for all time periods, above the random rate line in the PR plots and in the top left triangle for the ROC plots. Longer time periods correspond to a small number of Type A clusters, due to the elimination from the dataset of clusters that already had strong aftershocks. This affects the capability of both the training set and the test set to accurately describe the characteristics of the clusters, and thus, the reliability of results. Therefore, the analysis was stopped at $T_i$ = 0.75 days (18 h) so that we have at least three Type A clusters both in the training set and in the test set. The best performance for both ROC and PR graphs is at 6 h. We hypothesize that this is because, by including more data in the training set and using a more balanced dataset, we were able to better model the complexity of Greek seismicity, allowing the decision trees to converge to a more stable result. The good performances' short time intervals after the mainshock are noteworthy for the seismic risk mitigation assessment.

Figure 10c,d illustrates the characteristics of some features, selected because of their different performances, to illustrate the whole method procedure. They are calculated for different time intervals and are the normalized cumulative source area ($S$), the normalized radiated energy ($Q$), the cumulative variation of magnitude between each occurrence ($V_m$), and the number of events ($N_2$). From one feature to another, different time intervals $T_i$ were needed to achieve the best performance. For shorter intervals following the mainshock, not all of the features were considered reliable or could be computed. Six hours after the mainshock, the feature $Q$, $S$, $Z$, and $N_2$ are considered reliable, but only two of them supply high-performance results. Comparing feature performances in Figure 10c,d, it can be seen that the features $S$ and $Q$ produced the best performances, the feature $N_2$ produced the poorest, and the feature $V_m$ had intermediate results. These differences are mainly related to the smaller recall (True Positive Rate) of these features, very low especially for $N_2$ (blue dots). At 6 h after the mainshock, the NESTOREv1.0 Bayesian performances coincides with $Q$ and $S$ feature ones, with a True Positive Rate of 1 (all Type A clusters correctly classified), a False Positive Rate of 0.095 (90.5% of Type B clusters correctly classified), and a Precision of 0.75 (75% of the clusters classified as Type A were actually A). This result corresponds in Figure 11 to two Type B clusters wrongly classified: one in the third row and second column, correctly classified for longer time periods, and the cluster in the fourth row and fourth column, automatically outlined in yellow by NESTOREv1.0 as an outlier, because it supplies a wrong classification in all the analyzed time periods.

In order to evaluate the best value of the threshold for future application of the method to Greek seismicity (by using the near-real-time classification module), we used the ones obtained during the autotest. Since, using all the data, we have no independent test set to evaluate the performances, we stopped our analysis at $T_i$ = 18 h, as in the test shown in Figure 10. Table 2 shows the values of the thresholds for the training set of the autotest at these time intervals. It is noticeable that the larger training set eliminates the poorly performing feature $N_2$ from the classification at 6 and 12 h.

**Table 2.** Values of the thresholds of the features obtained by the training procedure on the whole dataset. Inh. = inherited threshold value.

| **Features** | **Thresholds** | | |
|---|---|---|---|
| | **Th (6 h)** | **Th (12 h)** | **Th (18 h)** |
| S | 0.053 | 0.053 | 0.084 |
| Z | 0.026 | 0.026 | 0.026 |
| SLCum | | 0.056 | Inh. |
| QLCum | | 2.318 | 2.318 |
| SLCum2 | | | 0.090 |
| QLCum2 | | | 2.79 |
| Q | 0.012 | 0.012 | 0.013 |
| Vm | | 0.035 | 0.450 |
| $N_2$ | | | 2.5 |

Table 3 shows the values of $p_u$ and $p_o$ that are used to evaluate $p_n$ of Equation (4): if the cluster is under the threshold, $p_u$ is used; otherwise, $p_o$ is used.

**Table 3.** Values of the probability of being Type A under and over the threshold for the whole dataset.

| **Features** | **Thresholds** | | | | | |
|---|---|---|---|---|---|---|
| | $p_u$ **(6 h)** | $p_o$ **(6 h)** | $p_u$ **(12 h)** | $p_o$ **(12 h)** | $p_u$ **(18 h)** | $p_o$ **(18 h)** |
| S | 0.02 | 0.85 | 0.00 | 0.83 | 0.02 | 0.88 |
| Z | 0.06 | 0.64 | 0.03 | 0.62 | 0.02 | 0.58 |
| SLCum | | | 0.02 | 0.82 | 0.02 | 0.78 |
| QLCum | | | 0.02 | 0.69 | 0.02 | 0.64 |
| SLCum2 | | | | | 0.02 | 1.00 |
| QLCum2 | | | | | 0.00 | 0.67 |
| Q | 0.02 | 0.92 | 0.02 | 0.90 | 0.02 | 0.88 |
| Vm | | | 0.02 | 0.75 | 0.02 | 0.78 |
| $N_2$ | | | | | 0.02 | 0.63 |

Figure 12 shows a comparison between the features $Q$ and $N_2$ for the 6 h time interval. The clusters are ordered in time, so the circles with cluster numbers from 1 to 46 are the ones from the 1995–2015 training set. It is noticeable how the A clusters can be clearly discriminated from the B ones using the $Q$ feature, while $N_2$ shows mixed classes. In particular, several Type B clusters show a number of events $N_2$ equal to 2 at 6 h, while there are Type A clusters with a smaller or equal number of events. The bad performances of the $N_2$ feature in Figure 12 can be explained with the attempt of the algorithm to discriminate the two classes setting high values of the threshold (in this case, 3.50). This choice supplied poor results for the testing of Figure 10c,d because half of the Type A clusters in the test set are under the threshold.

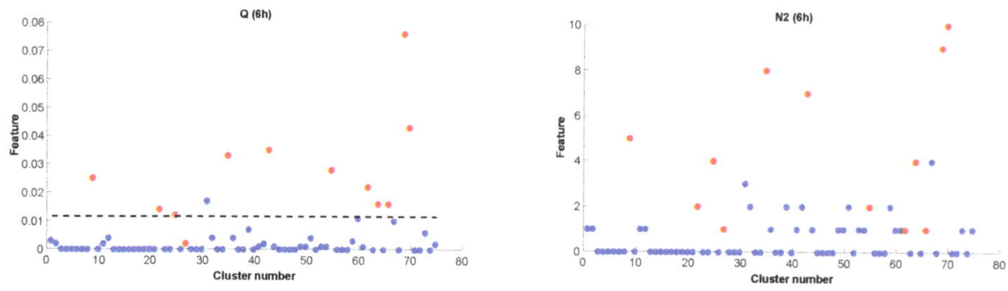


**(a)** **(b)**

**Figure 12.** Values of the features Q (**a**) and N2 (**b**) at the 6 h (6 h) time interval for the whole 1995–2022 dataset. Red circles: Type A clusters; blue circles: Type B clusters; black dashed line: the threshold obtained by NESTOREv1.0.

## 6. Discussion

In analyzing Greek seismicity from the perspective of Type A and B cluster analysis, several interesting results emerged that distinguish the seismicity of the area from that of other regions of the world. The first interesting result is that the percentage of Type A clusters in Greece is very low, even considering the smallest time interval analyzed after the mainshock (6 h). In fact, for a time interval of 6 h, the number of Type B clusters is about five times higher than the number of Type A clusters. This number is very high when compared with Italy, northeastern Italy and western Slovenia, and California, where NESTORE has already been applied [18,25,27], where the number of Type B clusters is between 1.5 and 2 times the number of Type A clusters. Moreover, there are no correlations between the cluster type and the focal mechanism, focal depth, magnitude, and location of the mainshock, as observed in some cases in other regions [27].

Previous studies in California [18], Italy [27], and northeastern Italy and western Slovenia [25], corresponding to very different seismotectonic regions, have shown good performance of classifiers based on the number of events (feature $N_2$) shortly after the mainshock [18]. Classifiers based on the features $Q$ and $S$ perform well in Italy and western Slovenia, while in California, they provide reliable results only some days after the mainshock. Conversely, feature $N_2$ gives the worst results in Greece, while features $Q$ and $S$ give the best results. The difference in performance between features $Q$ and $N_2$ in Greece can be clearly seen in Figure 12. The main difference is related to a large number of Type A clusters with a similar number of aftershocks as B clusters. A further comparison of the $N_2$ feature at 6 and 18 h, shows that the performance of the $N_2$ feature improves at longer time periods, since the Type A clusters, characterized by an early strong aftershock and therefore removed from the dataset, are precisely those with a low number of events. These strong early aftershock clusters are not very productive in terms of the number of aftershocks, but they are still productive in terms of the energy of the aftershocks and can therefore be discriminated from B clusters using the features $Q$ and $S$, which are related to the magnitude of the aftershocks. This fact and the low percentage of A clusters make us hypothesize that there may be fewer high-energy earthquakes in Greece for the same total energy radiation. This hypothesis is beyond the scope of this paper and should be verified in future work.

Another interesting feature is $QLcum$, which corresponds to the deviation of $Q$ from the long-term trend. This feature gives good short-term results after the mainshock for California as well as for northern Italy and western Slovenia, while it requires longer time intervals for Italian seismicity. The numerical values of this feature can only be compared with the application for California, since the interval start time and completeness requirements have changed from previous work. However, it is interesting to note that the thresholds of the other features defined in both Greece and California are similar, with variations within 25%, while the threshold of the *QLcum* feature in Greece is about 12 times higher than that in California. This could be related to the strong temporal variations in the radiated energy in both the Greek Type A and Type B clusters.

For this NESTOREv1.0 application, the performance is good at 6 h and deteriorates over time for longer periods. This trend is explained by the fact that shorter time intervals have a higher percentage of Type A clusters than longer time intervals, where performance is affected by the imbalance of A and B classes, resulting in a reduced ability of the classification system to distinguish between classes. In addition, the effects of background seismicity and activation of nearby fault segments, especially in case of large earthquakes, may reduce the reliability of the features. Importantly, the improved performance shortly after the mainshock is a notable advantage for the application of the algorithm for risk mitigation purposes.

## 7. Conclusions

The NESTORE machine learning algorithm, implemented in the NESTOREv1.0 software package [26], was applied to Greek seismicity to forecast the occurrence of a strong earthquake after an intense mainshock. We used the AUTH earthquake catalogue between 1995 and 2022 over a large area of Greece, consisting of the Gulf of Corinth, the Ionian Islands, the northern Aegean Sea, Thessaly and central Greece, Crete, and the Peloponnese, in order to obtain a long time period and a large area for analysis, and thus to analyze a sufficiently large number of clusters. Using a window-based approach, in which a cluster is defined as all events occurring within a temporal and spatial window around the main earthquake, we tested several laws for cluster detection and found that Uhrhammer's (1986) [69] law was the most appropriate for identifying clusters in Greece.

NESTORE classifies clusters into two classes, Type A or Type B, depending on the magnitude of the strongest aftershock. The algorithm analyzes seismicity features at increasing time intervals from the mainshock using a training procedure based on single-node decision trees (one threshold for each feature) and found statistically validated thresholds for the features to discriminate the two typologies. After training, a testing procedure estimates the probability for each feature to be a Type A cluster on an independent test set. The estimated probabilities from the different features are combined using a Bayesian approach to obtain the NESTORE response, which takes into account the different degrees of reliability of each feature.

The NESTOREv1.0 cluster identification module is independent of the other two. It allows the user to choose the equations for the radius and time interval of the cluster. This approach allows fitting to different regions for which different equations should be used. However, if a more accurate cluster evaluation procedure is required for a particular region, this module can be modified without affecting the following two modules.

The training and testing modules can be applied to clusters whose magnitude of completeness is at least equal to the magnitude of the mainshock minus 2. The modules require a dataset of tens of clusters for reliable training and testing. Thus, the success of the application of the NESTORE algorithm is influenced by the earthquake catalogue: if the completeness magnitude is too large, and thus, the number of clusters that can be analyzed is too small, the algorithm cannot be successfully applied. In addition, the performance of the features can be affected by the quality of the catalogue used and the magnitudes and the epicenters of the earthquakes. For this reason, a well-covered seismological network is important. To avoid too few clusters or problems related to changes in seismicity over time, the use of data over a period longer than 10–20 years is strongly recommended to cover the variability of seismicity features. Considering these catalogue property requirements, the algorithm has been shown to be robust enough to be applied in different seismotectonic environments. Crucial to this are the training procedure, which allows the algorithm to automatically adapt to the study area, and the clustering approach, which allows different region-specific equations as input.

In our work, we carried out the analysis by NESTOREv1.0 on 75 clusters reported in the AUTH earthquake catalogue from 1995 to 2022, using a training set from 1995 to 2015 and a test set in the following 7 years. In particular, by using ROC and Precision–Recall plots, we show that NESTOREv1.0 provided good performances in terms of Type A clusters forecasting. The best performance was obtained for a time interval of 0.25 days (6 h) after the o-mainshock. Notably, 100% of Type A clusters were forecasted correctly, the percentage of Type B clusters misclassified as Type A clusters was less than 10%, and the percentage of correct classifications was 92%. This makes the method particularly attractive for application in the field of seismic risk mitigation, as it allows estimating the probability of a future hazardous earthquake occurring after an initial strong event.

Our understanding of the SSLE preparation process can benefit from a detailed examination of the features and time periods in which they are relevant to the cluster classification. In particular, the features $S$ and $Q$, both depending on the earthquake's magnitude, perform well shortly after the mainshock, while $N_2$, depending on the number of earthquakes, performs poorly. Interestingly, in a previous application of the code to California, Italy, and northeastern Italy and western Slovenia, in [18,25,27], $N_2$ performed best, while $Q$ and $S$ feature performances depended on the analyzed region.

It is important to remark that NESTORE performs well independently on different regional characteristics because is based on a region-dependent training and because it is based on different features of seismicity. In our opinion, such an approach based on multiple features is pivotal to develop a robust algorithm able to work in different regions.

**Supplementary Materials:** The following supporting information can be downloaded at: https://www.mdpi.com/article/10.3390/e25050797/s1. Figure S1. Confusion Matrix [76,77].

**Author Contributions:** Conceptualization, S.G.; methodology, S.G.; software, S.G. and P.B.; validation, S.G., P.B. and E.-A.A.; formal analysis, S.G., P.B. and E.-A.A.; investigation, S.G., P.B. and E.-A.A.; resources, S.G. and P.B.; data curation, E.-A.A.; writing—original draft preparation, E.-A.A.; writing—review and editing, S.G., P.B. and F.V.; visualization, E.-A.A., S.G. and P.B.; supervision, S.G. and F.V.; project administration, S.G. and F.V.; funding acquisition, S.G. and F.V. All authors have read and agreed to the published version of the manuscript.

**Funding:** Funded by a grant from the Italian Ministry of Foreign Affairs and International Cooperation and Co-funded by the Erasmus+ programme of the European Union (EU).

**Data Availability Statement:** The NESTOREv1.0 toolbox is available for free download from GitHub at the address https://github.com/StefaniaGentili/NESTORE and the reproducibility package is available on Zenodo https://zenodo.org/account/settings/github/repository/StefaniaGentili/NESTORE. The catalogue used in this paper is the AUTH earthquake catalogue, available online: http://geophysics.geo.auth.gr/the_seisnet/WEBSITE_2005/station_index_en.html (last accessed on 28 July 2022).

**Conflicts of Interest:** The authors declare no conflict of interest. The funders had no role in the design of the study; in the collection, analyses, or interpretation of data; in the writing of the manuscript; or in the decision to publish the results.

## References

1. Jalayer, F.; Ebrahimian, H. Seismic risk assessment considering cumulative damage due to aftershocks. *Earthq. Eng. Struct. Dyn.* **2017**, *46*, 369–389. [CrossRef]
2. Raghunandan, M.; Liel, A.B.; Luco, N. Aftershock collapse vulnerability assessment of reinforced concrete frame structures. *Earthq. Eng. Struct. Dyn.* **2015**, *44*, 419–439.
3. DeVries, P.M.R.; Viégas, F.; Wattenberg, M.; Meade, B.J. Deep learning of aftershock patterns following large earthquakes. *Nature* **2018**, *560*, 632–634. [CrossRef]
4. Anyfadi, E.-A.; Avgerinou, S.-E.; Michas, G.; Vallianatos, F. Universal Non-Extensive Statistical Physics Temporal Pattern of Major Subduction Zone Aftershock Sequences. *Entropy* **2022**, *24*, 1850. [CrossRef]
5. Papazachos, B.C.; Papazachou, C. *The Earthquakes of Greece*; Ziti Publications: Thessaloniki, Greece, 2003.
6. Avgerinou, S.-E.; Anyfadi, E.-A.; Michas, G.; Vallianatos, F. A Non-Extensive Statistical Physics View of the Temporal Properties of the Recent Aftershock Sequences of Strong Earthquakes in Greece. *Appl. Sci.* **2023**, *13*, 1995. [CrossRef]

7. Tsapanos, T. Seismicity and Seismic Hazard Assessment in Greece. In *Earthquake Monitoring and Seismic Hazard Mitigation in Balkan Countries*; Springer: Dordrecht, The Netherlands, 2008.
8. Stiros, S.C. The AD 365 Crete Earthquake and Possible Seismic Clustering During the Fourth to Sixth Centuries AD in the Eastern Mediterranean: A Review of Historical and Archaeological Data. *J. Struct. Geol.* **2001**, *23*, 545–562. [CrossRef]
9. Altinok, Y.; Alpar, B.; Özer, N.; Gazioglu, C. 1881 and 1949 earthquakes at the Chios-Cesme Strait (Aegean Sea) and their relation to tsunamis. *Nat. Hazards Earth Syst. Sci.* **2005**, *5*, 717–725. [CrossRef]
10. Kouskouna, V.; Makropoulos, K. Historical earthquake investigations in Greece. *Ann. Geophys.* **2004**, *47*, 723–731.
11. Stiros, S.C.; Pirazzoli, P.A.; Laborel, J.; Laborel-Deguen, F. The 1953 earthquake in Cephalonia (Western Hellenic Arc): Coastal uplift and halotectonic faulting. *Geophys. J. Int.* **1994**, *117*, 834–849. [CrossRef]
12. Lekidis, V.A.; Karakostas, C.Z.; Dimitriu, P.P.; Margaris, B.N.; Kalogeras, I.; Theodulidis, N. The Aigio (Greece) seismic sequence of June 1995: Seismological, strong motion data and effects of the earthquakes on structures. *J. Earthq. Eng.* **1999**, *3*, 349–380. [CrossRef]
13. Mavroulis, S.; Fountoulis, I.; Lekkas, L. Environmental effects caused by the Andravida (08-06-2008, ML = 6.5, NW Peloponnese, Greece) earthquake. In Proceedings of the 11th IAEG, Auckland, New Zealand, 5–10 September 2010.
14. Saltogianni, V.; Gianniou, M.; Taymaz, T.; Yolsal-Çevikbilen, S.; Stiros, S. Fault slip source models for the 2014 Mw 6.9 Samothraki-Gökçeada earthquake (North Aegean trough) combining geodetic and seismological observations. *J. Geoph. Res. Solid Earth* **2015**, *120*, 8610–8622. [CrossRef]
15. Vallianatos, F.; Pavlou, K. Scaling properties of the Mw7.0 Samos (Greece), 2020 aftershock sequence. *Acta Geophys.* **2021**, *69*, 1067–1084. [CrossRef]
16. Michas, G.; Pavlou, K.; Avgerinou, S.E.; Anyfadi, E.A.; Vallianatos, F. Aftershock patterns of the 2021 Mw 6.3 Northern Thessaly (Greece) earthquake. *J. Seism.* **2022**, *26*, 201–225. [CrossRef]
17. Vallianatos, F.; Karakonstantis, A.; Michas, G.; Pavlou, K.; Kouli, M.; Sakkas, V. On the patterns and scaling properties of the 2021–2022 Arkalochori earthquake sequence (Central Crete, Greece) Based on Seismological, Geophysical and Satellite Observations. *Appl. Sci.* **2022**, *12*, 7716. [CrossRef]
18. Gentili, S.; Di Giovambattista, R. Forecasting strong subsequent earthquakes in California clusters by machine learning. *Phys. Earth Planet. Inter.* **2022**, *327*, 106879. [CrossRef]
19. Vorobieva, I.A.; Panza, G.F. Prediction of the Occurrence of Related Strong Earthquakes in Italy. *Pure Appl. Geophys.* **1993**, *141*, 25–41. [CrossRef]
20. Shcherbakov, R.; Zhuang, J.; Zoeller, G.; Ogata, Y. Forecasting the magnitude of the largest expected earthquake. *Nat. Commun.* **2019**, *10*, 4051. [CrossRef]
21. Helmstetter, A.; Sornette, D. Båth's law derived from the Gutenberg-Richter law and from aftershock properties. *Geophys. Res. Lett.* **2003**, *30*, 2069. [CrossRef]
22. Gulia, L.; Wiemer, S.; Vannucci, G. Pseudoprospective evaluation of the foreshock traffic light system in Ridgecrest and implications for aftershock hazard assessment. *Seismol. Res. Lett.* **2020**, *91*, 2828–2842. [CrossRef]
23. Rodríguez-Pérez, Q.; Zúñiga, F.R. Båth's law and its relation to the tectonic environment: A case study for earthquakes in Mexico. *Tectonophysics* **2016**, *687*, 66–77. [CrossRef]
24. Tahir, M.; Grasso, J.R.; Amorese, D. The largest aftershock: How strong, how far away, how delayed? *Geophys. Res. Lett.* **2012**, *39*, L04301. [CrossRef]
25. Gentili, S.; Di Giovambattista, R. Forecasting strong aftershocks in earthquake clusters from northeastern Italy and western Slovenia. *Phys. Earth Planet. Inter.* **2020**, *303*, 106483. [CrossRef]
26. Gentili, S.; Brondi, P.; Di Giovambattista, R. NESTOREv1.0: A MATLAB Package for Strong Forthcoming Earthquake Forecasting. *Seismol. Res. Lett.* **2023**. [CrossRef]
27. Gentili, S.; Di Giovambattista, R. Pattern recognition approach to the subsequent event of damaging earthquakes in Italy. *Phys. Earth Planet. Inter.* **2017**, *266*, 1–17. [CrossRef]
28. Papanikolaou, D.I. *The geology of Greece*; Springer: Berlin/Heidelberg, Germany, 2021. [CrossRef]
29. Frank, A.B.; Frei, R.; Triantaphyllou, M.; Vassilakis, E.; Kristiansen, K.; Frei, K.M. Isotopic range of bioavailable strontium on the Peloponnese peninsula, Greece: A multi-proxy approach. *Sci. Total Environ.* **2021**, *774*, 145181. [CrossRef]
30. Tsodoulos, I.M.; Koukouvelas, I.K.; Pavlides, S. Tectonic geomorphology of the easternmost extension of the Gulf of Corinth (Beotia, Central Greece). *Tectonophysics* **2008**, *453*, 211–232. [CrossRef]
31. Mariolakos, I.; Fountoulis, I.; Kranis, H. Geology and tectonics: Sterea Hellas area. *Eng. Geol. Environ. Mar. Koukis Tsiambaos Stournaras* **2001**, *5*, 3971–3986.
32. Alexakis, D.; Astaras, T.; Sarris, A.; Vouzaxakis, K.; Karimali, L. Reconstructing the neolithic landscape of Thessaly through a GIS and geological approach. *Reconstr. Anc. Landsc. Veg.* **2008**. [CrossRef]
33. Anders, B.; Reischmann, B.T.; Poller, U.; Kostopoulos, D. Age and origin of granitic rocks of the eastern Vardar Zone, Greece: New constraints on the evolution of the Internal Hellenides. *J. Geol. Soc.* **2005**, *162*, 857–870. [CrossRef]
34. Piippo, S.; Sadeghi, M.; Koivisto, E.; Skyttä, P.; Baker, T. Semi-automated geological mapping and target generation from geochemical and magnetic data in Halkidiki region, Greec. *Ore Geol. Rev.* **2022**, *142*, 104714. [CrossRef]
35. Higgins, M.D. *Geology of the Greek Islands*; University of California Press: Berkeley, CA, USA, 2009; pp. 392–396.

36. Evelpidou, N. Modelling of erosional processes in the Ionian Islands (Greece), Geomatics. *Nat. Hazards Risk* **2012**, *3*, 293–310. [CrossRef]
37. Kassaras, I.; Kapetanidis, V.; Ganas, A.; Tzanis, A.; Kosma, C.; Karakonstantis, A.; Valkaniotis, S.; Chailas, S.; Kouskouna, V.; Papadimitriou, P. The New Seismotectonic Atlas of Greece (v1.0) and Its Implementation. *Geosciences* **2020**, *10*, 447. [CrossRef]
38. Bathrellos, G.D.; Skilodimou, H.D.; Kelepertsis, A.; Alexakis, D.; Chrisanthaki, I.; Archonti, D. Environmental research of groundwater in the urban and suburban areas of Attica region, Greece. *Env. Geol.* **2008**, *56*, 11–18. [CrossRef]
39. Tavoularis, N.; Papathanassiou, G.; Ganas, A.; Argyrakis, P. Development of the Landslide Susceptibility Map of Attica Region, Greece, Based on the Method of Rock Engineering System. *Land* **2021**, *10*, 148. [CrossRef]
40. Valkanou, K.; Karymbalis, E.; Papanastassiou, D.; Soldati, M.; Chalkias, C.; Gaki-Papanastassiou, K. Assessment of Neotectonic Landscape Deformation in Evia Island, Greece, Using GIS-Based Multi-Criteria Analysis. *ISPRS Int. J. Geo-Inf.* **2021**, *10*, 118. [CrossRef]
41. Chatzipetros, A.; Pavlides, S.; Foumelis, M.; Sboras, S.; Galanakis, D.; Pikridas, C.; Bitharis, S.; Kremastas, E.; Chatziioannou, A.; Papaioannou, I. The northern Thessaly strong earthquakes of March 3 and 4, 2021, and their neotectonic setting. *Bull. Geol. Soc. Greece* **2021**, *58*, 222–255. [CrossRef]
42. Kassaras, I.; Kapetanidis, V.; Ganas, A.; Karakonstantis, A.; Papadimitriou, P.; Kaviris, G.; Kouskouna, V.; Voulgaris, N. Seismotectonic analysis of the 2021 Damasi-Tyrnavos (Thessaly, Central Greece) earthquake sequence and implications on the stress field rotations. *J. Geodyn.* **2022**, *150*, 101898. [CrossRef]
43. Nikolakopoulos, K.G.; Koukouvelas, I.K.; Lampropoulou, P. UAV, GIS, and Petrographic Analysis for Beachrock Mapping and Preliminary Analysis in the Compressional Geotectonic Setting of Epirus, Western Greece. *Minerals* **2022**, *12*, 392. [CrossRef]
44. Aristotle University of Thessaloniki Seismological Network. Available online: http://geophysics.geo.auth.gr/the_seisnet/WEBSITE_2005/station_index_en.html (accessed on 28 July 2022).
45. Bountzis, P.; Papadimitriou, E.; Tsaklidis, G. Identification and Temporal Characteristics of Earthquake Clusters in Selected Areas in Greece. *Appl. Sci.* **2022**, *12*, 1908. [CrossRef]
46. Papadimitriou, E.; Bonatis, P.; Bountzis, P.; Kostoglou, A.; Kourouklas, C.; Karakostas, V. The Intense 2020–2021 Earthquake Swarm in Corinth Gulf: Cluster Analysis and Seismotectonic Implications from High Resolution Microseismicity. *Pure Appl. Geophys.* **2022**, *179*, 3121–3155. [CrossRef]
47. Kapetanidis, V.; Deschamps, A.; Papadimitriou, P.; Matrullo, E.; Karakonstantis, A.; Bozionelos, G.; Kaviris, G.; Serpetsidaki, A.; Lyon-Caen, H.; Voulgaris, N.; et al. The 2013 earthquake swarm in Helike, Greece: Seismic activity at the root of old normal faults. *Geophys. J. Int.* **2015**, *202*, 2044–2073. [CrossRef]
48. Kaviris, G.; Elias, P.; Kapetanidis, V.; Serpetsidaki, A.; Karakonstantis, A.; Plicka, V.; De Barros, L.; Sokos, E.; Kassaras, I.; Sakkas, V.; et al. The Western Gulf of Corinth (Greece) 2020–2021 Seismic Crisis and Cascading Events: First Results from the Corinth Rift Laboratory Network. *Seism. Rec.* **2021**, *1*, 85–95. [CrossRef]
49. Vassilakis, E.; Kaviris, G.; Kapetanidis, V.; Papageorgiou, E.; Foumelis, M.; Konsolaki, A.; Petrakis, S.; Evangelidis, C.P.; Alexopoulos, J.; Karastathis, V.; et al. The 27 September 2021 Earthquake in Central Crete (Greece)-Detailed Analysis of the Earthquake Sequence and Indications for Contemporary Arc-Parallel Extension to the Hellenic Arc. *Appl. Sci.* **2022**, *12*, 2815. [CrossRef]
50. Becker, D.; Meier, T.; Bohnhoff, M.; Harjes, H.P. Seismicity at the convergent plate boundary offshore Crete, Greece, observed by an amphibian network. *J. Seism.* **2010**, *14*, 369–392. [CrossRef]
51. Sboras, S.; Chatzipetros, A.; Pavlides, S.B. North Aegean Active Fault Pattern and the 24 May 2014, Mw 6.9 Earthquake. In *Active Global Seismology: Neotectonics and Earthquake Potential of the Eastern Mediterranean Region*, 1st ed.; Çemen, I., Yılmaz, Y., Eds.; American Geophysical Union: Washington, DC, USA, 2017.
52. Koukouvelas, I.; Aydin, A. Fault structure and related basins of the North Aegean Sea and its surroundings. *Tectonics* **2002**, *21*, 10-1–10-17. [CrossRef]
53. Sboras, S. The Greek Database of Seismogenic Sources: Seismotectonic Implications for North Greece. Ph.D. Thesis, University of Ferrara, Ferrara, Italy, 2011.
54. Kokinou, E.; Kamperis, E.; Alves, T.M. Structural decoupling in a convergent forearc setting (Southern Crete, Eastern Mediterranean). *Geol. Soc. Am. Bull.* **2012**, *124*, 1352–1364. [CrossRef]
55. Keilis-Borok, V.I.; Kossobokov, V.G. Premonitory activation of earthquake flow: Algorithm M8. *Phys. Earth Planet. Inter.* **1990**, *61*, 73–83. [CrossRef]
56. Keilis-Borok, V.I.; Rotwain, I.M. Diagnosis of Time of Increased Probability of strong earthquakes in different regions of the world: Algorithm CN. *Phys. Earth Planet. Inter.* **1990**, *61*, 57–72. [CrossRef]
57. Vorobieva, I.A. Prediction of a subsequent large earthquake. *Phys. Earth Planet. Inter.* **1999**, *111*, 197–206. [CrossRef]
58. Dascher-Cousineau, K.; Brodsky, E.E.; Lay, T.; Goebel, T. What controls variations in aftershock productivity? *J. Geophys. Res. Solid Earth* **2020**, *125*, e2019JB018111. [CrossRef]
59. Seydoux, L.; Balestriero, R.; Poli, P.; Hoop, M.D.; Campillo, M.; Baraniuk, R. Clustering earthquake signals and background noises in continuous seismic data with unsupervised deep learning. *Nat. Commun.* **2020**, *11*, 3972. [CrossRef] [PubMed]
60. Wang, L.; Hare, B.M.; Zhou, K.; Stöcker, H.; Scholten, O. Identifying lightning structures via machine learning. *Chaos Solitons Fractals* **2023**, *170*, 113346. [CrossRef]

61. Van Stiphout, T.; Zhuang, J.; Marsan, D. Seismicity Declustering, Community Online Resource for Statistical Analysis. 2012. Available online: http://www.corssa.org (accessed on 28 February 2023). [CrossRef]
62. Jang, J.S.R.; Sun, C.T.; Mizutani, E. Neuro-fuzzy and soft computing-a computational approach to learning and machine intelligence. *IEEE Trans. Autom. Control* **1997**, *42*, 1482–1484. [CrossRef]
63. Breiman, L.; Friedman, J.; Olshen, R.A.; Stone, C.J. *Classification and Regression Trees*; Routledge: New York, NY, USA, 1984. [CrossRef]
64. Gentili, S.; Brondi, P.; Rossi, G.; Venturini, E.; Di Giovambattista, R. *The Italy-Japan Project "Analysis of Seismic Sequences for Strong Aftershock Forecasting"*; Centro di Ricerche Sismologiche-Istituto Nazionale di Oceanografia e di Geofisica Sperimentale (CRS-OGS): Udine, Italy, 2021.
65. Bailer-Jones, C.A.L.; Smith, K. Combining probabilities. *Gaia DPAC*. GAIA-C8-TN-MPIA-CBJ-053. 2011. Available online: https://www.mpia.de/3432751/probcomb_TN.pdf (accessed on 11 May 2022).
66. Gardner, J.K.; Knopoff, L. Is the sequence of earthquakes in Southern California, with aftershocks removed, Poissonian? *Bull. Seismol. Soc. Am.* **1974**, *64*, 1363–1367. [CrossRef]
67. Lolli, B.; Gasperini, P. Aftershocks hazard in Italy Part I: Estimation of time-magnitude distribution model parameters and computation of probabilities of occurrence. *J. Seismol.* **2003**, *7*, 235–257. [CrossRef]
68. Gentili, S.; Bressan, G. The partitioning of radiated energy and the largest aftershock of seismic sequences occurred in the northeastern Italy and western Slovenia. *J. Seism.* **2008**, *12*, 343–354. [CrossRef]
69. Uhrhammer, R. Characteristics of northern and central California seismicity. *Earthq. Notes* **1986**, *57*, 21.
70. Kagan, Y.Y. Seismic moment distribution revisited: I. Statistical results. *Geophys. J. Int.* **2002**, *148*, 520–541. [CrossRef]
71. Knopoff, L. The magnitude distribution of declustered earthquakes in Southern California. *Proc. Nat. Acad. Sci. USA* **2000**, *97*, 11880–11884. [CrossRef]
72. Ganas, A.; Mouzzakiotis, E.; Moshou, A.; Karastathis, V. Left-lateral shear inside the North Gulf of Evia Rift, Central Greece evidenced by relocated earthquake sequences and moment tensor inversion. *Tectonophysics* **2016**, *682*, 237–248. [CrossRef]
73. Wiemer, S. A Software Package to Analyze Seismicity: ZMAP. *Seismol. Res. Lett.* **2001**, *72*, 373–382. [CrossRef]
74. Gholamy, A.; Kreinovich, V.; Kosheleva, O. Why 70/30 or 80/20 Relation Between Training and Testing Sets: A Pedagogical Explanation. *Dep. Tech. Rep. UTEP-CS-18-09* **2018**, 1209. Available online: https://scholarworks.utep.edu/cs_techrep/1209 (accessed on 11 May 2023).
75. Xu, Y.; Goodacre, R. On Splitting Training and Validation Set: A Comparative Study of Cross-Validation, Bootstrap and Systematic Sampling for Estimating the Generalization Performance of Supervised Learning. *J. Anal. Test.* **2018**, *2*, 249–262. [CrossRef] [PubMed]
76. Di Giovambattista, R.; Tyupkin, Y.S. Burst of aftershocks as a manifestation of instability of the earth crust in an area of strong earthquake preparation. In Proceedings of the European Seismological Commission (ESC), XXVIII General Assembly, Genoa, Italy, 1–6 September 2002.
77. Gutenberg, B.; Richter, C.F. Earthquake magnitude, intensity, energy, and acceleration: (Second paper). *Bull. Seismol. Soc. Am.* **1956**, *46*, 105–145. [CrossRef]

**Disclaimer/Publisher's Note:** The statements, opinions and data contained in all publications are solely those of the individual author(s) and contributor(s) and not of MDPI and/or the editor(s). MDPI and/or the editor(s) disclaim responsibility for any injury to people or property resulting from any ideas, methods, instructions or products referred to in the content.

*entropy*

MDPI

*Article*

# Aftershocks and Fluctuating Diffusivity

**Sumiyoshi Abe [1,2,3,*], Norikazu Suzuki [4] and Dmitrii A. Tayurskii [2]**

1 Department of Physics, College of Information Science and Engineering, Huaqiao University, Xiamen 361021, China
2 Institute of Physics, Kazan Federal University, Kazan 420008, Russia
3 Department of Natural and Mathematical Sciences, Turin Polytechnic University in Tashkent, Tashkent 100095, Uzbekistan
4 College of Science and Technology, Nihon University, Chiba 274-8501, Japan
* Correspondence: suabe@sf6.so-net.ne.jp

**Abstract:** The Omori-Utsu law shows the temporal power-law-like decrease of the frequency of earthquake aftershocks and, interestingly, is found in a variety of complex systems/phenomena exhibiting catastrophes. Now, it may be interpreted as a characteristic response of such systems to large events. Here, hierarchical dynamics with the fast and slow degrees of freedom is studied on the basis of the Fokker-Planck theory for the load-state distribution to formulate the law as a relaxation process, in which diffusion coefficient in the space of the load state is treated as a fluctuating slow variable. The evolution equation reduced from the full Fokker-Planck equation and its Green's function are analyzed for the subdynamics governing the load state as the fast degree of freedom. It is shown that the subsystem has the temporal translational invariance in the logarithmic time, not in the conventional time, and consequently the aging phenomenon appears.

**Keywords:** Omori-Utsu law for aftershocks; slow relaxation; Fokker-Planck theory with hierarchical dynamics; fluctuating diffusivity; logarithmic time and aging

**Citation:** Abe, S.; Suzuki, N.; Tayurskii, D.A. Aftershocks and Fluctuating Diffusivity. *Entropy* **2023**, *25*, 989. https://doi.org/10.3390/e25070989

Academic Editor: Georgios Michas

Received: 6 April 2023
Revised: 7 June 2023
Accepted: 9 June 2023
Published: 28 June 2023

**Copyright:** © 2023 by the authors. Licensee MDPI, Basel, Switzerland. This article is an open access article distributed under the terms and conditions of the Creative Commons Attribution (CC BY) license (https://creativecommons.org/licenses/by/4.0/).

## 1. Introduction

Given a system, it is essential for understanding its properties out of equilibrium to look at how the system responds to a sudden disturbance and returns to a typical state such as the equilibrium one. Therefore, the relaxation phenomena are of major interest for a stable system. Most celebrated may be the exponential relaxation, to the type of which the mechanical stress relaxation of Maxwell [1] and the Drude-Lorentz-Debye dielectric(-paramagnetic) relaxation [2,3] belong. However, it is also known that there exist systems of another type characterized by the stretched exponential function exhibiting temporal decays of disturbances slower than the exponential type. Such an exotic type has been identified already in the 19th century [4]. It is termed the Kohlrausch-Williams-Watts relaxation and is widely observed in disordered systems such as glass-forming liquids (Reference [5] for a review) and polymers (see Reference [6], for example). Furthermore, it has been found [7,8] that there are soft matters obeying even slower power-law stress relaxation.

The power-law relaxation phenomenon has actually been long known in seismology. Omori [9] has discovered in the 19th century that the temporal decrease of aftershocks following a large earthquake obeys a power law. Later, this empirical law has been modified and made more precise by Utsu [10]. Accordingly, today it is commonly referred to as the Omori-Utsu law for earthquake aftershocks. It is as follows. Suppose that a main shock has occurred at $t = 0$. Then, the law states that the number of aftershocks occurring during the later time interval between $t$ and $t + \Delta t$, $\Delta N(t) = N(t + \Delta t) - N(t)$, is

$\Delta N(t) \sim (c+t)^{-p}\ \Delta t$ with $p$ and $c$ being positive constants. In the continuous approximation, it is written in the form of differential as

$$\frac{d\,N(t)}{d\,t} = n_0\,\phi(t), \tag{1}$$

where $n_0$ is a positive constant, and $\phi(t)$ is the function describing the power-law relaxation

$$\phi(t) = \frac{1}{(1+t/\tau)^{p}} \tag{2}$$

with $\tau = c$. The value of the exponent $p$ is generally considered to be close to unity but actually ranges between about 0.5 and 1.5, depending on datasets.

An interesting point is that the Omori-Utsu law has its analogs outside seismology. A couple of examples have been observed in connection with the World Wide Web. It has been reported [11] that the download rate of an article uploaded online obeys the Omori-Utsu-like law. It has also been shown [12] that the pattern of the Internet traffic after a heavily congested state also obeys it. Furthermore, analogous phenomena have been discovered in the behaviors of the financial markets. In Reference [13], it has been found that the dynamics of the stock market after a large crash is well described by the Omori-Utsu-like law. In addition, it has also been demonstrated to be the case concerning the crash of the currency exchange rate [14]. These facts suggest that the Omori-Utsu law may serve as a key for understanding complex systems with catastrophes, in general.

In this paper, we develop a discussion about the Omori-Utsu law for occurrence of earthquake aftershocks based on physical kinetics. In particular, a generalization of the Fokker-Planck theory, in which the diffusion coefficient is not a constant but a fluctuating variable that may describe the effects originating from heterogeneity of the crust and a complex landscape of the stress distribution at faults, and see how such an theory can explain the features of the temporal pattern of aftershocks. For this purpose, "a modified version" of the approach recently proposed in Reference [15] is employed for representing the dynamical hierarchy. It is shown that the relaxation function in Equation (2) can be obtained for the subsystem defined through elimination of the fluctuating diffusivity. This, in turn, gives information on the dynamics underlying the aftershock phenomenon. Then, we study the reduction of the generalized Fokker-Planck equation and analyze associated Green's function characterizing the transition of the load state. It is found that Green's function has the temporal translational invariance not in the conventional time but in the logarithmic time. As a result, the system exhibits the phenomenon of aging, that is, the system has its own clock. This is in conformity with the discovery in real seismicity reported in References [16,17]. On the other hand, time evolution is however still Markovian since Green's function obeys the Chapman-Kolmogorov equation, in contrast to the non-Markovian nature of aftershocks [18] (see also Reference [17]). Thus, the present theory can partially describe the properties of aftershocks.

## 2. Omori-Utsu Law and Hierarchical Fokker-Planck Equation with Fluctuating Diffusivity

The concept of Brownian motion has been discussed in the context of seismology in References [19,20] (see also other works cited therein). There, recurrence of earthquakes has been modeled as a problem of passage times [21]. The basic random variable considered there, which experiences random perturbations, is termed "load state" [20]. In the context of statistical mechanics, it may be thought of as the mean field of stresses distributed over faults in a geographical region under consideration. We shall denote its realization by $\xi$. Thus, $\xi$ *should not be confused with a spatial coordinate.*

Our point is as follows. We hypothesize that the load state in the spatial regime of aftershocks undergoes the heterogeneity described by the fluctuating diffusivity, implying that the diffusion coefficient $D$ itself is regarded as the realization of a random variable, not a fixed constant. On the other hand, such exotic diffusion with fluctuating diffusivities has been observed in laboratories [22,23], where the heterogeneities of the systems are essential. To theoretically describe it, a kinetic approach has been developed in Reference [15] on the basis of the Fokker-Planck theory endowed with the hierarchical dynamics characterized by largely separated time scales. In view of that approach, the load-state variable $\xi$ is the fast degree of freedom, whereas the diffusivity variable $D$ should be the slow degree of freedom.

Thus, we consider a 2-tuple of the dynamical random variables $\mathbf{X} = (X_1, X_2)^T$, which obeys a general stochastic differential equation [24]: $d\,\mathbf{X} = \tilde{\mathbf{K}}\; d\,t + \tilde{G}\; d\,\mathbf{W}$. $\tilde{\mathbf{K}}$ is a drift term and $\tilde{G}$ is a $2 \times 2$ matrix. Both of them depend on $X_1$, $X_2$ and time $t$. $d\,\mathbf{W} = (d\,W_1, d\,W_2)^T$ is assumed to be the Wiener noise satisfying Itô's rule: $d\,W_i\, d\,W_j = \delta_{ij}\, d\,t$. This implies that $d\,W_1$ and $d\,W_2$ are mutually independent, but a possible correlation between them may effectively be taken into account by a nondiagonal $\tilde{G}$. Now, $X_1$ is taken to be the random variable of the load state whose realization is $\xi$ and $X_2$ is the logarithm of the nonnegative random variable of diffusivity with its realization $D$. Since the diffusivity variable is dimensioned, its logarithm contains a constant eliminating the dimensionality. However, without loss of generality, such a constant can be set equal to unity. Therefore, the realization $\mathbf{x}$ of $\mathbf{X}$ is denoted by $\mathbf{x} = (x_1,\; x_2)^T = (\xi,\; \ln D)^T$.

Our interest is in determining $\tilde{\mathbf{K}}$ and $\tilde{G}$ that lead to the Omori-Utsu law since these quantities carry information on the dynamics underlying the law. In this respect, the Fokker-Planck theory may give a clue. In what follows, we examine this point based on a modification of the method presented in Reference [15].

Let $\tilde{P}\,(x_1,\; x_2,\; t)$ be the probability distribution normalized in the whole plane: $-\infty < x_1 < \infty$, $-\infty < x_2 < \infty$. Physically, this infinite support can be approximated to be a finite one for the well-localized system/phenomenon. See the later discussion about Equation (31). This distribution obeys the Fokker-Planck equation [24] associated with the above-mentioned stochastic differential equation in the following general form:

$$\frac{\partial \tilde{P}}{\partial t} = -\sum_{i=1,2} \frac{\partial}{\partial x_i}\left(\tilde{K}_i\, \tilde{P}\right) + \sum_{i,j=1,2} \frac{\partial^2}{\partial x_i\, \partial x_j}\left(\tilde{\sigma}_{ij}\, \tilde{P}\right), \tag{3}$$

where $\tilde{\sigma} = (\tilde{\sigma}_{ij})$ is a positive matrix (i.e., being symmetric and having only positive eigenvalues) given by $\tilde{\sigma} = (1/2)\, \tilde{G}\, \tilde{G}^{\,T}$, and both $\tilde{K}_i$ and $\tilde{\sigma}_{ij}$ are the functions of $x_1$, $x_2$ and $t$. Mathematically, the drift term has the dependency on calculi since the underlying stochastic process mentioned above is multiplicative. Here, Itô calculus has been employed.

To be explicit, it is convenient to directly employ the pair $(\xi,\; D)$. Accordingly, $\partial/\partial\, x_1 = \partial/\partial\,\xi$, $\partial/\partial\, x_2 = D\,\partial/\partial\, D$ and $P\,(\xi,\; D,\; t) = (1/D)\tilde{P}\,(\xi,\; lnD,\; t)$ with the prefactor $1/D$ being the Jacobian. $P\,(\xi,\; D,\; t)$ is normalized in the upper-half plane since $D$ is nonnegative. Then, Equation (1) is rewritten as

$$\begin{aligned}\tfrac{\partial P}{\partial t} &= -\tfrac{\partial}{\partial \xi}(K_1\, P) - D\tfrac{\partial}{\partial D}(K_2\, P) \\ &\quad + \tfrac{\partial^2}{\partial \xi^2}(\sigma_{11}\, P) + 2\tfrac{\partial^2}{\partial \xi\, \partial D}(\sigma_{12}\, P) + \tfrac{\partial^2}{\partial D^2}(\sigma_{22}\, P)\end{aligned}\;, \tag{4}$$

which is "a modification" of Equation (13) in Reference [15]. In this equation, $K$'s and $\sigma$'s may depend on $\xi$ and $D$ as well as $t$ and are related to $\tilde{K}$'s and $\tilde{\sigma}$'s in Equation (3) as follows: $K_1 = \tilde{K}_1$, $K_2 = D\left(\tilde{K}_2 + \tilde{\sigma}_{22}\right)$, $\sigma_{11} = \tilde{\sigma}_{11}$, $\sigma_{12} = D\,\tilde{\sigma}_{12}$, $\sigma_{22} = D^2\,\tilde{\sigma}_{22}$. As mentioned, $\xi$ and $D$ are the fast and slow degrees of freedom, respectively. To implement such a hierarchy, the method analogous to the Born-Oppenheimer approximation may be applied [15]. That is, *the fast degree of freedom is strongly influenced by the slow degree of freedom, whereas the slow degree of freedom is not affected by the fast degree of freedom*. This leads to

$$K_2 = K_2(D),\; \sigma_{22} = \sigma_{22}(D), \tag{5}$$

provided that time dependence in these quantities are ignored since they are relevant to the slow degree of freedom. In addition, the joint probability distribution should be factorized as follows:

$$P\,(\xi,\,D,\,t) = p\,(\xi,\,t|D)\,\Pi(D), \tag{6}$$

where $p\,(\xi,\,t|D)$ is the conditional probability distribution given the value of $D$ and $\Pi(D)$ is the marginal probability distribution of $D$, time dependence of which is ignored as in Equation (5). Then, Equation (4) becomes

$$\begin{aligned}\Pi(D)\frac{\partial\, p(\xi,\,t|D)}{\partial\, t} = & -\Pi(D)\frac{\partial}{\partial\,\xi}[K_1(\xi,\,D,\,t)\,p(\xi,\,t|D)] \\ & -\frac{\partial}{\partial\, D}[K_2(D)\,p(\xi,\,t|D)\,\Pi(D)] \\ & +\Pi(D)\frac{\partial^2}{\partial\,\xi^2}[\sigma_{11}(\xi,\,D,\,t)\,p\,(\xi,\,t|D)] \\ & +2\frac{\partial}{\partial\, D}\left\{\Pi(D)\,\frac{\partial}{\partial\,\xi}[\sigma_{12}(\xi,\,D,\,t)\,p\,(\xi,\,t|D)]\right\} \\ & +\frac{\partial^{\,2}}{\partial\, D^2}[\sigma_{22}(D)\,p\,(\xi,\,t|D)\,\Pi(D)].\end{aligned} \tag{7}$$

As shown in Reference [15] with the present modification, this equation can be separated as follows:

$$\begin{aligned}\frac{\partial\, p\,(\xi,\,t|D)}{\partial\, t} = & -\frac{\partial}{\partial\,\xi}[K_1(\xi,\,D,\,t)\,p\,(\xi,\,t|D)] \\ & +\frac{\partial^2}{\partial\,\xi^2}[\sigma_{11}(\xi,\,D,\,t)\,p\,(\xi,\,t|D)]\end{aligned} \tag{8}$$

for the fast degree of freedom and

$$\begin{aligned} & -\frac{\partial}{\partial\, D}[K_2(D)\,p\,(\xi,\,t|D)\,\Pi(D)] \\ & +2\frac{\partial}{\partial\, D}\left\{\Pi(D)\frac{\partial}{\partial\,\xi}[\sigma_{12}(\xi,\,D,\,t)\,p\,(\xi,\,t|D)\,]\right\} \\ & +\frac{\partial^{\,2}}{\partial\, D^{\,2}}[\sigma_{22}(D)\,p\,(\xi,\,t|D)\,\Pi(D)] = 0\end{aligned} \tag{9}$$

for the rest. Equation (9) is immediately integrated to be

$$\begin{aligned} & -K_2(D)\,p\,(\xi,\,t|D)\,\Pi(D) + 2\,\Pi(D)\frac{\partial}{\partial\,\xi}[\sigma_{12}(\xi,\,D,\,t)\,p\,(\xi,\,t|D)\,] \\ & +\frac{d\,[\sigma_{22}(D)\,\Pi(D)]}{d\,D}\,p\,(\xi,\,t|D) + \sigma_{22}(D)\,\Pi(D)\frac{\partial\, p\,(\xi,\,t|D)}{\partial\, D} \\ & = f\,(\xi,\,t).\end{aligned} \tag{10}$$

Here, $f\,(\xi,\,t)$ is a certain function. This equation is further separated as follows:

$$-\,K_2(D)\,\Pi(D) + \frac{d\,[\sigma_{22}(D)\,\Pi(D)]}{d\,D} = 0 \tag{11}$$

for the slow degree of freedom and

$$\begin{aligned} & 2\,\Pi(D)\frac{\partial}{\partial\,\xi}[\sigma_{12}(\xi,\,D,\,t)\,p\,(\xi,\,t|D)\,] + \sigma_{22}(D)\,\Pi(D)\frac{\partial\, p\,(\xi,\,t|D)}{\partial\, D} \\ & = f\,(\xi,\,t).\end{aligned} \tag{12}$$

for the coupling between the fast and slow degrees of freedom. Actually, it is possible to set

$$f\,(\xi,\,t) = 0. \tag{13}$$

This is because the integration of Equation (12) over $-\infty < \xi < \infty$ leads to the fact that such an integration of $f\,(\xi,\,t)$ vanishes, provided that $\sigma_{12}(\xi,\,D,\,t)\,p\,(\xi,\,t|D) \to 0 \;\; (\xi \to \pm\infty)$ is assumed. [There is still a possibility that $f(\xi,\,t)$ is an integrable odd function of $\xi$, but such a case turns out to be ruled out]. Thus, the equation for the coupling is given by

$$2\;\frac{\partial}{\partial\,\xi}[\sigma_{12}(\xi,\,D,\,t)\,p\,(\xi,\,t|D)] + \sigma_{22}(D)\;\;\frac{\partial\, p\,(\xi,\,t|D)}{\partial\, D} = 0. \tag{14}$$

As required, the slow degree of freedom does not depend on the fast degree of freedom in Equation (11). Equations (8), (11) and (14) characterize the hierarchical dynamics. Let us analyze these for the Omori-Utsu law.

Firstly, we discuss Equation (8) for the fast degree of freedom. Since the external loading on the region under consideration such as the tectonic one [19] is negligible, we impose the condition on the drift term in Equation (8) that it is actually absent:

$$K_1 = 0. \tag{15}$$

Therefore, we have

$$\frac{\partial\, p\,(\xi,\, t|D)}{\partial\, t} = \frac{\partial^2}{\partial\, \xi^2}[\sigma_{11}(\xi,\, D,\, t)\, p\,(\xi,\, t|D)]. \tag{16}$$

Following the Brownian model of seismicity proposed in References [19,20], we take the Gaussian solution to Equation (16):

$$p\,(\xi,\, t|D) = \frac{1}{\sqrt{4\pi D\, t}} \exp\left(-\frac{\xi^2}{4\, D\, t}\right), \tag{17}$$

which corresponds to the initial condition $p\,(\xi,\, 0|D) = \delta\,(\xi)$, representing a main shock at $t = 0$. Although a more general initial condition may actually be employed, here we consider this simplest one. Substituting Equation (17) into Equation (16), we have

$$\sigma_{11} = D, \tag{18}$$

as expected.

Let us see how Equation (17) can give rise to the Omori-Utsu law. Recall that the relaxation function in a symmetric random walk model is given by the characteristic function of the distribution [25–27]. In the present case, it is the characteristic function of the marginal distribution of $\xi$ obtained by the integration of the joint distribution $P\,(\xi,\, D,\, t)$ over $D$. That is,

$$\phi_{\,k}(t) = \int_0^\infty d\, D\, \chi\,(k,\, t|D)\, \Pi(D). \tag{19}$$

In this equation, $\chi\,(k,\, t|D)$ stands for the characteristic function of the conditional distribution

$$\chi\,(k,\, t|D) = \int_{-\infty}^{\infty} d\,\xi\, e^{\,i\,k\,\xi}\, p\,(\xi,\, t|D), \tag{20}$$

which is calculated for Equation (17) to be

$$\chi\,(k,\, t|D) = \exp\left(-D\, k^2 t\right). \tag{21}$$

Let us examine the case when the fluctuating diffusivity obeys the gamma distribution

$$\Pi(D) = \frac{1}{\Gamma\,(p)} \frac{D^{\,p-1}}{D_0^{\,p}} \exp\left(-\frac{D}{D_0}\right)\ (p > 0), \tag{22}$$

where $\Gamma\,(p)$ is the gamma function and $D_0$ is a constant given by $D_0 = \langle D \rangle / p$ with $\langle Q \rangle \equiv \int_0^\infty dD \int_{-\infty}^\infty d\xi\ Q\, P\,(\xi,\, D,\, t)$, Equations (6), (17) and (22). Then, from Equations (19)–(21), we obtain the relaxation function for the Omori-Utsu law

$$\phi_k(t) = \frac{1}{(1 + t/\tau_k)^{\,p}}, \tag{23}$$

where $\tau_k$ is given by

$$\tau_k = \frac{1}{D_0\, k^2}. \tag{24}$$

Thus, Equation (2) is in fact obtained for each mode $k$. This is our first result.

Secondly, let us analyze Equation (11). Using Equation (22), that equation leads to the following relation:

$$K_2(D) = \frac{d\,\sigma_{22}(D)}{d\,D} + \left(\frac{p-1}{D} - \frac{1}{D_0}\right)\sigma_{22}(D). \tag{25}$$

$\sigma_{22}$ will be determined later.

Thirdly, substituting Equation (17) into Equation (14), we have the following equation for the coupling between the fast and slow degrees of freedom:

$$2\frac{\partial\,\sigma_{12}(\xi,\, D,\, t)}{\partial\,\xi} - \frac{\xi}{D\,t}\sigma_{12}(\xi,\, D,\, t) + \left(\frac{\xi^2}{4D^2 t} - \frac{1}{2D}\right)\sigma_{22}(D) = 0. \tag{26}$$

This equation has the solution

$$\sigma_{12}(\xi,\, D) = \frac{\xi}{4D}\sigma_{22}(D), \tag{27}$$

which shows that the coupling is fixed in time.

Finally, as in Reference [15], we determine $\sigma_{22}$ by the positivity of $\sigma$. Using Equations (5), (18) and (27), this matrix is written as

$$\sigma = \begin{pmatrix} D & [\xi/(4D)]\,\sigma_{22}(D) \\ [\xi/(4D)]\,\sigma_{22}(D) & \sigma_{22}(D) \end{pmatrix}. \tag{28}$$

Since the positivity implies that the eigenvalues of this $2 \times 2$ matrix is positive, it follows that

$$\mathrm{tr}\,\sigma = D + \sigma_{22}(D) > 0, \tag{29}$$

$$\det\sigma = \sigma_{22}(D)\left[D - \frac{\xi^2}{16\,D^2}\sigma_{22}(D)\right] > 0. \tag{30}$$

Equation (29) is natural, whereas Equation (30) needs some considerations. Clearly, in order for the quantity inside the square brackets in Equation (30) to be positive, $\xi$ cannot arbitrarily be large. This, in turn, imposes a constraint on the time scale for the validity of the present theory [15]. Let such a time scale be denoted by $T$. The diffusion property suggests that $\xi^2 \sim \langle \xi^2 \rangle = 2p\, D_0\, T$, where $\langle \xi^2 \rangle = \int_{-\infty}^{\infty} d\,\xi \int_0^{\infty} dD\, \xi^2\, P\,(\xi,\, D,\, T)$ is the variance of $\xi$ in terms of the distribution in Equation (6) with Equations (17) and (22) and $\langle \xi \rangle = 0$ (see the last paragraph in this section). This may indicate that the corresponding scale $S$ is a constant satisfying

$$\sqrt{2\,p\,D_0 T} << S. \tag{31}$$

In other words, the value of $\xi$ should be well localized in this way. Thus, Equation (30) may hold up to $S$ (or correspondingly $T$). In terms of such a scale, we have

$$\sigma_{22}(D) = \frac{16\,D^3}{S^2} \tag{32}$$

as the solution. Substituting this equation into Equations (25) and (27), we obtain

$$K_2(D) = (p+2)\frac{16\,D^2}{S^2} - \frac{16\,D^3}{D_0\,S^2}, \tag{33}$$

$$\sigma_{12}(\xi,\, D) = \frac{4\,\xi D^2}{S^2}, \tag{34}$$

respectively.

Consequently, we find that the present theory based on the Fokker-Planck equation with fluctuating diffusivity describes the Omori-Utsu law if $K$'s and $\sigma$'s are given by Equations (15), (18) and (32)–(34).

Closing this section, we present the explicit form of the marginal distribution of $\xi$, which is denoted here by $\hat{p}\,(\xi,\, t)$. As seen in Equations (19) and (20), it is given by the inverse Fourier transformation of Equation (23):

$$\hat{p}\,(\xi,\, t) = \frac{1}{2\pi}\int_{-\infty}^{\infty} dk\, e^{-i\,k\,\xi}\,\phi_k(t). \tag{35}$$

Then, using the Formula (9.6.25) in Reference [28], we have

$$\hat{p}\,(\xi,\, t) = \frac{1}{2^{\,p-1/2}\sqrt{\pi}\,\Gamma\,(p)\,\sqrt{D_0\,t}}\left(\frac{|\xi|}{\sqrt{D_0\,t}}\right)^{p-1/2} K_{p-1/2}\left(\frac{|\xi|}{\sqrt{D_0\,t}}\right), \tag{36}$$

where $K_{\,\nu}(z)$ is the modified Bessel function. The $\xi$-dependence appears only in the combined form $|\xi|/\sqrt{D_0\,t}$, implying the normal diffusion: the variance of $\xi$ linearly grows in time $t$. Therefore, the present model offers an example of the non-Gaussian normal diffusion, the phenomenon of which is currently attracting much attention [22,23]. It is however noted that this diffusion process takes place in *the space of the load state*, not in the conventional space.

## 3. Subdynamics, Logarithmic Time and Aging

Now, we address ourselves to studying the subdynamics obtained by reduction of the fluctuating diffusivity.

Let us derive the evolution equation for the marginal distribution $\hat{p}\,(\xi,\, t)$. It is should be noted that such an equation is necessarily specific to the initial condition. Equation (17) satisfies

$$\frac{\partial\, p\,(\xi,\, t|D)}{\partial\, t} = -\frac{1}{2\,t}\left(\xi\frac{\partial}{\partial\,\xi}+1\right)p\,(\xi,\, t|D). \tag{37}$$

With this form, multiplying the both sides by $\Pi(D)$ and integrating over $D$, we have the following equation for the marginal distribution:

$$\frac{\partial\, \hat{p}\,(\xi,\, t)}{\partial\, t} = -\frac{1}{2\,t}\left(\xi\frac{\partial}{\partial\,\xi}+1\right)\hat{p}\,(\xi,\, t), \tag{38}$$

which describes how the subsystem evolves in time.

Here, we wish to make a comment on the fact that the operator in Equation (37) does not have explicit dependence on $D$ and accordingly the marginal distribution satisfies a closed form as in Equation (38). In fact, upon deriving that equation, we do not have to assume the explicit form of $\Pi(D)$ in Equation (22). Actually, this feature has its origin in the scaling property of the conditional distribution in Equation (17):

$$p\,(\xi,\, t|D) = \frac{1}{t^{1/2}}\overline{p}\,(\xi/t^{1/2}|D), \tag{39}$$

where $\overline{p}\,(x|D)$ is the Gaussian scaling function $\overline{p}\,(x|D) = (4\pi D)^{-1/2}\exp\left[-x^2/(4D)\right]$. In this context, the scale invariance of Equation (38) should be noted: it does not change its form under the individual rescaling transformations of $\xi$ and $t$. In Section 4, a further discussion will be made about the relation between the scaling property and derivability of a local equation of evolution for a marginal distribution in a closed form.

To see the property of the subdynamics, let us analyze Green's function $G(\xi,\ t:\xi',\ t')$ associated with Equation (38), which is the solution of the equation

$$\left[\frac{\partial}{\partial t}+\frac{1}{2t}\left(\xi\frac{\partial}{\partial \xi}+1\right)\right]\ G(\xi,\ t:\xi',\ t')=\delta(\xi-\xi')\,\delta(t-t'), \tag{40}$$

satisfying the condition

$$\lim_{t\to t'+0}G(\xi,\ t:\xi',\ t')=\delta(\xi-\xi'). \tag{41}$$

The explicit form of the solution is found to be given by

$$G(\xi,\ t:\xi',\ t')=\frac{1}{t^{1/2}}\delta\left(\xi/t^{1/2}-\xi'/t'^{1/2}\right)\theta(t-t'), \tag{42}$$

where $\theta(s)$ is the Heaviside step function: $\theta(s)=0\ (s<0), 1/2\ (s=0), 1\ (s>0)$.

From Equation (42), we obtain three important results, which are as follows.

Firstly, *the transition from one value of the load state to another is deterministic* because of the delta-function nature of Green's function. This is due to the fact that Equation (40) does not depend on $D_0$: no remnants of the diffusivity are contained. The functional form in Equation (36) is kept unchanged under time evolution (as discussed in Section 4, this comes from the scaling property of the conditional distribution that still depends on $D$).

Secondly, Green's function clearly satisfies the Chapman-Kolmogorov equation [24]

$$G(\xi,\ t:\xi',\ t')=\int_{-\infty}^{\infty}d\,\xi''\ G(\xi,\ t:\xi'',\ t'')\,G(\xi'',\ t'':\xi',\ t')\ (t>t''>t'). \tag{43}$$

Therefore, the evolution process is Markovian.

Thirdly, the evolution is not stationary since the dependence of Green's function on $t$ and $t'$ cannot be expressed in terms only of the difference $t-t'$ and so the temporal translational invariance is violated. This is actually clear since the operator in Equation (40) has explicit time dependence. However, it is of significance to rewrite Equation (42) in the following form:

$$G(\xi,\ t:\xi',\ t')=\delta\left(\xi-\xi'\,e^{(\ln t-\ln t')/2}\right)\theta(\ln t-\ln t'). \tag{44}$$

This implies that *the temporal translational invariance is restored in terms of the logarithmic time.*

Finally, let us examine the above-mentioned second and third results in view of the known properties of aftershocks. The second result is unsatisfactory since it has been reported in References [17,18] that processes of aftershocks are generally non-Markovian. Clearly, Markovianity of the present theory comes from that of Equation (3) and the hierarchical structure although subdynamics of a Markovian dynamics is not necessarily Markovian, in general. This point needs further investigations. On the other hand, the third result is intriguing since it shows how the subdynamics experiences *slowing down* in terms of the logarithmic time. This captures an element of *criticality*. Let us note that Green's function is actually a tri-variate function $G(\xi,\ t:\xi',\ t')\equiv g(\xi,\ \xi',\ t/t')$, as seen in Equation (44). Rewriting as $t\to t+t_w$ and $t'\to t_w$ with the waiting time $t_w$, this becomes

$$G(\xi,\ t+t_w:\xi',\ t_w)=g(\xi,\ \xi',\ t/t_w+1), \tag{45}$$

which depends on not only $t$ but also the waiting time, showing nonstationarity of the evolution. The dependence on the waiting time here is specific: the larger the waiting time is, the slower the evolution in terms of the conventional time $t$ becomes. That is, the subsystem exhibits the aging phenomenon (not in the two-time correlation function but in Green's function), implying that the subsystem has its own "internal clock". We mention that such a phenomenon has been discovered for the event-event correlation of real aftershocks [16,17]. It is also noted that the logarithmic time and the aging phenomenon appear in glassy dynamics [29].

**4. Concluding Remarks**

We have presented a theoretical approach to describing the Omori-Utsu law for earthquake aftershocks. Assuming fluctuating diffusivity effectively representing the system heterogeneity, we have examined the Fokker-Planck theory with the hierarchical structure, in which the load-state and diffusivity variables are the fast and slow degrees of freedom, respectively. In this way, we have extracted the information about the dynamics underlying the law that can be used in the stochastic process of aftershocks. Then, we have studied the evolution equation for the load state that is reduced from the Fokker-Planck equation. We have analyzed Green's function of that equation and have observed how the logarithmic time and the aging phenomenon naturally appear.

An additional point we make here is concerned with Equation (38) or, more fundamentally, Equation (37). As mentioned, these equations have the invariance under the individual rescaling transformations of the load-state variable and time. This symmetry makes the equations independent of the diffusivity (i.e., $D_0$) and leads to the deterministic transition between the load states. We have claimed that this symmetry has its origin in the scaling property of the conditional distribution in Equation (39). To see this somewhat in a wider context, let us look at, as an explicit example, the symmetric Lévy distribution indexed by $\alpha \in (0,\ 2)$ [30]:

$$L_\alpha(\xi,\ t|D_*) = \frac{1}{2\pi}\int_{-\infty}^{\infty} dk\ \exp\left(-i\,k\,\xi - D_*\,t\,|k|^{\alpha}\right), \tag{46}$$

where $D_*$ stands for a generalized diffusion coefficient. The Gaussian case corresponds to the limit $\alpha \to 2^-$. This decays as a power law, $L_\alpha(\xi,\ t|D_*) \sim 1/|\xi|^{1+\alpha}$ and therefore its second moment is divergent. Accordingly, the diffusion property should be characterized not by the standard deviation but by e.g., the half width. We note that the conditional distribution in Equation (46) has the scaling property

$$L_\alpha(\xi,\ t|D_*) = \frac{1}{t^{1/\alpha}}\overline{L}_\alpha(\xi/t^{1/\alpha}|D_*) \tag{47}$$

with the Lévy scaling function $\overline{L}_\alpha(x) = (2\pi)^{-1}\int_{-\infty}^{\infty} dk\ \exp(-i\,k\,x - D_*\,|k|^{\alpha})$. This implies that the half width grows in time as $\sim t^{1/\alpha}$, exhibiting superdiffusion faster than normal diffusion $\sim \sqrt{t}$. Then, from Equation (47), it follows that

$$\frac{\partial L_\alpha(\xi,\ t|D_*)}{\partial t} = -\frac{1}{\alpha\,t}\left(\xi\,\frac{\partial}{\partial \xi} + 1\right) L_\alpha(\xi,\ t|D_*), \tag{48}$$

which generalizes Equation (37). This equation still does not explicitly contain the (generalized) diffusion coefficient and therefore is invariant under the rescaling transformations of $\xi$ and $t$. It is however known that, in order to obtain the Lévy distribution as a solution of the Fokker-Planck equation, the operator $\partial^2/\partial\ \xi^2$ should be fractionalized [30,31] and replaced by e.g. Riesz's fractional Laplacian. In general, not limited to the example in Equation (46), appearance of the deterministic transition is related to the scaling property of the conditional distribution.

*Note added.* After completing the present work, we have noticed Reference [32]. There, the authors discuss the Fokker-Planck theory with the hierarchical structure to the biological process experienced by the cell in connection with decision making. It shows how the theory can shed new light on its application to information theory.

**Author Contributions:** S.A. has formulated the problem and the all authors have equally contributed to the theoretical work. S.A. has organized the paper and the other two authors have agreed to publish it. All authors have read and agreed to the published version of the manuscript.

**Funding:** S.A. acknowledges the support from the Program of Fujian Province. The work of S.A. and D.A.T. has been supported by the Kazan Federal University Strategic Academic Leadership Program (PRIORITY-2030). N.S. is indebted to a Grant-in-Aid for Start-up Research of College of Science and Technology, Nihon University.

**Institutional Review Board Statement:** Not applicable.

**Data Availability Statement:** Not applicable.

**Conflicts of Interest:** The authors declare no conflict of interest.

## References

1. Christensen, R.M. *Theory of Viscoelasticity*, 2nd ed.; Academic Press: Cambridge, MA, USA, 1982.
2. de Groot, S.R.; Mazur, P. *Non-Equilibrium Thermodynamics*; Dover: Mineola, NY, USA, 1984.
3. Pottier, N. *Nonequilibrium Statistical Physics*; Oxford University Press: Oxford, UK, 2010.
4. Cardona, M.; Chamberlin, R.V.; Marx, W. The history of the stretched exponential function. *Ann. Phys.* **2007**, *16*, 842–845. [CrossRef]
5. Sillescu, H. Heterogeneity at the glass transition: A review. *J. Non-Cryst. Solids* **1999**, *243*, 81–108. [CrossRef]
6. Meng, F.; Pritchard, R.H.; Terentjev, E.M. Stress relaxation, dynamics, and plasticity of transient polymer networks. *Macromolecules* **2016**, *49*, 2843–2852. [CrossRef]
7. Kapnistos, M.; Lang, M.; Vlassopoulos, D.; Pyckhout-Hintzen, W.; Richter, D.; Cho, D.; Chang, T.; Rubinstein, M. Unexpected power-law stress relaxation of entangled ring polymers. *Nat. Mater.* **2008**, *7*, 997–1002. [CrossRef] [PubMed]
8. Bonfanti, A.; Kaplan, J.L.; Charras, G.; Kabla, A. Fractional viscoelastic models for power-law materials. *Soft Matter* **2020**, *16*, 6002–6020. [CrossRef]
9. Omori, F. On the after-shocks of earthquakes. *J. Coll. Sci. Imp. Univ. Jpn.* **1894**, *7*, 111–200.
10. Utsu, T. A statistical study on the occurrence of aftershocks. *Geophys. Mag.* **1961**, *30*, 521–605.
11. Johansen, A. Response time of internauts. *Physica A* **2001**, *296*, 539–546. [CrossRef]
12. Abe, S.; Suzuki, N. Omori's law in the Internet traffic. *Europhys. Lett.* **2003**, *61*, 852–855. [CrossRef]
13. Lillo, F.; Mantegna, R.N. Power-law relaxation in a complex system: Omori law after a financial market crash. *Phys. Rev. E* **2003**, *68*, 016119. [CrossRef] [PubMed]
14. Usmanova, V.; Lysogorskiy, Y.V.; Abe, S. Aftershocks following crash of currency exchange rate: The case of RUB/USD in 2014. *Europhys. Lett.* **2018**, *121*, 48001. [CrossRef]
15. Abe, S. Fokker-Planck approach to non-Gaussian normal diffusion: Hierarchical dynamics for diffusing diffusivity. *Phys. Rev. E* **2020**, *102*, 042136. [CrossRef]
16. Abe, S.; Suzuki, N. Aging and scaling of earthquake aftershocks. *Physica A* **2004**, *332*, 533–538. [CrossRef]
17. Abe, S.; Suzuki, N. Aftershocks in modern perspectives: Complex earthquake network, aging, and non-Markovianity. *Acta Geophys.* **2012**, *60*, 547–561. [CrossRef]
18. Abe, S.; Suzuki, N. Violation of the scaling relation and non-Markovian nature of earthquake aftershocks. *Physica A* **2009**, *388*, 1917–1920. [CrossRef]
19. Kagan, Y.Y.; Knopoff, L. Random stress and earthquake statistics: Time dependence. *Geophys. J. R. Astron. Soc.* **1987**, *88*, 723–731. [CrossRef]
20. Matthews, M.V.; Ellsworth, W.L.; Reasenberg, P.A. A Brownian model for recurrent earthquakes. *Bull. Seism. Soc. Am.* **2002**, *92*, 2233–2250. [CrossRef]
21. Redner, S. *A Guide to First-Passage Processes*; Cambridge University Press: Cambridge, UK, 2001.
22. Wang, B.; Kuo, J.; Bae, S.C.; Granick, S. When Brownian diffusion is not Gaussian. *Nat. Mater.* **2012**, *11*, 481–485. [CrossRef]
23. Bhattacharya, S.; Sharma, D.K.; Saurabh, S.; De, S.; Sain, A.; Nandi, A.; Chowdhury, A. Plasticization of poly(vinylpyrrolidone) thin films under ambient humidity: Insight from single-molecule tracer diffusion dynamics. *J. Phys. Chem. B* **2013**, *117*, 7771–7782. [CrossRef]
24. Gardiner, C. *Stochastic Methods*, 4th ed.; Springer: Heidelberg, Germany, 2009.
25. Gomi, S.; Yonezawa, F. Anomalous relaxation in the fractal time random walk model. *Phys. Rev. Lett.* **1995**, *74*, 4125–4128. [CrossRef]
26. Weron, K.; Kotulski, M. On the Cole-Cole relaxation function and related Mittag-Leffler distribution. *Physica A* **1996**, *232*, 180–188. [CrossRef]
27. Metzler, R.; Klafter, J. From stretched exponential to inverse power-law: Fractional dynamics, Cole-Cole relaxation processes, and beyond. *J. Non-Cryst. Solids* **2002**, *305*, 81–87. [CrossRef]
28. Abramowitz, M.; Stegun, I.A. (Eds.) *Handbook of Mathematical Functions*; Dover: Mineola, NY, USA, 1965.
29. Fischer, K.H.; Hertz, J.A. *Spin Glasses*; Cambridge University Press: Cambridge, UK, 1991.

30. Metzler, R.; Klafter, J. The random walk's guide to anomalous diffusion: A fractional dynamics approach. *Phys. Rep.* **2000**, *339*, 1–77. [CrossRef]
31. Zaslavsky, G.M. *Hamiltonian Chaos and Fractional Dynamics*; Oxford University Press: Oxford, UK, 2005.
32. Barua, A.; Hatzikirou, H. Cell decision making through the lens of Bayesian learning. *Entropy* **2023**, *25*, 609. [CrossRef] [PubMed]

**Disclaimer/Publisher's Note:** The statements, opinions and data contained in all publications are solely those of the individual author(s) and contributor(s) and not of MDPI and/or the editor(s). MDPI and/or the editor(s) disclaim responsibility for any injury to people or property resulting from any ideas, methods, instructions or products referred to in the content.

*entropy*

Article

# Viscoelastic Slider Blocks as a Model for a Seismogenic Fault

Charlotte A. Motuzas [1,2,*] and Robert Shcherbakov [1,2,*]

1 Department of Earth Sciences, Western University, London, ON N6A 5B7, Canada
2 Department of Physics and Astronomy, Western University, London, ON N6A 3K7, Canada
* Correspondence: cmotuzas@uwo.ca (C.A.M.); rshcherb@uwo.ca (R.S.)

**Abstract:** In this work, a model is proposed to examine the role of viscoelasticity in the generation of simulated earthquake-like events. This model serves to investigate how nonlinear processes in the Earth's crust affect the triggering and decay patterns of earthquake sequences. These synthetic earthquake events are numerically simulated using a slider-block model containing viscoelastic standard linear solid (SLS) elements to reproduce the dynamics of an earthquake fault. The simulated system exhibits elements of self-organized criticality, and results in the generation of avalanches that behave similarly to naturally occurring seismic events. The model behavior is analyzed using the Epidemic-Type Aftershock Sequence (ETAS) model, which suitably represents the observed triggering and decay patterns; however, parameter estimates deviate from those resulting from natural aftershock sequences. Simulated aftershock sequences from this model are characterized by slightly larger $p$-values, indicating a faster-than-normal decay of aftershock rates within the system. The ETAS fit, along with realistic simulated frequency-size distributions, supports the inclusion of viscoelastic rheology to model the seismogenic fault dynamics.

**Keywords:** earthquake physics; ETAS model; friction; viscoelasticity; slider-block model; power–law

**Citation:** Motuzas, C.A.; Shcherbakov, R. Viscoelastic Slider Blocks as a Model for a Seismogenic Fault. *Entropy* **2023**, *25*, 1419. https://doi.org/10.3390/e25101419

Academic Editor: Georgios Michas

Received: 15 September 2023
Revised: 2 October 2023
Accepted: 4 October 2023
Published: 6 October 2023

**Copyright:** © 2023 by the authors. Licensee MDPI, Basel, Switzerland. This article is an open access article distributed under the terms and conditions of the Creative Commons Attribution (CC BY) license (https://creativecommons.org/licenses/by/4.0/).

## 1. Introduction

Despite the complex nature of earthquake dynamics, simple models may be used to understand many aspects of earthquake behavior. In particular, these models aim to explain and represent the physical mechanisms behind the generation of earthquakes [1,2]. This is accomplished by analyzing and replicating patterns observed in aftershock sequences and other seismic activity in accordance with statistical observations and known properties of the Earth's crust. These models are designed to examine earthquake behavior from the perspective of either statistical seismology or physics, and allow for a more comprehensive understanding of the physics involved with common earthquake patterns [3,4]. A better understanding of these characteristics provides additional resources for risk estimation and forecasting efforts, which serve to mitigate the damage resulting from future earthquake events [5–8].

When considering a traditional earthquake model, earthquakes result from interactions between tectonic plates within the Earth's crust. This system, defined by networks of constantly moving plates and their corresponding dynamics, can be classified as an incredibly complex nonlinear system, exhibiting self-organized criticality [9,10]. Plainly, as the plates are constantly driven forward and interacting with one another, the system hovers at a state extremely close to instability, and at a certain stress threshold, critical instabilities appear in the form of abrupt slippage along a fault [3,11,12]. In this context, earthquakes are viewed as the outward expression of these sudden slippages, as segments of the rock walls on either side of the fault are suddenly displaced in a series of "avalanches" [12,13]. The resulting earthquake dynamics within a certain region (a single fault or a system of faults) then depend on the physical properties of the surrounding rock medium, particularly its elastic, viscous, and frictional responses.

Due to the complexity of this system, it is impossible to design a model that represents the Earth's crust in its entirety. Instead, models are designed to be much simpler analogs, which still exhibit similar system dynamics on a much smaller scale. The slider-block model, first introduced by Burridge and Knopoff in 1967, is an example of one such model used to represent the behavior of a single seismogenic fault within a surrounding elastic medium [14]. This slider-block model focuses on the interaction between two opposing walls of a fault, and how the corresponding friction and elastic responses play a role in the occurrence of earthquakes [11,12,14,15]. The original formulation represents a chain of blocks of equal mass, connected in series using elastic springs. These blocks are placed on a surface with uniform friction, which acts in opposition to a constant driving force. This system, much like the Earth's system of tectonic plates, is persistently driven toward instability. This instability results from the constant driving force, alongside the linear and nonlinear forces acting on each block due to the elastic springs and velocity weakening friction force [12,16]. Each block eventually approaches critical slipping points, resulting in abrupt displacements of previously stuck blocks, or "avalanches" affecting a series of neighboring blocks, mimicking earthquakes propagation along a fault.

Many studies incorporating this slider-block concept have been carried out, utilizing two-dimensional arrays of slider blocks, cellular automata, and a variety of nonlinear velocity-dependant friction laws, all in an attempt to discern more information regarding the dynamics of seismogenic faults [1,3,17–19]. Those studies examined the sizes of generated avalanche events, and determined that models similar to the Burridge and Knopoff model produce size distributions similar to those commonly associated with seismic activity. In particular, for slider-block models and other models that display aspects of self-organized criticality, frequency-size distributions often follow power–law-type functional form with relevant finite-size effects. [1,12]. It should be stated that although slider-block models are not considered as systems that exhibit strictly self-organized critical behaviour, due to the presence of tuning parameters, these models still provide meaningful methods through which this behavior can be studied.

Despite the simple premise, the slider-block model can be a useful tool for simulating the behavior of earthquake faults as complex, chaotic systems, while simultaneously examining the properties of self-organized criticality within the field of seismology [2,3,20]. However, this model does have its limitations. In particular, large earthquakes are often followed by a series of aftershocks [21,22], proportional to the initial shock, which decrease in magnitude and frequency according to several well-studied patterns, such as the Omori–Utsu law [23], or the Epidemic-Type Aftershock Sequence (ETAS) model [24,25]. Purely elastic slider-block models often lack these aftershock patterns, reducing the applicability of these models to real-life earthquake forecasting or risk assessments. It is because of this that recent studies have introduced slider-block models with additional components or processes to better resemble seismologic observations, primarily through the introduction of physical properties such as viscoelasticity [19,26–31]. The addition of viscoelastic components to existing slider-block models serves to recreate the physical properties of the Earth's crust along a seismogenic fault, in the hopes of recreating realistic aftershock sequences following a sufficiently large event.

Although the linearly elastic properties of a rock medium strongly influence earthquake dynamics over short timescales, nonlinear properties of the Earth's crust are thought to influence many aspects of observed seismic activity [32,33]. The lithosphere, the lower regions of the Earth's crust, the upper mantle, and regions along active fault zones exhibit the greatest deviation from linear elasticity [34–37]. In these regions, the addition of nonlinear viscoelasticity may be responsible for the presence of the temporal clustering, or aftershocks, generated independently of the initial driving forces behind the initial avalanche [19,28,38].

In this paper, we investigate the influence of viscoelasticity on a one-dimensional slider-block model, specifically through the use of standard linear solid (SLS) viscoelastic components, composed of a Maxwell element connected in parallel to an elastic spring.

Using computer simulation to depict the motion of each slider-block over a set time interval, a catalog of avalanche events is then analyzed using the ETAS model to determine the model parameters and the quantitative behaviour of the model. The viscoelastic slider-block model reproduces frequency–magnitude behavior and temporal clustering similar to that of natural seismic activity. Reasonable values are obtained through ETAS parameter estimates, which indicate the generation of realistic, rapidly decaying aftershock sequences. The purpose of this model is to investigate the behavior of simulated earthquake catalogs. This is performed to better understand the influence of different physical properties on seismic activity, using metrics like ETAS model parameters to compare simulated events to real-life seismicity.

The paper is organized as follows. In Section 2, the model is formulated and the governing equations are derived. In Section 3, the model simulations are presented and the obtained results are described. And finally, in Section 4, the results are discussed and future research directions are outlined.

## 2. Viscoelastic Slider-Block Model

The model consists of a one-dimensional chain of slider blocks, each of equal mass $m$ and connected to neighboring blocks by the SLS elements. This model is illustrated in Figure 1. These SLS components contain a viscous dashpot and an elastic spring, with characteristic parameters $\eta$ and $k_{\mathrm{d}}$, respectively, connected in series. This is known as a Maxwell element. This Maxwell element is connected in parallel to a second elastic spring, with parameter $k$, to form the SLS component [39]. This component was chosen to model the observed nonlinear viscoelastic response of common natural materials, specifically with regard to the behavior of the Earth's crust. Each block is then individually attached to an upper plate via separate SLS components, characterized by parameters $k_{\mathrm{pd}}$ and $\eta_{\mathrm{p}}$ within the Maxwell element, and parameter $k_{\mathrm{p}}$ for the elastic spring placed in parallel. The chain of blocks is then placed on a conveyor belt, which moves at a constant driving velocity, $-v_{\mathrm{dr}}$.

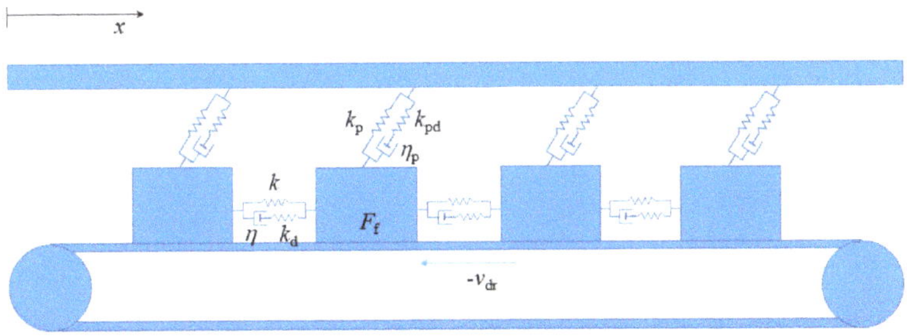


**Figure 1.** Schematic illustration of the model with $N = 4$ blocks, with model parameters $\{k, k_{\mathrm{d}}, \eta, k_{\mathrm{p}}, k_{\mathrm{pd}}, \eta_{\mathrm{p}}\}$. The slider blocks are interconnected by SLS elements and are driven by the conveyor belt. The positive $x$ direction is opposite the direction of the driving velocity, $-v_{dr}$.

This model also relies on a nonlinear slip–stick friction law, dependant on the velocity of the blocks relative to the moving conveyor belt. The chosen friction law originates from the 1989 Carlson and Langer slider-block model, in which the magnitude of the friction force ranges between $f_{\mathrm{s}}$ and $-f_{\mathrm{s}}$. This friction force takes the form [11]:

$$F_f(v_i) = f_{\mathrm{s}}\phi\left(\frac{v_i + v_{\mathrm{dr}}}{v_0}\right), \tag{1}$$

where $v_i$ is the velocity of a single block $i$; $v_0$ is a chosen reference velocity; and the velocity-dependant component, $\phi$, is chosen such that the friction force vanishes at high velocity. In this model, $\phi$ is defined as follows [11]:

$$\phi(z) = \begin{cases} (-\infty; 1], & z = 0, \\ \dfrac{\text{sign}(z)}{1 + \delta|z|}, & z > 0. \end{cases} \tag{2}$$

To simulate the motion of $N$ blocks in a linear array, the equation of motion for a single block $i$ can be written as

$$\begin{aligned} m\frac{d^2x_i}{dt^2} &= -k\,(2x_i - x_{i-1} - x_{i+1}) - k_\text{p}\,x_i + F_\text{M}^{(i-1)}(t) + F_\text{M}^{(i+1)}(t) \\ &\quad + F_\text{Mp}(t) - f_\text{s}\,\phi\left(\frac{v_i + v_\text{dr}}{v_0}\right), \end{aligned} \tag{3}$$

where $F_\text{Mp}(t)$ is the force exerted by the Maxwell element connecting the block to the top plate, and $F_\text{M}^{(i-1)}(t)$ and $F_\text{M}^{(i+1)}(t)$ are the forces due to the two Maxwell elements connecting the block to its neighboring blocks. When the forces acting on the block from both the upper driving plate and the nearest neighbor blocks exceed that of the velocity-weakening frictional force, the block abruptly slips. These sudden displacements may trigger subsequent slippages of neighboring blocks, resulting in an avalanche in the system.

The force exerted by the Maxwell elements in both SLS components can be defined as follows. Assume that $x$ represents the total displacement of the Maxwell element and is defined by $x = x_\text{d} + x_\text{s}$, where $x_\text{d}$ is the displacement in the dashpot and $x_\text{s}$ is the displacement in the spring. It is possible to show that the force due to the Maxwell element satisfies the following ODE [39]:

$$\frac{dF}{dt} = -k_\text{d}\frac{dx}{dt} - \frac{k_\text{d}}{\eta}F. \tag{4}$$

This was used to define the forces exerted by the Maxwell elements in the SLS component connected to the top plate, and on either side of block $i$, by blocks $(i-1)$ and $(i+1)$.

$$\frac{dF_\text{Mp}}{dt} = -k_\text{pd}\frac{dx_i}{dt} - \frac{k_\text{pd}}{\eta_\text{p}}F_\text{Mp}, \tag{5}$$

$$\frac{dF_\text{M}^{(i-1)}}{dt} = -k_\text{d}\frac{d(x_i - x_{i-1})}{dt} - \frac{k_\text{d}}{\eta}F_\text{M}^{(i-1)}, \tag{6}$$

$$\frac{dF_\text{M}^{(i+11)}}{dt} = -k_\text{d}\frac{d(x_i - x_{i+1})}{dt} - \frac{k_\text{d}}{\eta}F_\text{M}^{(i+1)}. \tag{7}$$

Equations (6) and (7) can be combined into a single ordinary differential equation, where the force exerted by the Maxwell elements of both neighbor blocks is equal to $F_\text{M} = F_\text{M}^{(i-1)} + F_\text{M}^{(i+1)}$, and is defined by:

$$\frac{dF_\text{M}}{dt} = -k_\text{d}\frac{d(2x_i - x_{i-1} - x_{i+1})}{dt} - \frac{k_\text{d}}{\eta}F_\text{M}. \tag{8}$$

A chain of $N$ slider blocks can then be described by the system of ODE equations for each block $i$, which can then be solved numerically to simulate the motion of each block within a set time interval. The linearly elastic interactions between neighboring blocks and the upper plate result in an instantaneous transfer of stress within the system, while the presence of the viscous dashpots allows for a delay in transfer that enables the further potential triggering of slipping events [19,28,38,39].

Performing the nondimensionalization of this system of equations allows for a further analysis of the model behavior, while reducing the number of independent parameters. The following dimensionless variables can be introduced for this purpose:

$$\tau = \frac{t}{\sqrt{\frac{m}{k_{\mathrm{pd}}}}}, \quad u_i = \frac{x_i k_{\mathrm{pd}}}{f_{\mathrm{s}}}, \quad \tilde{F}_{\mathrm{M}} = \frac{F_{\mathrm{M}} k_{\mathrm{pd}}}{f_{\mathrm{s}} k_{\mathrm{d}}}, \quad \tilde{F}_{\mathrm{Mp}} = \frac{F_{\mathrm{Mp}}}{f_{\mathrm{s}}}, \tag{9}$$

where

$$V_i = \dot{u}_i = \frac{du_i}{d\tau} = \frac{\sqrt{m k_{\mathrm{pd}}}}{f_{\mathrm{s}}} \frac{dx_i}{dt} = \frac{\sqrt{m k_{\mathrm{pd}}}}{f_{\mathrm{s}}} \dot{x}_i = \frac{\sqrt{m k_{\mathrm{pd}}}}{f_{\mathrm{s}}} v_i\,. \tag{10}$$

These nondimensional variables can be used to rewrite the equations of motion (3) with (5) and (8) in the following form:

$$\begin{aligned}
\frac{d^2 u_i}{d\tau^2} &= -\omega\,(2u_i - u_{i-1} - u_{i+1}) - \omega_{\mathrm{p}}\, u_i + \omega_{\mathrm{f}} \tilde{F}_{\mathrm{M}}(\tau) + \tilde{F}_{\mathrm{Mp}}(\tau) - \phi(z)\,, \\
\frac{d\tilde{F}_{\mathrm{M}}}{d\tau} &= -\frac{d(2u_i - u_{i-1} - u_{i+1})}{d\tau} - \omega_{\mathrm{d}}\, \tilde{F}_{\mathrm{M}}\,, \\
\frac{d\tilde{F}_{\mathrm{Mp}}}{d\tau} &= -\frac{du_i}{d\tau} - \omega_{\mathrm{pd}}\, \tilde{F}_{\mathrm{Mp}}\,,
\end{aligned} \tag{11}$$

where the new dimensionless parameters are given as follows:

$$\begin{aligned}
\omega &= \frac{k}{k_{\mathrm{pd}}}, \qquad \omega_{\mathrm{p}} = \frac{k_{\mathrm{p}}}{k_{\mathrm{pd}}}, \qquad \omega_{\mathrm{f}} = \frac{k_{\mathrm{d}}}{k_{\mathrm{pd}}}, \\
\omega_{\mathrm{d}} &= \frac{k_{\mathrm{d}}}{\eta} \sqrt{\frac{m}{k_{\mathrm{pd}}}}, \qquad \omega_{\mathrm{pd}} = \frac{k_{\mathrm{pd}}}{\eta_{\mathrm{p}}} \sqrt{\frac{m}{k_{\mathrm{pd}}}}\,.
\end{aligned} \tag{12}$$

The parameters $\omega$ and $\omega_{\mathrm{p}}$ describe the elastic coupling of the system, and $\omega_{\mathrm{f}}$, $\omega_{\mathrm{pd}}$, and $\omega_{\mathrm{d}}$ dictate the viscous response of the SLS elements. The variable representing the velocity of each block $i$ within the velocity-weakening friction law, $z$, is given by

$$z = \frac{v_i + v_{\mathrm{dr}}}{v_0} = \delta\,(V_i + \nu), \tag{13}$$

where

$$\delta = \frac{1}{v_0}\frac{B}{A} = \frac{1}{v_0}\frac{f_{\mathrm{s}}}{k_{\mathrm{pd}}}\sqrt{\frac{k_{\mathrm{pd}}}{m}}, \qquad \nu = v_{\mathrm{dr}} \frac{k_{\mathrm{pd}}}{f_{\mathrm{s}}} \sqrt{\frac{m}{k_{\mathrm{pd}}}}\,. \tag{14}$$

To simulate the model using this system of equations, one can use the following switch algorithm, which allows for transitions between the stick and slip states [40,41]. This is evaluated at each time step within the numerical solution process, where $F_{\mathrm{SLS}}^{(i)}$ is the force applied by both the upper plate and neighboring blocks (Algorithm 1).

Finally, for the model to produce suitable results, boundary conditions must be assigned to prevent irregular behavior on either side of the slider-block chain. For this model, the displacement of the end blocks was set to zero. Additionally, the initial condition for each slider block is the "stuck" position, with a velocity of zero relative to the conveyor belt. To provide the required spatial heterogeneity in the model, the initial displacement of each block was generated according to uniformly distributed random numbers within a limited interval.

**Algorithm 1** The switch algorithm to simulate the numerical integration of slider blocks assembled in a chain.

**procedure** SLIDERBLOCK($u_i, u_{i-1}, u_{i+1}, V_i, V_{i-1}, V_{i+1}, \omega, \omega_p, \omega_f, \omega_d, \omega_{pd}, \nu, dV$)

  $F_{SLS}^{(i)} \leftarrow -\omega\,(2u_i - u_{i-1} - u_{i+1}) - \omega_p\, u_i + \omega_f \tilde{F}_M(\tau) + \tilde{F}_{Mp}(\tau)$ ▷ compute the force

  **if** $|V_i + \nu| > dV$ **then** ▷ slip phase

    $\frac{du_i}{d\tau} \leftarrow V_i$

    $\frac{dV_i}{d\tau} \leftarrow F_{SLS}^{(i)} - \phi(z)$

  **else if** $\left|F_{SLS}^{(i)}\right| > 1$ **then** ▷ stick to slip transition

    $\frac{du_i}{d\tau} \leftarrow V_i$

    $\frac{dV_i}{d\tau} \leftarrow F_{SLS}^{(i)} - \mathrm{sign}\left(F_{SLS}^{(i)}\right)$

  **else** ▷ stick phase

    $\frac{du_i}{d\tau} \leftarrow -\nu$

    $\frac{dV_i}{d\tau} \leftarrow -(V_i + \nu)$

  $\frac{d\tilde{F}_M}{d\tau} \leftarrow -(2V_i - V_{i-1} - V_{i+1}) - \omega_d\, \tilde{F}_M$

  $\frac{d\tilde{F}_{Mp}}{d\tau} \leftarrow -V_i - \omega_{pd}\, \tilde{F}_{Mp}$

  **return** $\left(\frac{du_i}{d\tau}, \frac{dV_i}{d\tau}, \frac{d\tilde{F}_M}{d\tau}, \frac{d\tilde{F}_{Mp}}{d\tau}\right)$ ▷ return the slider block state

Within the simulation itself, an analysis was performed, following an initial transient regime to allow for the system to fall into a statistically steady state. During the steady state, a procedure was conducted to identify, examine, and record the properties of the observed block movement that form clusters or avalanches. The properties of each cluster, including the number of involved blocks, the velocity at which the slippage events occurred, and the area over which the total displacement occurred, were used to quantify the size of each avalanche.

It is the size and frequency regarding these events that form the basis of the following analysis. Particularly, the triggering and decay of avalanche events following a sufficiently large event was analyzed using the ETAS model [25]. This model describes aftershock sequences as a clustering of seismic activity; each earthquake triggers a subsequent proportional increase in the rate of earthquake events depending on the frequency and magnitude of past earthquakes. The variation of event occurrence rate can be described by the following equation [25]:

$$\lambda_\omega(t) = \mu + K \sum_{i:t_i<t}^{N_t} \frac{e^{\alpha(m_i - m_0)}}{\left(\frac{t-t_i}{c} + 1\right)^p}, \tag{15}$$

where $\lambda_\omega(t)$ is the event rate with respect to time, with reference magnitude $m_0$ and the model parameters $\omega = \{\mu, \alpha, K, c, p\}$. In this model, the rate is a superposition of a constant background activity rate $\mu$, alongside contributions from each previous event. The parameter $c$ describes the rate of aftershocks in the beginning stages of an aftershock sequence, the parameter $p$ describes the speed at which the aftershock rate decays, and both the parameters $K$ and $\alpha$ describe the productivity of an aftershock sequence.

The parameters of simulated aftershock sequences according to the ETAS model were estimated using the maximum likelihood estimation (MLE) method, utilizing the corresponding log-likelihood function:

$$\log(L) = -\mu(T_e - T_s) - \frac{Kc}{p-1}\sum_{i=1}^{k} e^{\alpha(m_i - m_0)}\left[\left(\frac{T_s - t_i}{c}+1\right)^{1-p} - \left(\frac{T_e - t_i}{c}+1\right)^{1-p}\right]$$
$$-\frac{Kc}{p-1}\sum_{i=k+1:T_s\leq t_i\leq T_e}^{N_{T_e}} e^{\alpha(m_i-m_0)}\left[1-\left(\frac{T_e-t_i}{c}+1\right)^{1-p}\right] \tag{16}$$
$$+\sum_{j=1}^{n}\log\left[\mu + K\sum_{i:t_i<t_{k+j}}^{N_{t_{k+j}}}\frac{e^{\alpha(m_i-m_0)}}{\left(\frac{t_{k+j}-t_i}{c}+1\right)^{p}}\right],$$

where $t_i$ is the event times within a time interval $[T_0, T_e]$ in a given catalog with $N_{T_e}$ events and the time interval $[T_s, T_e]$, with $T_s > T_0$ encompassing all events within the fitting time interval. $k$ is the number of events in the interval $[T_0, T_s]$ and $n$ is the number of event in $[T_s, T_e]$.

## 3. Results

### 3.1. Model Simulation

Simulations of the above viscoelastic slider-block model were performed through the numerical integration of a system of ODE equations for a linear array of $N = 100$ slider blocks. Each simulation was permitted to complete an initial transient regime, after which statistics and observations were collected regarding the behavior of the system while in a steady state. Each simulation had a transient regime length of $\tau = 10,000$, followed by a steady state regime with length $\tau = 50,000$.

The simulation described in this section is characterized by the dimensionless model parameters $\{\omega, \omega_{\mathrm{p}}, \omega_{\mathrm{f}}, \omega_{\mathrm{d}}, \omega_{\mathrm{pd}}, \nu, \delta\} = \{8, 2.5, 1.5, 0.03, 0.01, 0.001, 10\}$. The degree of elastic coupling within the system is defined by the initial three parameters, while parameters $\omega_{\mathrm{d}}$ and $\omega_{\mathrm{pd}}$ (12) define the magnitude of the viscous response of the SLS components connected between slider blocks and to the upper plate, respectively. The magnitude of the friction between slider blocks and the conveyor belt is determined by the parameter $\delta$ (14), and the driving velocity of the conveyor belt is determined by the parameter $\nu$ (14).

These dimensionless parameters $\{\omega, \omega_{\mathrm{p}}, \omega_{\mathrm{f}}, \omega_{\mathrm{d}}, \omega_{\mathrm{pd}}, \nu, \delta\}$ define the amplitudes of the corresponding forces acting on each slider block. Values were chosen to reproduce realistic behaviour observed for natural seismicity, and later on, these values were varied to enhance the viscoelastic effect, and to observe how the frequency–magnitude behaviour changes accordingly. For this, we used small values of $\omega_{\mathrm{p}}$ and $\omega_{\mathrm{pd}}$, as they are entered as a dumping parameter for forces $F_{\mathrm{M}}$ and $F_{\mathrm{Mp}}$ in (11). The smaller values result in longer viscous effects associated with the slippage of each block. Parameter $\nu$ was chosen to be small as it represents the driving or loading velocity.

The velocity of each slider block $v_i$, determined at each time step of the numerical solution, is shown in Figure 2. The sharp, velocity spikes represent the sudden slippage of a slider block that occurs once the nonlinear friction force is overcome. Sufficiently large sudden displacements of a single slider block acts as a trigger for neighboring blocks to undergo similar displacements, generating an avalanche. The size, $s$, of each avalanche is determined by the sum displacement of each block involved in the avalanche. Figure 3 displays the displacement resulting from the above velocity spikes. Figure 4 displays the distribution of these events in time.

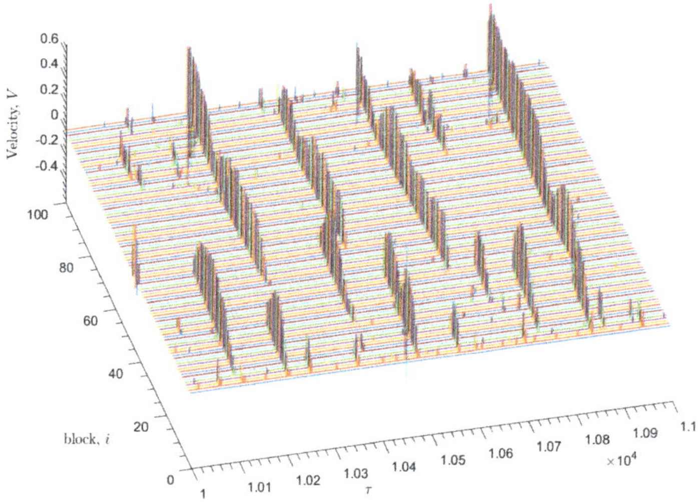


**Figure 2.** Nondimensional steady-state velocity measurements with time $\tau$ for $N = 100$ slider blocks for the simulation of the model with parameters $\omega = 8$, $\omega_p = 2.5$, $\omega_f = 1.5$, $\omega_d = 0.03$, $\omega_{pd} = 0.01$, $\nu = 0.001$, $\delta = 10$, recorded following an initial transient regime. Sharp velocity spikes spanning multiple slider blocks within a sufficiently small time interval are counted as avalanche events. Velocity is measured in arbitrary units.

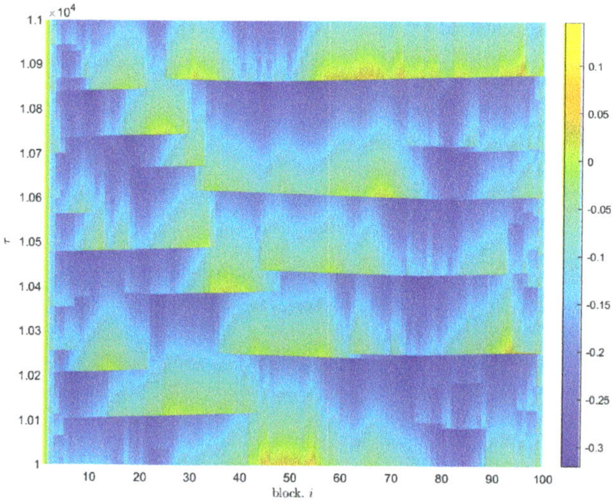


**Figure 3.** Pseudocolor plot displaying nondimensional displacement measurements with time $\tau$, for $N = 100$ slider blocks, for an iteration of the model with the same parameters as in Figure 2. The magnitude of displacement is represented by the adjacent color bar, with arbitrary units.

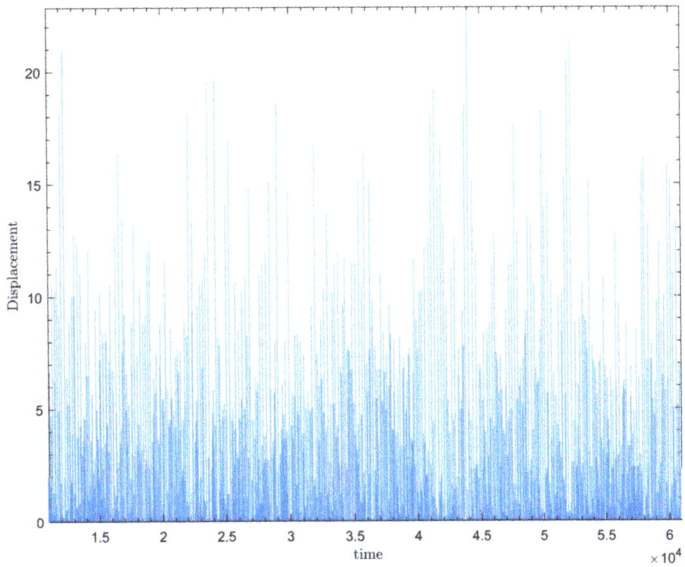


**Figure 4.** Plot depicting the evolution of event sizes in time for $N = 100$ slider blocks, and for the simulation of the model with the same parameters as in Figure 2.

The simulation collects characteristics of the avalanches, including the block at which the sequence was initiated and the duration of the avalanche, as defined by the period of time in which any block within the event has a velocity relative to the conveyor belt greater than some small cutoff value. This cutoff value allows for a distinction between avalanche events. In both Figures 2 and 3, regular instances of both negative velocity and negative or "backward" displacement may be observed alongside movement in the positive direction.

For the model simulations performed in this work, a single run with the specified transient and steady time intervals took several hours to run on a PC computer. Increasing the system size can be a challenging task; however, the implementation of the parallel version of the model can help to speed up the computations.

### 3.2. Avalanche Statistics and Model Fitting

The viscoelastic slider-block model was capable of generating frequency-size distributions, resembling those observed in natural seismicity [21,22,42]. The frequency-size distributions associated with five parameter variations can be observed in Figure 5, alongside an associated power–law fit $P(s) \sim s^{-\gamma}$, where $P(s)$ is the probability distribution function for the sum displacement $s$, and $\gamma$ is the scaling exponent. Each simulation was performed over $\tau = 200,000$, following an initial transient regime.

When compared to naturally occurring aftershock sequences, the temporal clustering observed in this model deviated slightly. The slider-block avalanche sequences were described using parameters obtained from the fitting of the ETAS model. The ETAS model is used to describe the rate of aftershock generation, Equation (15), as a direct response to the triggering and decay of seismic activity following prior earthquake events. Parameters were obtained using maximum likelihood estimation. A comparison between the cumulative number of simulated avalanche events for one simulation with the given model parameters during a specific time interval and the corresponding ETAS fit is shown in Figure 6.

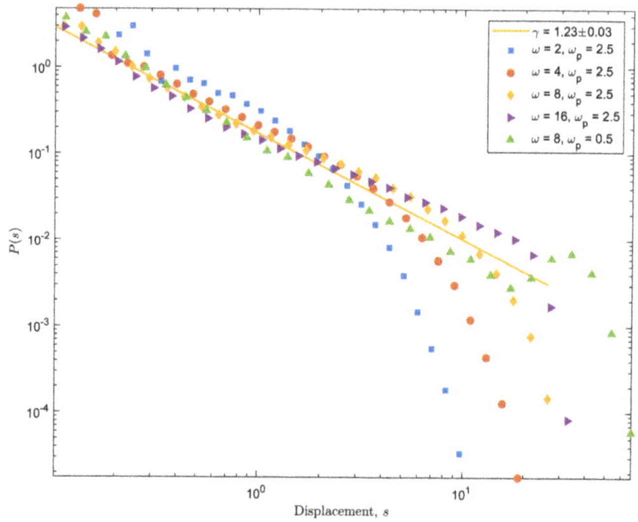


**Figure 5.** Frequency-size distribution of simulated avalanches for the simulations of the model with $N = 100$ blocks, with parameters $\omega_\mathrm{f} = 1.5$, $\omega_\mathrm{d} = 0.03$, $\omega_\mathrm{pd} = 0.01$, $\nu = 0.001$, $\delta = 10$, and varying parameters $\omega$ and $\omega_\mathrm{p}$. The corresponding parameters for each simulation run are in the legend, with the associated symbol. The straight solid line corresponds to the power–law fit $P(s) \sim s^{-\gamma}$ to the data with $\omega = 8$ and $\omega_\mathrm{p} = 2.5$.

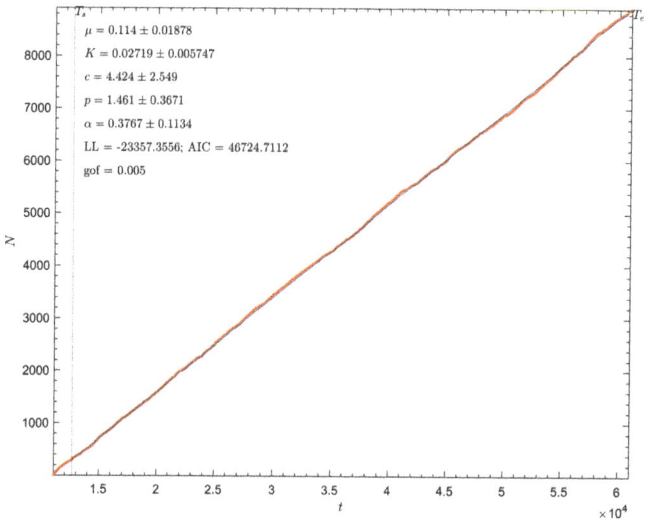


**Figure 6.** ETAS model fitting for cumulative simulated avalanche events for an iteration of an $N = 100$ linear array, with parameters $\omega = 8$, $\omega_\mathrm{p} = 2.5$, $\omega_\mathrm{f} = 1.5$, $\omega_\mathrm{d} = 0.03$, $\omega_\mathrm{pd} = 0.01$, $\nu = 0.001$, $\delta = 10$. The red points represent the cumulative number of events during the given nondimensional time interval. The blue line represents the corresponding ETAS fit using MLE (15). All parameters are displayed within a 95% confidence interval.

As the model avalanches do allow for a successful ETAS model fitting, temporal clustering does exist within the model; however, across repeated simulations, the estimated $p$-value took a value of approximately 1.46, and an $\alpha$-value of 0.377. These values, respectively, indicate that the aftershock rate of decay is relatively high, and that the aftershock rate stays relatively constant during this time interval, then drops abruptly. When reviewed in contrast to Omori's law, this behavior indicates that these simulated aftershock sequences may not be as prolific as naturally occurring aftershock sequences. This $\alpha$-value is lower than typically observed in real-life earthquake events, indicating that generated aftershock sequences are not as vigorous as those observed in real world; however, a $p$-value of 1.46 is a physically reasonable value, and indicates similarity between simulated and real earthquake behavior [43,44].

## 4. Discussion and Conclusions

The objective of this work was to use the viscoelastic slider-block model to replicate the conditions of a seismogenic fault within a medium similar to that of the Earth's crust. This model is characterized by the introduction of SLS elements, and subsequently, the introduction of nonlinear viscous processes involved with the redistribution of stress. Simulations were performed to determine the role of viscous processes in the generation of avalanche events that follow well-known laws regarding frequency-size distribution and aftershock decay rates.

The simulations provided favorable results regarding frequency-size distributions, resulting in consistent power–law scaling [45–47]. All simulations reproduced frequency-size relations similar to those generated by natural earthquake dynamics. Using this model, and other similar slider-block models exhibiting elements of self-organized criticality, simulations can produce realistic results and can re-affirm the hypothesis that naturally occurring frequency-size distributions of events may partially be the result of SOC within the Earth's crust [9,31].

Moreover, the ETAS model fitting produced favorable results, despite relatively high $p$-values and a low productivity of aftershock sequences reflected in the $\alpha$ parameter. Using the input model parameters outlined in the previous section, model fitting returned $p$-values of approximately 1.46, demonstrating a comparatively high, but still physically reasonable, decay rate of aftershocks. The results of this work confirm that to observe an Omori-like decay rates for aftershock sequences, one needs to consider the viscous effects governed by linear or power–law rheology as was demonstrated by similar slider-block and cellular automata models [19,29,30].

In this paper, initial conditions and physical parameters were varied to observe how frequency–magnitude statistics change as the properties of the model change. In future works, a full analysis of the parameter space of the model will be completed, both to determine the effects of the input parameters on the frequency–magnitude distribution, but also to determine the relationship between input parameters and corresponding ETAS model parameters. This will help deduce the physical meaning of the ETAS parameters, and may provide a theoretical basis to support the application of the ETAS model to natural seismicity. In particular, future model parameter space exploration should further examine how variations in viscoelasticity influence the generation of simulated earthquake sequences.

The ETAS model describes the behavior of aftershock sequences as a direct consequence of previous seismic activity. This places an emphasis on the redistribution of stress within a system following any seismic activity, as events are generated not through the direct constant application of force (plate tectonics, conveyor belt), but through previous trigger activity and instability [28,48]. The introduction of nonlinear behavior within a medium provides a mechanism through which this redistribution of stress may be delayed, allowing for the generation of aftershocks. The presence of temporal clustering within the system described by this model supports the theory that nonlinear viscoelasticity influences the generation of aftershocks, and that viscous responses within the Earth's crust and upper mantle may contribute to the observed aftershock dynamics.

**Author Contributions:** Conceptualization, C.A.M. and R.S.; methodology, C.A.M. and R.S.; software, C.A.M. and R.S.; validation, C.A.M. and R.S.; manuscript writing, C.A.M.; manuscript revision and editing, C.A.M. and R.S.; visualization, C.A.M. and R.S.; supervision, R.S.; project administration, R.S.; funding acquisition, R.S. All authors have read and agreed to the published version of the manuscript.

**Funding:** This research was funded by the NSERC Discovery grant.

**Institutional Review Board Statement:** Not applicable.

**Data Availability Statement:** No new data were created in this work.

**Acknowledgments:** We would like to thank the two reviewers for their constructive and helpful comments that helped to improve the presentation and clarify results.

**Conflicts of Interest:** The authors declare no conflict of interest.

## References

1. Kawamura, H.; Hatano, T.; Kato, N.; Biswas, S.; Chakrabarti, B.K. Statistical physics of fracture, friction, and earthquakes. *Rev. Mod. Phys.* **2012**, *84*, 839–884. [CrossRef]
2. Shcherbakov, R.; Turcotte, D.L.; Rundle, J.B. Complexity and Earthquakes. In *Earthquake Seismology*, 2nd ed.; Kanamori, H., Ed.; Elsevier: Amsterdam, The Netherlands, 2015; Volume 4 of *Treatise on Geophysics*; pp. 627–653. [CrossRef]
3. Rundle, J.B.; Turcotte, D.L.; Shcherbakov, R.; Klein, W.; Sammis, C. Statistical physics approach to understanding the multiscale dynamics of earthquake fault systems. *Rev. Geophys.* **2003**, *41*, 1019. [CrossRef]
4. Rundle, J.B.; Stein, S.; Donnellan, A.; Turcotte, D.L.; Klein, W.; Saylor, C. The complex dynamics of earthquake fault systems: New approaches to forecasting and nowcasting of earthquakes. *Rep. Prog. Phys.* **2021**, *84*, 076801. [CrossRef] [PubMed]
5. Shcherbakov, R.; Turcotte, D.L.; Rundle, J.B.; Tiampo, K.F.; Holliday, J.R. Forecasting the Locations of Future Large Earthquakes: An Analysis and Verification. *Pure Appl. Geophys.* **2010**, *167*, 743–749. [CrossRef]
6. Tiampo, K.F.; Shcherbakov, R. Seismicity-based earthquake forecasting techniques: Ten years of progress. *Tectonophysics* **2012**, *522*, 89–121. [CrossRef]
7. Shcherbakov, R.; Zhuang, J.; Zöller, G.; Ogata, Y. Forecasting the magnitude of the largest expected earthquake. *Nat. Commun.* **2019**, *10*, 4051. [CrossRef]
8. Shcherbakov, R. Statistics and Forecasting of Aftershocks During the 2019 Ridgecrest, California, Earthquake Sequence. *J. Geophys. Res.* **2021**, *126*, e2020JB020887. [CrossRef]
9. Bak, P.; Tang, C. Earthquakes as a Self-Organized Critical Phenomenon. *J. Geophys. Res.* **1989**, *94*, 15635–15637. [CrossRef]
10. Turcotte, D.L. Self-organized criticality. *Rep. Prog. Phys.* **1999**, *62*, 1377–1429. [CrossRef]
11. Carlson, J.M.; Langer, J.S. Mechanical Model of an Earthquake Fault. *Phys. Rev. A* **1989**, *40*, 6470–6484. [CrossRef]
12. Carlson, J.M.; Langer, J.S.; Shaw, B.E. Dynamics of Earthquake Faults. *Rev. Mod. Phys.* **1994**, *66*, 657–670. [CrossRef]
13. de Arcangelis, L.; Godano, C.; Grasso, J.R.; Lippiello, E. Statistical physics approach to earthquake occurrence and forecasting. *Phys. Rep.* **2016**, *628*, 1–91. [CrossRef]
14. Burridge, R.; Knopoff, L. Model and Theoretical Seismicity. *Bull. Seismol. Soc. Am.* **1967**, *57*, 341–371. [CrossRef]
15. Huang, J.; Narkounskaia, G.; Turcotte, D.L. A Cellular-Automata, Slider-Block Model for Earthquakes. II. Demonstration of Self-Organized Criticality for a 2-D System. *Geophys. J. Int.* **1992**, *111*, 259–269. [CrossRef]
16. Carlson, J.M.; Langer, J.S.; Shaw, B.E.; Tang, C. Intrinsic-Properties of a Burridge-Knopoff Model of an Earthquake Fault. *Phys. Rev. A* **1991**, *44*, 884–897. [CrossRef]
17. Olami, Z.; Feder, H.J.S.; Christensen, K. Self-organized criticality in a continuous, nonconservative cellular automaton modeling earthquakes. *Phys. Rev. Lett.* **1992**, *68*, 1244–1247. [CrossRef]
18. Shcherbakov, R.; Davidsen, J.; Tiampo, K.F. Record-breaking avalanches in driven threshold systems. *Phys. Rev. E* **2013**, *87*, 052811. [CrossRef]
19. Zhang, X.; Shcherbakov, R. Power-law rheology controls aftershock triggering and decay. *Sci. Rep.* **2016**, *6*, 36668. [CrossRef]
20. Turcotte, D.L. *Fractals and Chaos in Geology and Geophysics*, 2nd ed.; Cambridge University Press: Cambridge, UK, 1997.
21. Shcherbakov, R.; Turcotte, D.L.; Rundle, J.B. Aftershock statistics. *Pure Appl. Geophys.* **2005**, *162*, 1051–1076. [CrossRef]
22. Shcherbakov, R.; Goda, K.; Ivanian, A.; Atkinson, G.M. Aftershock Statistics of Major Subduction Earthquakes. *Bull. Seismol. Soc. Am.* **2013**, *103*, 3222–3234. [CrossRef]
23. Utsu, T.; Ogata, Y.; Matsu'ura, R.S. The Centenary of the Omori Formula for a Decay Law of Aftershock Activity. *J. Phys. Earth* **1995**, *43*, 1–33. [CrossRef]
24. Ogata, Y.; Utsu, T.; Katsura, K. Statistical Features of Foreshocks in Comparison with Other Earthquake Clusters. *Geophys. J. Int.* **1995**, *121*, 233–254. [CrossRef]
25. Ogata, Y. Statistical-Models For Earthquake Occurrences And Residual Analysis For Point-Processes. *J. Am. Stat. Assoc.* **1988**, *83*, 9–27. [CrossRef]

26. Gran, J.D.; Rundle, J.B.; Turcotte, D.L. A possible mechanism for aftershocks: Time-dependent stress relaxation in a slider-block model. *Geophys. J. Int.* **2012**, *191*, 459–466. [CrossRef]
27. Amendola, A.; Dragoni, M. Dynamics of a two-fault system with viscoelastic coupling. *Nonlinear Proc. Geophys.* **2013**, *20*, 1–10. [CrossRef]
28. Jagla, E.A. Aftershock production rate of driven viscoelastic interfaces. *Phys. Rev. E* **2014**, *90*, 042129. [CrossRef] [PubMed]
29. Jagla, E.A.; Landes, F.P.; Rosso, A. Viscoelastic Effects in Avalanche Dynamics: A Key to Earthquake Statistics. *Phys. Rev. Lett.* **2014**, *112*, 174301. [CrossRef]
30. Sakaguchi, H.; Okamura, K. Aftershocks and Omori's law in a modified Carlson-Langer model with nonlinear viscoelasticity. *Phys. Rev. E* **2015**, *91*, 052914. [CrossRef]
31. Baró, J.; Davidsen, J. Universal avalanche statistics and triggering close to failure in a mean-field model of rheological fracture. *Phys. Rev. E* **2018**, *97*, 033002. [CrossRef]
32. Freed, A.M.; Lin, J. Delayed triggering of the 1999 Hector Mine earthquake by viscoelastic stress transfer. *Nature* **2001**, *411*, 180–183. [CrossRef]
33. Barbot, S.; Fialko, Y. A unified continuum representation of post-seismic relaxation mechanisms: Semi-analytic models of afterslip, poroelastic rebound and viscoelastic flow. *Geophys. J. Int.* **2010**, *182*, 1124–1140. [CrossRef]
34. Pollitz, F.F.; Wicks, C.; Thatcher, W. Mantle flow beneath a continental strike-slip fault: Postseismic deformation after the 1999 Hector Mine earthquake. *Science* **2001**, *293*, 1814–1818. [CrossRef] [PubMed]
35. Freed, A.M.; Burgmann, R. Evidence of power-law flow in the Mojave desert mantle. *Nature* **2004**, *430*, 548–551. [CrossRef] [PubMed]
36. Hetland, E.A.; Hager, B.H. Postseismic and interseismic displacements near a strike-slip fault: A two-dimensional theory for general linear viscoelastic rheologies. *J. Geophys. Res.* **2005**, *110*, B10401. [CrossRef]
37. Burgmann, R.; Dresen, G. Rheology of the lower crust and upper mantle: Evidence from rock mechanics, geodesy, and field observations. *Annu. Rev. Earth Planet. Sci.* **2008**, *36*, 531–567. [CrossRef]
38. Petrillo, G.; Lippiello, E.; Landes, F.P.; Rosso, A. The influence of the brittle-ductile transition zone on aftershock and foreshock occurrence. *Nat. Commun.* **2020**, *11*, 3010. [CrossRef]
39. Lakes, R.S. *Viscoelastic Materials*; Cambridge University Press: Cambridge, UK, 2009.
40. Leine, R.I.; Van Campen, D.H.; De Kraker, A.; Van den Steen, L. Stick-slip vibrations induced by alternate friction models. *Nonlin. Dyn.* **1998**, *16*, 41–54. [CrossRef]
41. Leine, R.I.; Nijmeijer, H. *Dynamics and Bifurcations of Non-Smooth Mechanical Systems*; Springer: Berlin/Heidelberg, Germany, 2004.
42. Kothari, S.; Shcherbakov, R.; Atkinson, G. Statistical Modeling and Characterization of Induced Seismicity Within the Western Canada Sedimentary Basin. *J. Geophys. Res.* **2020**, *125*, e2020JB020606. [CrossRef]
43. Shcherbakov, R.; Yakovlev, G.; Turcotte, D.L.; Rundle, J.B. Model for the Distribution of Aftershock Interoccurrence Times. *Phys. Rev. Lett.* **2005**, *95*, 218501. [CrossRef]
44. Zhang, L.Z.; Werner, M.J.; Goda, K. Variability of ETAS Parameters in Global Subduction Zones and Applications to Mainshock-Aftershock Hazard Assessment. *Bull. Seismol. Soc. Am.* **2020**, *110*, 191–212. [CrossRef]
45. Shcherbakov, R.; Van Aalsburg, J.; Rundle, J.B.; Turcotte, D.L. Correlations in aftershock and seismicity patterns. *Tectonophysics* **2006**, *413*, 53–62. [CrossRef]
46. Bhattacharya, P.; Phan, M.; Shcherbakov, R. Statistical Analysis of the 2002 Mw 7.9 Denali Earthquake. *Bull. Seismol. Soc. Am.* **2011**, *101*, 2662–2674. [CrossRef]
47. Burroughs, S.M.; Tebbens, S.F. The upper-truncated power law applied to earthquake cumulative frequency-magnitude distributions: Evidence for a time-independent scaling parameter. *Bull. Seismol. Soc. Am.* **2002**, *92*, 2983–2993. [CrossRef]
48. Jagla, E.A. Creep dynamics of viscoelastic interfaces. *Europhys. Lett.* **2014**, *105*, 46003. [CrossRef]

**Disclaimer/Publisher's Note:** The statements, opinions and data contained in all publications are solely those of the individual author(s) and contributor(s) and not of MDPI and/or the editor(s). MDPI and/or the editor(s) disclaim responsibility for any injury to people or property resulting from any ideas, methods, instructions or products referred to in the content.

entropy

MDPI

*Article*

# Defining the Scale to Build Complex Networks with a 40-Year Norwegian Intraplate Seismicity Dataset

**Claudia Pavez-Orrego [1,*] and Denisse Pastén [2]**

[1] SINTEF Industry, S.P. Andersens Vei 15B, 7031 Trondheim, Norway
[2] Departamento de Física, Facultad de Ciencias, Universidad de Chile, Las Palmeras 3425, Ñuñoa, Santiago 7800003, Chile; denisse.pasten.g@gmail.com
* Correspondence: claudia.pavez.orrego@sintef.no

**Abstract:** We present a new complex network-based study focused on intraplate earthquakes recorded in southern Norway during the period 1980–2020. One of the most recognized limitations of spatial complex network procedures and analyses concerns the definition of adequate cell size, which is the focus of this approach. In the present study, we analyze the influence of observational errors of hypocentral and epicentral locations of seismic events in the construction of a complex network, looking for the best cell size to build it and to develop a basis for interpreting the results in terms of the structure of the complex network in this seismic region. We focus the analysis on the degree distribution of the complex networks. We observed a strong result of the cell size for the slope of the degree distribution of the nodes, called the critical exponent $\gamma$. Based on the Abe–Suzuki method, the slope ($\gamma$) showed a negligible variation between the construction of 3- and 2-dimensional complex networks. The results were also very similar for a complex network built with subsets of seismic events. These results suggest a weak influence of observational errors measured for the coordinates latitude, longitude, and depth in the outcomes obtained with this particular methodology and for this high-quality dataset. These results imply stable behavior of the complex network, which shows a structure of hubs for small values of the cell size and a more homogeneous degree distribution when the cell size increases. In all the analyses, the $\gamma$ parameter showed smaller values of the error bars for greater values of the cell size. To keep the structure of hubs and small error bars, a better range of the side sizes was determined to be between 8 to 16 km. From now on, these values can be used as the most stable cell sizes to perform any kind of study concerning complex network studies in southern Norway.

**Keywords:** complex networks; intraplate seismicity; earthquake distribution

**Citation:** Pavez-Orrego, C.; Pastén, D. Defining the Scale to Build Complex Networks with a 40-Year Norwegian Intraplate Seismicity Dataset. *Entropy* **2023**, *25*, 1284. https://doi.org/10.3390/e25091284

Academic Editor: Georgios Michas

Received: 13 July 2023
Revised: 17 August 2023
Accepted: 24 August 2023
Published: 31 August 2023

**Copyright:** © 2023 by the authors. Licensee MDPI, Basel, Switzerland. This article is an open access article distributed under the terms and conditions of the Creative Commons Attribution (CC BY) license (https://creativecommons.org/licenses/by/4.0/).

## 1. Introduction

The analysis of the distribution and physical behavior of earthquakes in Norway has been approached from different points of view, which include, for example, imaging studies [1–13], the physics of earthquakes [14], seismic hazards [15,16], and tectonics and seismology [14,17–19]. Recently, a whole new area regarding complex networks and fractality has been investigated, which aims to build a deeper understanding of the connections between seismic events in time and space [20].

Complex networks are able to show non-trivial behavior of physical systems through the analysis of their topological features. They can be categorized into many different types, such as random complex networks [21], small-world behavior [22], which indicates the need to take just a small number of steps to go from one node to another one, and scale-free behavior [23], which shows the structural organization of a system. Complex networks have been developed and applied in the study of different systems that show non-trivial topological behavior, such as biology [24–26], economics [27,28] or social relationships [29,30]. The application of complex networks in geophysics has grown in recent years, providing a new perspective in the Earth sciences [31–41].

In particular, over the last twenty years, complex networks have been used to analyze the complex behavior of earthquake distribution in time and space. In this sense, many studies have been carried out. For example, building networks based on the spatial position of earthquakes, their epicenters or hypocenters [39,42–44], or following their time series [45,46], has included application of the visibility graph method (VG) to earthquakes in Italy, finding that the probability distribution of connectivity follows a power law. Moreover, Telesca et al. (2020) [47], applied a horizontal visibility graph (HVG) to determine changes in the reversibility of the time series measured for the Iquique earthquake (Chile, 2014), showing that they were reversible for the dataset without the aftershocks and irreversible for the catalog with aftershocks.Therefore, earthquakes have been studied through complex networks mainly in two different ways: (1) based on following temporal sequences or (2) based on the spatial distribution of seismic events. One of the most recognized approaches is that developed by Abe and Suzuki (2006) [39], whose method involves the building of spatial complex networks with earthquake datasets. The method is based on the size of a cubic or a plane cell: if the cell contains a hypocenter (cubic) or an epicenter (plane), a network node is defined. Then, the network can be based on the dimension of the cubes used to divide the three-dimensional space (latitude, longitude, and depth). In this sense, any attempt to define complex networks on these kinds of systems will require a suitable and stable cell size as the network topology and its physical properties depend on it.

In this study, we build a complex network with earthquakes recorded in southern Norway, using the above-described method of Abe and Suzuki, (2006) [39]. Our purpose is to define a cell size range independent of the measurement errors generally associated with hypocenter and epicenter estimations, with the aim of gaining a better understanding of scale effects in the complex network. This will be carried out by testing the stability of the values obtained for the critical exponent $\gamma$ in directed complex networks built with an earthquake dataset measured in southern Norway and with temporal and spatial subsets. We also analyze the influence of the measurement errors in the size of the chosen cell in the degree distribution. As a main result, we obtain a range of cubic cell sizes that show independence of the measurement errors and for which the construction of a complex network will be reliable.

## 2. Intraplate Seismicity in Southern Norway

Seismicity rates in Norway are the highest in northern Europe [17]. Seismic events occur periodically (NNSN report, 2018) [48], with low to intermediate magnitudes ($M_L \leq 4.0$). Natural earthquakes are strictly intraplate, both onshore and offshore, at the passive continental margin [17,49].

It is commonly accepted by the seismological community that this intraplate activity emerges due to a combination of stress-generating mechanisms, which include crustal to local scale (Olesen et al., 2013): (i) gravitational potential energy changes produced due to topographic loads, (ii) post-glacial isostatic adjustments, (iii) Mid-Atlantic ridge push (iv) Quaternary glacial erosion (v) flexural stresses through sedimentation [17], (vi) crustal density variations, and spatial coincidence with anomalous low-velocity zones of seismic waves in the upper mantle [50–52] (Figure 1).

Concerning the spatial earthquake location, the highest seismicity levels occur in the rifted continental margin as well as in the strongly faulted regions near to the rifts in the North Sea, and in the coast of south-western Norway [18,48]. This area, in which we focus this study, also presents high amounts of seismic events on the mainland in a zone that is highly influenced by the post-Caledonian faults [53]. Most of the seismic events are located in the upper 20 km of the crust [17,18] (Figure 2). Meanwhile, deep earthquakes occur mainly offshore, dominated by reverse faulting and the Mid-Atlantic Ridge push. On the other side, shallow earthquakes occur onshore, where normal faulting is dominant and the horizontal tension is coast perpendicular [54].

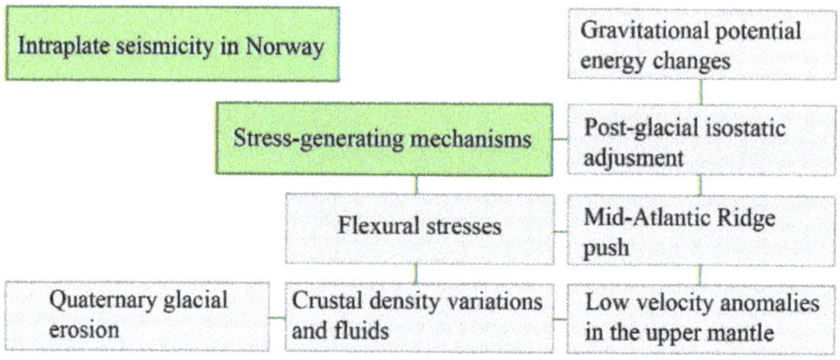


**Figure 1.** Summary of the main stress generating mechanisms for intraplate seismicity in Norway.

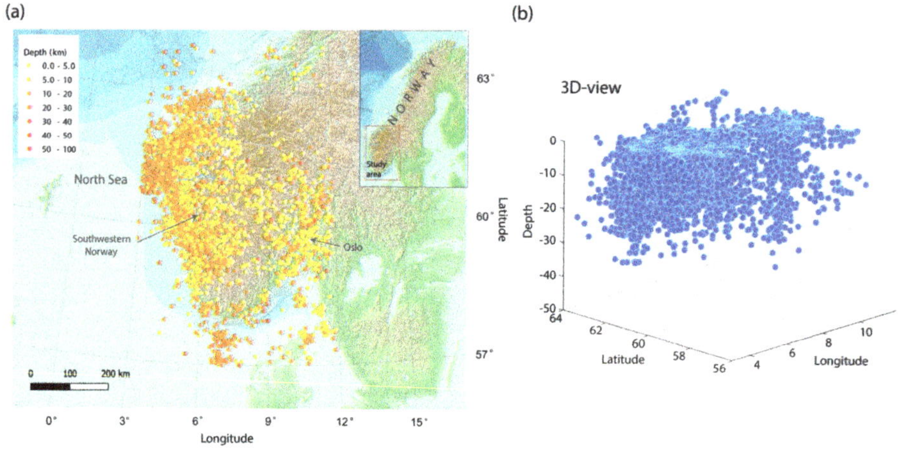


**Figure 2.** (**a**) Seismicity map with local events. Color legend shows earthquakes arranged by depth. The study area is marked with a red box in the regional inset. (**b**) 3D−view of hypocenters, showing in−depth earthquake distribution.

## 3. Data

The seismic catalog for the period 1980–2019 was downloaded from the EPOS Norway portal, the European Plate Observing System https://epos-no.uib.no/eposn-data-portal/ (accessed on 12 January 2023). To the present day, the database offers a list of 40 possible data services. This list includes access to the seismological data offered by the Norwegian National Seismic Network. The direct way to access the data is shown in Appendix A. The EPOS-N project is research-based, and is focused on understanding of the Earth's deformational processes, geohazards, and georesources. The project is collaborative, with many Norwegian institutions continuously contributing geological and geophysical datasets to the data portal: the University of Bergen (coordinator), the University of Oslo, the Geological Survey of Norway, NORSAR, the Norwegian Mapping Authority, and the Christian Michelsen Research Centre.

The original catalog consisted of 77,622 global seismic events. This catalog was downloaded as an Excel sheet (see Appendix A) and contains origin time and hypocentral location with their respective errors, magnitude and magnitude type, number of stations, rms, azimuthal gap, strike, dip, and rake. More details about the construction process of this catalog can be found in the 2018 and 2019 NNSN annual reports (NNSN annual report, 2018–2019) [48,55].

With the aim of selecting the relevant information for our study, the catalog was initially filtered for southern Norway, considering the area between 3° and 12° E and

57° and 64° N. Large hypocentral errors were observed in the catalog, which might be explained as a consequence of picking errors or large azimuthal gaps for events with a high number of observations. To avoid this, we consider a minimum number of recording stations equal to six. With this number, we also prevent incorrect low errors based on a low number of observations, where the location solutions might fit with small residuals. Based on the distribution of latitude, longitude, and depth errors, the last filtering consisted of the inspection of the histogram, from where out-of-average events—48, 81, and 100 km error locations for the y, x, z axes—were excluded from the curve (Figure 3). The spatial, error, and distribution filters resulted in a total of 6469 earthquakes for further analysis (Figures 2 and 3). Moreover, the catalog was filtered by its magnitude of completeness $M_c = 1.3$. After this process, the total number of events was 3739 (Figure 4a,b).

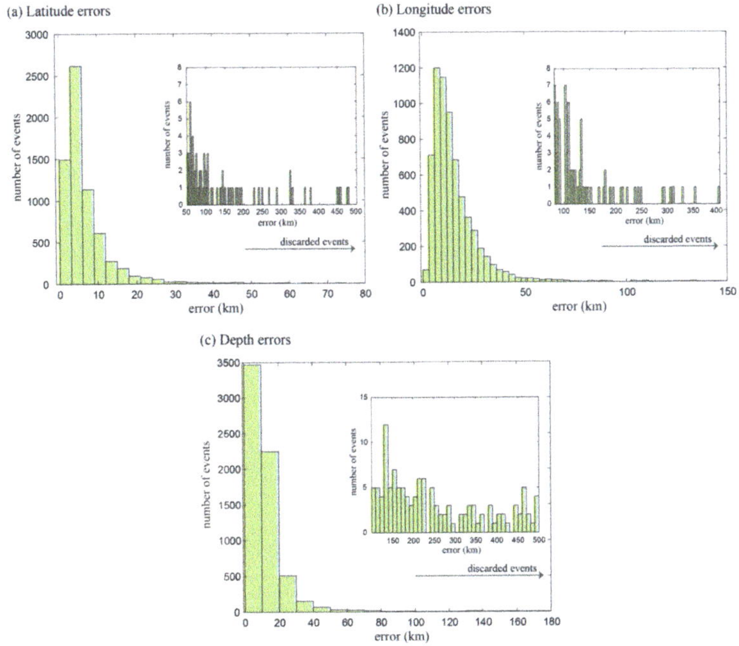


**Figure 3.** (**a**–**c**) Histograms with location errors in latitude, longitude and depth, respectively.

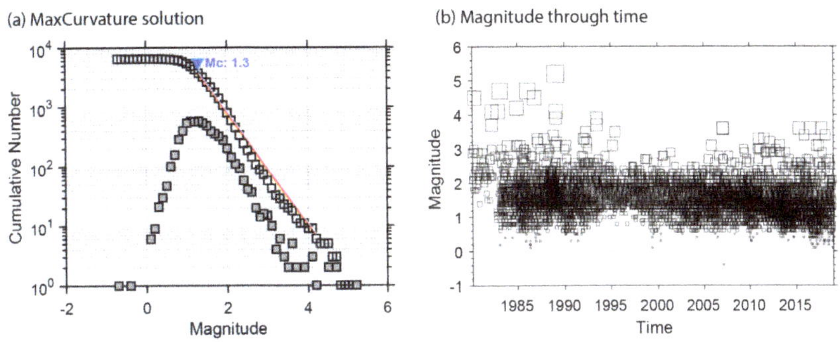


**Figure 4.** (**a**) Magnitude of completeness (blue triangle), calculated through the maximum curvature method. Gray and white squares represent the cumulative and discrete number of events, respectively. (**b**) Magnitude distribution vs. time.

## 4. Methodology

We built a complex network using the Abe and Suzuki [39] method applied over a seismic dataset measured in southern Norway during the period 1980–2020, i.e., we used their hypocenter coordinates and the time sequence as inputs. To perform the spatial analyses, we then used the spatial coordinates (longitude, latitude, depth) in kilometers. The latitude is represented by the angle $\theta$, and the longitude is represented by the angle $\phi$. This conversion is performed by using the following expressions:

$$d_i^{NS} = R(\theta_i - \theta_0), \tag{1}$$

$$d_i^{EW} = R(\phi_i - \phi_0)\cos(\theta_{av}), \tag{2}$$

$$d_i^z = z_i, \tag{3}$$

where $z_i$ is the depth and $\theta_{av}$ is the average latitude, $\theta_0$ and $\phi_0$ are the minimum values for the latitude and longitude, and $R$ is the radius of the Earth, assumed for this study to be 6370 km.

A complex network consists of **nodes** connected between them through **links**. Once we have converted the spatial coordinates (longitude, latitude, depth) into kilometers, we can build the complex network with the seismic dataset. To follow the method of Abe and Suzuki (2006) [39], we must define what a node is. For this, we divide each zone into cubic cells with a side size between $\Delta = 5$ km and $\Delta = 20$ km. Then, we check if one or more hypocenters are inside the cubic cell; if so, the cell is called a node. Then, we place the connections between the nodes following the temporal sequence of the seismic events. The direction of the connections between the nodes is defined through the temporal sequence of the seismic events in the region, as shown in Figure 5. Consequently, a directed network following the Abe–Suzuki method [39] is built.

Then, we place the connections between the nodes following the temporal sequence of the seismic events, so the order of events is preserved as in the standard time analysis [56–58]. The direction of the connections between nodes is defined through the temporal sequence of the seismic events in the region, as shown in Figure 5.

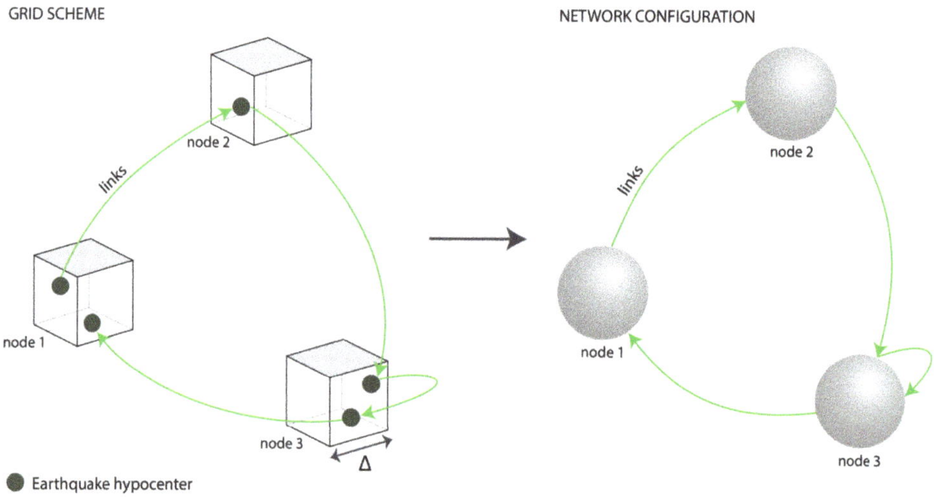


**Figure 5.** Schematic representation of the cubic cells representing nodes in the complex network. The cube side size is Δ.

A complex network can be characterized by different metrics. Among them, it is important here to mention the clustering coefficient ($C$), which measures the tendency of the nodes to form triangles (clusters) between them, the path length ($L$), defined as the average of the shortest path length in the network, and the degree $k_i$, which represents the number of connections of the node $i$. Good examples regarding metrics applications can be found, for example, in Watts and Strogatz (1998) [22], who defined the term "small world" for complex networks, following certain metric criteria imposed over $L$ and $C$. Also, Barabási and Albert, (1999) [23] defined "scale-free" behavior of a complex network when the probability of the degree $k_i$ is a power law in a log-log plot.

Now, once we have built the directed complex network for a one cubic cell side value, we focus on a complex network basic measure: the degree of the node. We analyze the behavior of the critical exponent $\gamma$, which corresponds to the slope of the degree distribution. For earthquake datasets, this probability is represented by a power law [39–43],

$$P(k) \sim k^{-\gamma}, \tag{4}$$

where $k$ is the degree of the nodes and $\gamma$ is the slope of the distribution in a log-log plot.

To understand the response of the $\gamma$ exponent in different scenarios, the analyses are carried out over both the prefiltered and the complete datasets (without and with $M_c$). For the first case, we analyze the entire catalog (6469 events), and the first half of it (3235 events). For the second case, we separate the dataset into two regions: the western zone (longitude 3.0° to 7.5° E, with 3111 seismic events) and the Oslo region (longitude 7.5° to 12.0° E, with 628 seismic events).

## 5. Results

### 5.1. Prefiltered Dataset

As a first result, we show the scale-free behavior for the degree distribution of the dataset without the magnitude of the completeness filter, and, for all the used sizes of $\Delta$, which vary from 5 to 20 km (Figure 6). We choose those values of the cell side size to have a sufficient number of cubic cells to analyze. As particular examples, Figure 7a–d show the scale-free behavior and the adjustment of the slopes for $\Delta = 5, 10, 15$ and 20 km, respectively. In order to calculate the best fit for each degree distribution, we considered the same $k_{min} = 1$ for all the cubic cell sizes, and we neglected between one to three points of the tails. We then computed their respective slopes, which represent the critical exponent $\gamma$. From Figure 7, it is possible to observe a better fit of the slope from $\Delta = 10$ km, being the best fit for $\Delta = 20$ km. The values of $\gamma$, together with their respective errors, are listed in Table 1.

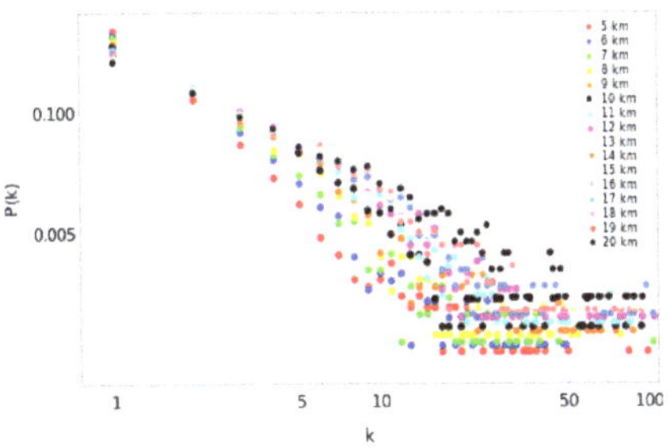


**Figure 6.** Log-log plot of the degree distribution for the studied complex networks. The complex networks were built with values of the side between $\Delta = 5$ km and $\Delta = 20$ km. The scale-free behavior is clear.

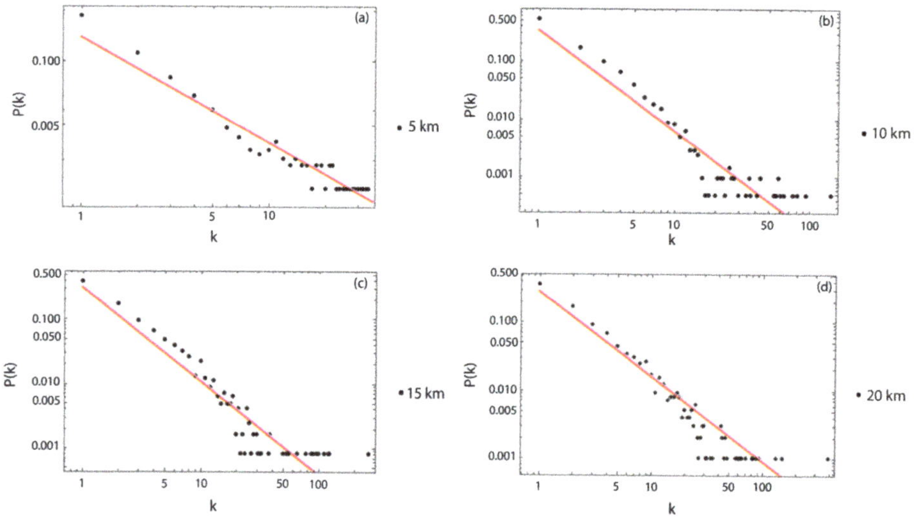


**Figure 7.** Examples of the scale-free behavior of the degree distribution of nodes with their respective linear fits. (**a**) Degree distribution for $\Delta = 5$ km with $\gamma = 2.1 \pm 0.1$. (**b**) Degree distribution for $\Delta = 10$ km with $\gamma = 1.7 \pm 0.1$. (**c**) Degree distributions for $\Delta = 15$ km with $\gamma = 1.45 \pm 0.08$. (**d**) Degree distribution for $\Delta = 20$ km with $\gamma = 1.24 \pm 0.05$.

After corroborating that all the complex networks show scale-free behavior, we focus on the $\gamma$ value. Figure 8a,b show the values of $\gamma$ and the number of nodes, respectively, for different values of the cubic cell side, considering a range of $\Delta$ from 5 to 20 km. Figures 7a–d and 8a show that smaller values of $\Delta$ have few nodes with a large degree, implying that the slope of the degree distribution is large (close to 2.0). On the other hand, values of $\Delta$ close to 20 km show a more homogeneous degree distribution, with a lower value of $\gamma$ (close to 1.2). We can interpret these results in dependence of the cubic cell size as follows: with side sizes close to 20 km, the number of contained hypocenters is higher, so the degree distribution is more homogeneous. The error bars from the linear fit of the degree distribution decrease as the cubic cell side grows. We can also notice from Table 1 that the error bars decrease from $\Delta = 14$ km.

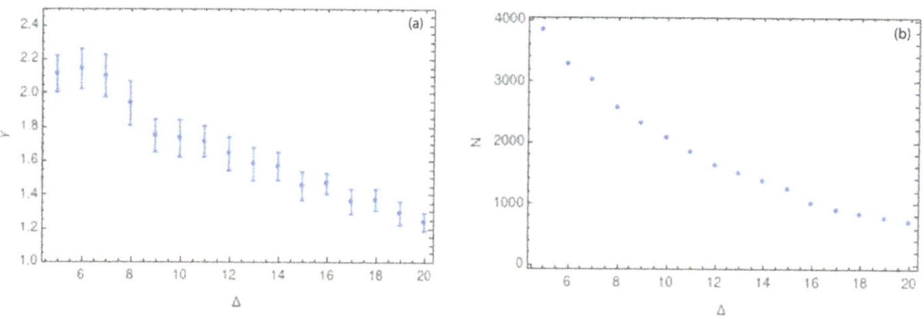


**Figure 8.** (**a**) Values of the critical exponent $\gamma$ for side sizes $\Delta$ between 5 and 20 km. (**b**) Values of the number of nodes $N$ for the same range of $\Delta$.

The earthquake dataset measured in southern Norway is of high quality, with associated errors for each hypocentral location. This allows us to include the average hypocentral errors in kilometers for all nodes in longitude, latitude, and depth. As an example, Figure 9 shows the average errors for $\Delta = 5, 9, 10, 11, 15$, and 20 km, respectively. Specifically, groups (a), (b) and (c) in Figure 9 show a comparison between the average hypocentral longitude, latitude, and depth in km, with their respective average errors.

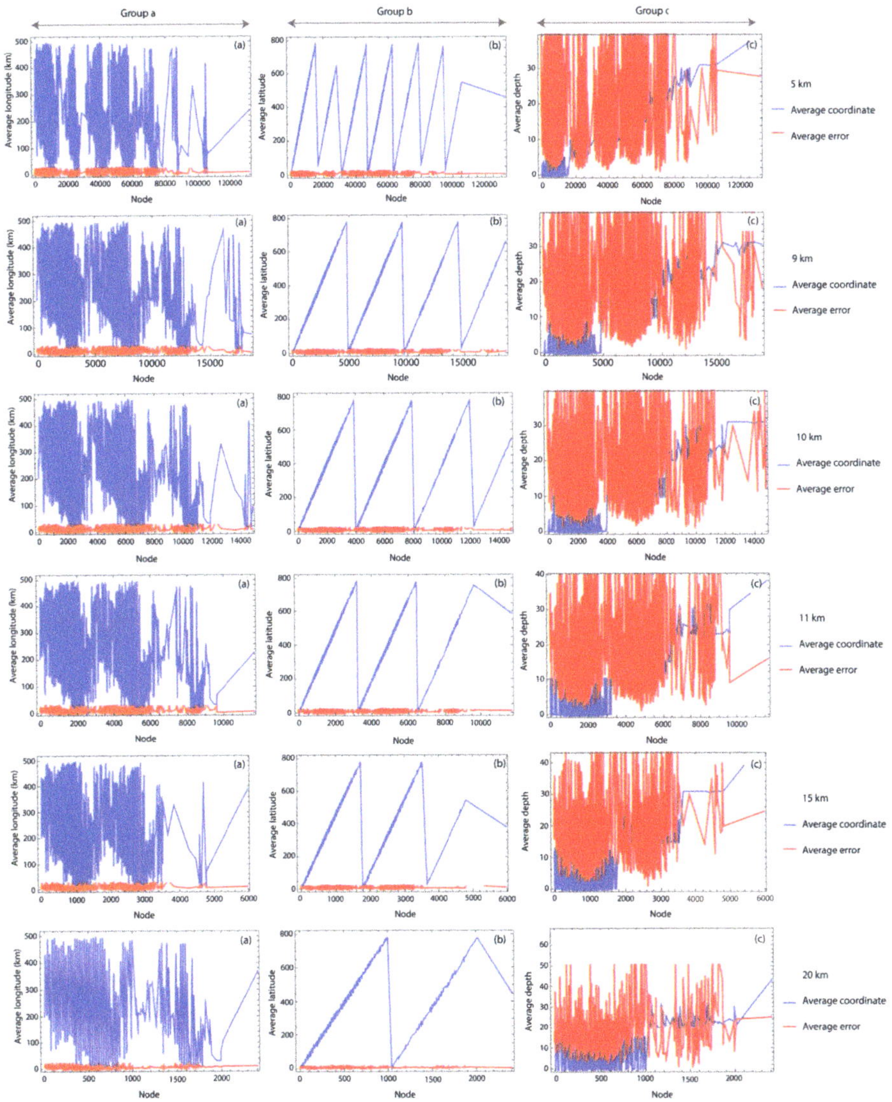


**Figure 9.** Values of the average coordinates in longitude (group (**a**)), latitude (group (**b**)), and depth (group (**c**)) and associated errors. These plots correspond to $\Delta$ equal to 5, 9, 10, 11, 15, and 20 km, from top to bottom. All averages are in km.

Figures 3 and 9 show the measurement errors in latitude, longitude, and depth. From Figure 3, we can observe that the largest errors in both latitude and longitude are concentrated in a few measurements, while most of the data have low errors (less than 15 km). Meanwhile, Figure 9 shows that the average error by node in the coordinates' longitude and latitude is negligible. Nevertheless, the errors in depth are considerable. The percentage of seismic events with errors greater than 20 km is less than the 4%, 23%, and 12% in latitude, longitude, and depth, respectively. As a consequence of this, we analyze the stability of the values of $\gamma$ for a 2-dimensional complex network, i.e., using only the coordinates of longitude and latitude. This result is shown in Figure 10a,b.

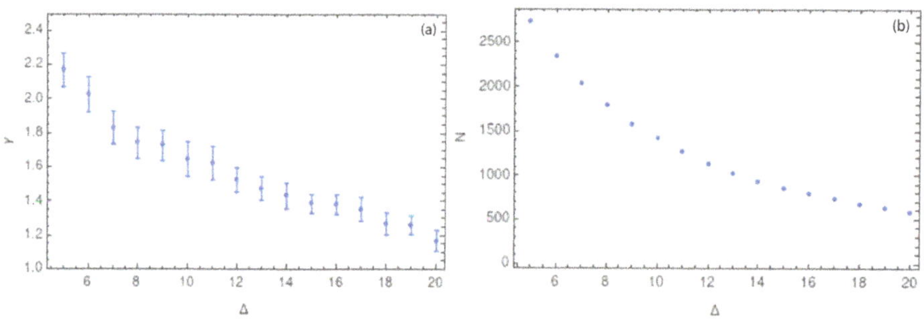


**Figure 10.** (**a**) Values of the critical exponent $\gamma$ in 2 dimensions for $\Delta$ between 5 and 20 km. (**b**) Values of the number of nodes $N$ for each value of $\Delta$ between 5 and 20 km.

Finally, we compare the results obtained in three dimensions (latitude, longitude, depth) and in two dimensions (latitude, longitude) in Figure 11. From this comparison, we observe a considerable similarity in the results for the value of $\gamma$ in 2D and 3D, showing that the distribution of degrees is not affected by the change in dimension. This point is interesting because it means that the complex network keeps its behavior with few nodes with a large degree for small values of $\Delta$, while the degree distribution evolves to a homogeneous distribution of the degree for larger values of $\Delta$. In addition, we observe a decrease in the error bars for $\Delta$ values larger than 14 km, which suggests that this range is the best for applying any kind of analyses in complex networks for this tectonic environment.

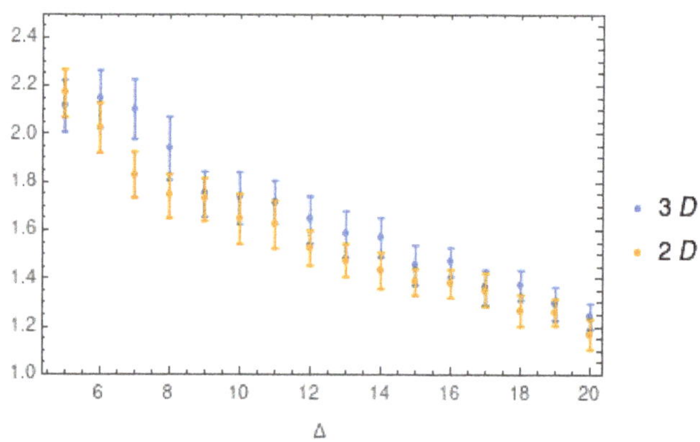


**Figure 11.** Values of the critical exponent $\gamma$ in 3 dimensions (blue dots) and for 2 dimensions (yellow dots) for $\Delta$ between 5 km and 20 km.

Prefiltered Subset

A relevant step during complex network analyses is to check the results' stability. To verify our first round of outcomes, we perform the same analysis by taking a subset of the entire catalog.

Figure 12a show the same behavior that we found for the first analysis: for small values of $\Delta$, the complex networks have few nodes with a larger degree (high values of $\gamma$), whilst larger values of $\Delta$ give small values of $\gamma$ (homogeneous degree distribution). We note how the error bars grow by using less seismic data.

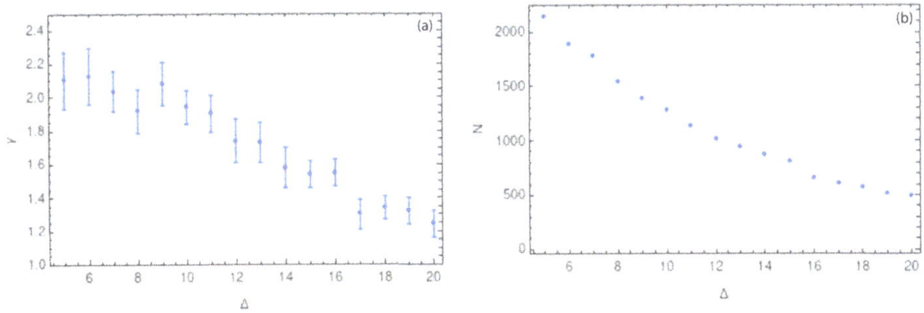


**Figure 12.** (**a**) Values of the critical exponent $\gamma$ in 3 dimensions for $\Delta$ between 5 and 20 km (**b**) Values of the number of nodes $N$ for each value of $\Delta$ between 5 and 20 km.

### 5.2. Complete Dataset

The analyses must then be complemented using the seismic dataset with the magnitude of completeness $M_c = 1.3$, which was computed in Section 3 (Figure 4). Figure 13 shows the scale-free behavior for the complex networks built with $M_c$.

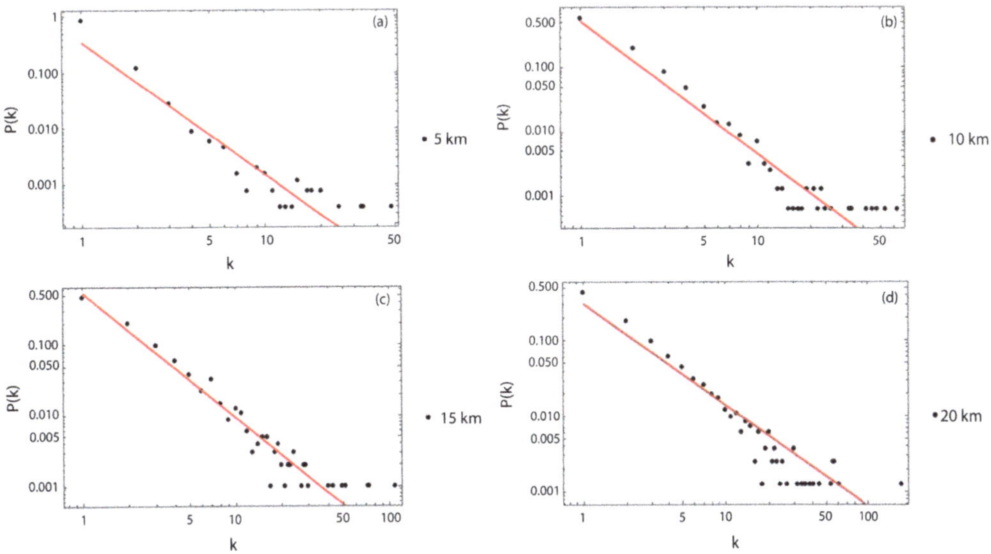


**Figure 13.** Examples of the scale-free behavior of the degree distribution of nodes with the respective linear fit for the dataset with the completeness magnitude. (**a**) Degree distribution for $\Delta = 5$ km. (**b**) Degree distribution for $\Delta = 10$ km. (**c**) Degree distribution for $\Delta = 15$ km. (**d**) Degree distribution for $\Delta = 20$ km.

Figure 14a shows the behavior of $\gamma$ vs. different cubic cell sizes. It is possible to observe the same behavior as for the previous results: the value of $\gamma$ decreases with growing cubic cells.

Table 1 shows an increase in the value of $\gamma$ for all the values of $\Delta$ in the case in which the completeness magnitude was considered, with respect to the previous results. Values of $\gamma$ are close to 2.0 for all the previous analyses and between 2.2 to 2.4 when $M_c$ is added.

In this case, the error bars are larger for small values of Δ, between 5 to 8 km. However, the general behavior of the complex network remains: a structure with hubs for small values of Δ (few nodes with large degree) and a more homogeneous degree distribution for larger values of Δ.

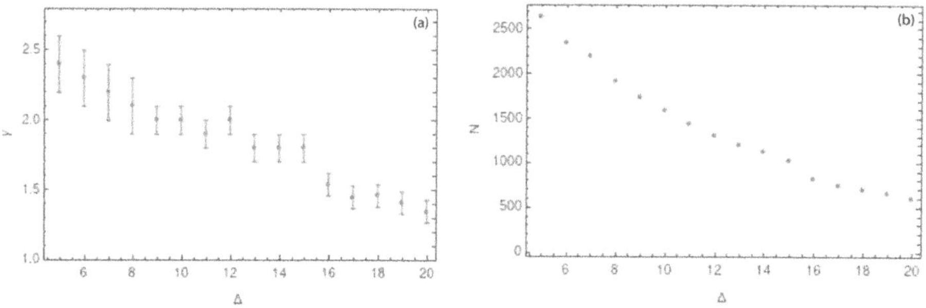

**Figure 14.** (**a**) Values of the critical exponent $\gamma$ in 3 dimensions for Δ between 5 and 20 km (**b**) Values of the number of nodes $N$ for each value of Δ between 5 and 20 km. Both figures use the $M_C \geq 1.3$.

**Table 1.** Values of the critical exponent $\gamma$ in 3D, 2D, for half of the seismic events and for the $M_c$, respectively.

| Δ | $\gamma$ 3D | $\gamma$ 2D | $\gamma$ Half Data | $M_C$ |
|---|---|---|---|---|
| 5 km | 2.1 ± 0.1 | 2.2 ± 0.1 | 2.1 ± 0.1 | 2.4 ± 0.2 |
| 6 km | 2.1 ± 0.1 | 2.0 ± 0.1 | 2.1 ± 0.1 | 2.3 ± 0.2 |
| 7 km | 2.1 ± 0.1 | 1.83 ± 0.09 | 2.0 ± 0.1 | 2.2 ± 0.2 |
| 8 km | 1.9 ± 0.1 | 1.74 ± 0.09 | 1.9 ± 0.1 | 2.1 ± 0.2 |
| 9 km | 1.7 ± 0.1 | 1.73 ± 0.09 | 2.1 ± 0.1 | 2.0 ± 0.1 |
| 10 km | 1.7 ± 0.1 | 1.6 ± 0.1 | 1.9 ± 0.1 | 2.0 ± 0.1 |
| 11 km | 1.7 ± 0.09 | 1.6 ± 0.1 | 1.9 ± 0.1 | 1.9 ± 0.1 |
| 12 km | 1.6 ± 0.1 | 1.52 ± 0.07 | 1.7 ± 0.1 | 2.0 ± 0.1 |
| 13 km | 1.6 ± 0.1 | 1.47 ± 0.07 | 1.7 ± 0.1 | 1.8 ± 0.1 |
| 14 km | 1.56 ± 0.08 | 1.43 ± 0.08 | 1.6 ± 0.1 | 1.8 ± 0.1 |
| 15 km | 1.45 ± 0.08 | 1.38 ± 0.06 | 1.54 ± 0.08 | 1.8 ± 0.1 |
| 16 km | 1.46 ± 0.06 | 1.38 ± 0.06 | 1.55 ± 0.08 | 1.54 ± 0.08 |
| 17 km | 1.36 ± 0.07 | 1.35 ± 0.07 | 1.30 ± 0.09 | 1.45 ± 0.08 |
| 18 km | 1.37 ± 0.06 | 1.27 ± 0.07 | 1.34 ± 0.07 | 1.46 ± 0.08 |
| 19 km | 1.29 ± 0.06 | 1.26 ± 0.06 | 1.32 ± 0.08 | 1.41 ± 0.08 |
| 20 km | 1.24 ± 0.05 | 1.17 ± 0.06 | 1.24 ± 0.08 | 1.35 ± 0.08 |

### 5.2.1. Southwestern Norway

The intraplate seismicity in Norway is varied in terms of causes (Figure 1). The seismicity patterns, if any, are still unknown, and the seismic events seem to behave in different ways according to the region, with large influence of regional variation in geological structures and stress fields [4,17]. For example, it has been established that seismicity in the southwestern coast is related to a high degree of weakness in the area, which strongly depends on the high degree of observed fracturing [17]. To determine if these behaviors are reflected in the complex network parameters, we analyzed 3111 seismic events located in the western zone (longitude 3.0° to 7.5° E). The results are shown in Figure 15.

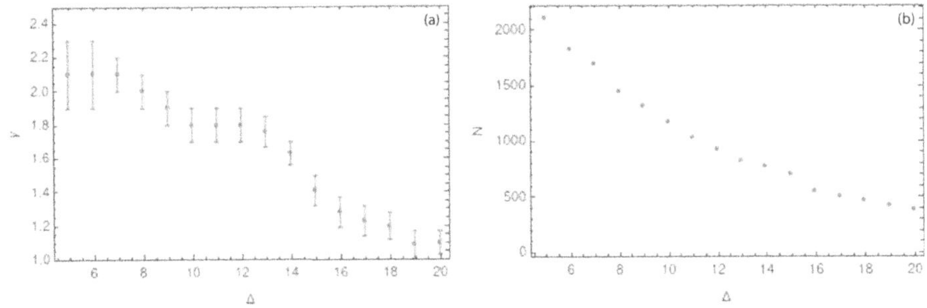


**Figure 15.** (**a**) Southwestern Norway: Values of the critical exponent $\gamma$ in 3 dimensions for $\Delta$ between 5 and 20 km (**b**) Values of the number of nodes $N$ for each value of $\Delta$ between 5 and 20 km. Both figures use the $M_c$.

### 5.2.2. Southeastern Norway

Seismicity in southeastern Norway is predominantly linked to the Oslo region. This and adjacent areas were exposed to stretching and rifting between 359 and 252 Ma ago (Late Carboniferous–Early Permian). The rifting process implied high levels of magmatism, volcanism, and seismic activity. Traces of this intense activity include, for example, the main bodies of igneous rocks, which can be found inside the Oslo Graben [59]. The rifting process, which stopped 65 Ma ago (Cretaceous), left behind several tectonic episodes. Some of these are related to the emplacement of large intrusive bodies, which created a set of extensional structures like normal faults and grabens [59]. Nowadays, the seismic activity in the area can be partially linked to these faults [16].

The Oslo region subset has 628 seismic events after filtering (longitude 7.5° to 12.0° E). In this case, the scale-free behavior of the degree distribution can be analyzed only for side sizes starting from 7 km due to the small number of degrees for the 5 and 6 km cases. Figure 16a show the values of $\gamma$.

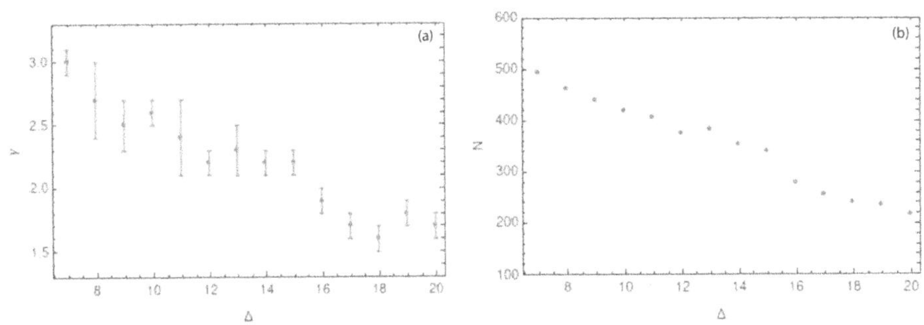


**Figure 16.** (**a**) Southeastern Norway: Values of the critical exponent $\gamma$ in 3 dimensions for $\Delta$ between 5 and 20 km (**b**) Values of the number of nodes $N$ for each value of $\Delta$ between 5 and 20 km. Both figures use the $M_c$.

If we compare both areas, it is possible to observe differences between these two complex networks (Table 2). In the case of southwestern Norway, the behavior is the same as that we have found before: a structure of hubs. However, for the case of southeastern Norway, we found larger values of $\gamma$, and the slope did not decrease as fast as in the case of the western region. In this sense, the southeastern region clearly shows a different complex network structure.

Table 2 shows the $\gamma$ values for the southwestern and the southeastern sub-catalogs in the first and second columns, respectively. It is easy to note that the values of the southwestern zone are smaller than the values for the southeastern zone. However, this might be because the quantity of data used to make the complex network analysis was smaller for the southeastern zone. To determine if the larger values of the critical exponent are influenced by the quantity of data, we repeated the same analysis for a small temporally sorted subset of data for the southwestern region. The results are shown in the third column of Table 2. The value of $\gamma$ for the 628 seismic events of the southwestern zone is very similar to the results for the total of 3111 seismic events in the same zone, but with larger error bars. This result suggests that the larger values of $\gamma$ obtained in the southeastern zone are due to the seismic environment of this zone and not due to the small quantity of data considered for this analysis. In fact, this result shows that the southeastern region has a smaller range of degrees than the southeastern region, which could suggest the presence of a larger number of hubs in the Oslo area.

**Table 2.** Values of the critical exponent $\gamma$ for southwestern Norway, southeastern Norway and for 628 seismic events in southwestern Norway.

| $\Delta$ | $\gamma$ **SW Norway** | $\gamma$ **SE Norway (Oslo)** | $\gamma$ **Subset SW Norway** |
|---|---|---|---|
| 5 km | $2.1 \pm 0.1$ | – | — |
| 6 km | $2.1 \pm 0.1$ | – | — |
| 7 km | $2.1 \pm 0.1$ | $3.0 \pm 0.1$ | $1.9 \pm 0.4$ |
| 8 km | $2.0 \pm 0.1$ | $2.7 \pm 0.2$ | $1.8 \pm 0.4$ |
| 9 km | $1.9 \pm 0.1$ | $2.5 \pm 0.2$ | $1.9 \pm 0.3$ |
| 10 km | $1.7 \pm 0.1$ | $2.6 \pm 0.2$ | $1.9 \pm 0.3$ |
| 11 km | $1.8 \pm 0.1$ | $2.4 \pm 0.2$ | $1.7 \pm 0.3$ |
| 12 km | $1.8 \pm 0.1$ | $2.2 \pm 0.2$ | $1.7 \pm 0.3$ |
| 13 km | $1.76 \pm 0.09$ | $2.3 \pm 0.2$ | $1.7 \pm 0.2$ |
| 14 km | $1.63 \pm 0.08$ | $2.2 \pm 0.1$ | $1.7 \pm 0.2$ |
| 15 km | $1.48 \pm 0.09$ | $2.2 \pm 0.1$ | $1.7 \pm 0.2$ |
| 16 km | $1.28 \pm 0.09$ | $1.9 \pm 0.1$ | $1.5 \pm 0.2$ |
| 17 km | $1.23 \pm 0.09$ | $1.7 \pm 0.1$ | $1.5 \pm 0.1$ |
| 18 km | $1.20 \pm 0.08$ | $1.7 \pm 0.1$ | $1.4 \pm 0.2$ |
| 19 km | $1.09 \pm 0.08$ | $1.8 \pm 0.1$ | $1.4 \pm 0.1$ |
| 20 km | $1.10 \pm 0.07$ | $1.8 \pm 0.1$ | $1.4 \pm 0.1$ |

## 6. Discussion and Conclusions

We used the method developed by Abe and Suzuki (2006) [39] to analyze the scale used to build nodes in a spatial earthquake complex network. To do so, we placed emphasis on the degree distribution of the nodes in the network, which is a well-known and extensively used measure in complex analyses. For earthquakes, it has been previously shown that the behavior of the degree distribution is a power law in a log-log plot of $P(k)$ versus $k$, where $k$ is the degree [39,43,45]. The slope of this power law is the critical exponent $\gamma$. In this study, our focal point is the analysis of $\gamma$, examining the reliability of the cell size that is used to define the nodes in this spatial complex network.

The analysis was performed using the hypocenters of seismic events for prefiltered and complete datasets, which, in this context, means without and with the magnitude of completeness, $M_c = 1.3$. Figure 8a show how the value of the critical exponent depends on the side size of the cubic cell, which is called $\Delta$. In this figure, error bars decrease with increasing $\Delta$, a fact that could suggest a better adjustment for larger values of $\Delta$. Figure 8b shows a decreasing number of nodes for larger $\Delta$ values.

In order to analyze the influence of the observational errors, Figure 9 shows the average error for each spatial coordinate: longitude, latitude, and depth. In groups (a) and (b) in Figure 9, the errors in longitude and latitude are negligible with respect to the change in the hypocentral longitude and latitude values in the studied area. However, the average error in depth is larger and considerable when compared to the average depth hypocentral values. To explore the influence of the error in depth on the obtained results, we also consider the complex network in two dimensions, i.e., (latitude, longitude). Figure 10a,b show the values of $\gamma$ and the number of nodes for values of $\Delta$ between 5 and 20 km, respectively. Here, we can observe the same behavior that we found for the 3-dimensional case: the error bars decrease as $\Delta$ grows. However, for the largest values of $\Delta$, the network structure is lost due to the homogenization of the degree distribution. Then, the best range of $\Delta$ is defined to be between 8 and 16 km.

Figure 11 shows the values of the critical exponent $\gamma$ for 2D and 3D. It is interesting to note how the $\gamma$ values do not seem to be affected by the change in dimension. This result is not conclusive, so we performed new analyses using a temporal subset of the data to verify if the quantity of data could affect the $\gamma$ value. Figure 12a shows the behavior of $\gamma$ for half of the seismic events, showing very similar results to those obtained before. Table 1 summarizes these three results, showing how the values of $\gamma$ are very similar between them for all the cases, presenting great stability under a change of dimension or the number of events considered. The values obtained for $\Delta = 5, 10$ and 20 km are close to 2, 1.4, and 1.2, respectively. These results suggest that the topology of the complex network changes with the values of $\Delta$: the complex network built with $\Delta = 5$ km shows a structure with central nodes; meanwhile, with $\Delta = 20$ km, the values of $\gamma$ are associated with a more homogeneous distribution of the degree in the complex network. Although these first results are not conclusive concerning the influence of measurement errors, they clearly show that nodes with smaller delta sizes can provide information about the complex network topology.

Since the previous results failed to provide clear information about the physics of intraplate seismic events, we performed a second analysis for the same dataset, but using the magnitude of completeness. Figure 14a and Table 1 summarize these results. In Table 1, it is possible to observe how the value of $\gamma$ increases when the magnitude of completeness is used. For $\Delta$ between 5 and 7 km, the value of $\gamma$ is greater than 2.0, and $\Delta$ between 9 and 15 km shows values of $\gamma$ close to 2.0. For $\Delta$ between 16 and 20 km, the values of $\gamma$ decrease and fluctuate between 1.35 and 1.54, showing again a more homogeneous distribution of the node degree. These results show the different behaviors of the complex network when the magnitude of completeness is used: the filtered low-magnitude events improve the complex network connection. When removed, the network has fewer nodes with a higher degree.

We have additionally compared the values obtained for Norway with the values computed for other regions of the planet, with different cubic cell sizes and using the same method. Table 3 summarizes these values. It is possible to observe that the values obtained by Abe and Suzuki [39] are smaller than the values obtained by Pastén et al. [40]. This difference may be because some results published by Abe and Suzuki do not consider the magnitude of completeness. However, the values of $\gamma$ seem to follow the same behavior: larger $\gamma$ values for smaller values of $\Delta$. The values for Japan and Chile show similar behavior. In the Chilean case, the results before the occurrence of a large earthquake are more similar to the results obtained for the case of Norway, but with values 5 km lower than those found for Norwegian seismicity. California and Iran show the greatest difference from the other seismic regions. Although our intention in presenting these results was to initiate discussions in terms of the comparison between the different $\gamma$ values, we finally conclude that there is no universality in the results. This means that some complex network parameters, for example, the one analyzed in this study, are strongly dependent on the seismicity type.

Finally, in order to understand the relevance of these results for the physical processes of intraplate seismicity in southern Norway, we added an extra analysis considering the spatial distribution of earthquakes in terms of longitude. This criterion is not random as the geological structures, the seismicity rates, the stress field, and the fracturing levels vary between the western and the eastern south Norwegian coasts, as explained before. The values of $\gamma$ are greater for the Oslo region than in the zone of southwestern Norway, showing behavior where there are few nodes with a larger degree and more nodes with a small degree, as a tree structure with main hubs. This corresponds to an advance in linking the behavior of these parameters of complex networks with the physics of intraplate seismicity.

**Table 3.** Different results for the $\gamma$ value obtained for different authors at several tectonic settings.

| Place | Δ (km) | $\gamma$ | Ref. |
|---|---|---|---|
| California | 5 | 1.61 | Abe and Suzuki 2006 [39] |
| | 10 | 1.33 | Abe and Suzuki 2006 [39] |
| | 20 | 1.28 | Abe et al. 2011 [43] |
| Japan | 5 | 2.5 | Abe and Suzuki 2006 [39] |
| | 10 | 2.22 | Abe and Suzuki 2006 [39] |
| | 20 | 1.40 | Abe et al. 2011 [43] |
| Iran | 20 | 2.01 | Abe et al. 2011 [43] |
| Chile | 20 | 1.35 | Abe et al. 2011 [43] |
| Chile (Illapel earthquake) | 5 | 3.0 (before earthquake) | Pastén et al. 2016 [40] |
| Chile (Illapel earthquake) | 10 | 2.2 (before earthquake) | Pastén et al. 2016 [40] |
| Chile (Illapel earthquake) | 5 | 3.6 (after earthquake) | Pastén et al. 2016 [40] |
| Chile (Illapel earthquake) | 10 | 2.2 (after earthquake) | Pastén et al. 2016 [40] |

This is the first complex network-related study that has been carried out with seismic data recorded in Norway, so this research constitutes the first step to associate measurements of complex networks with the underlying physics involved in the occurrence of earthquakes, especially in this area of the planet and for this intraplate seismicity environment. All the constructed networks show stable behavior, which is replicated for the prefiltered catalog, the complete catalog, and the subsets. Even though we could characterize the complex network with all the above-mentioned variations, we can conclude that it is not easy to determine the best range of delta values. As a tentative conclusion, it seems better to analyze the complex network using 5, 10, and 15 km cell sizes, as larger values homogenize the network. Now, we will expand this study by calculating the clustering coefficient, different measures of centrality, and their associated critical exponents.

**Author Contributions:** Conceptualization, C.P.-O. and D.P.; methodology, C.P.-O. and D.P.; software, D.P.; validation, C.P.-O. and D.P.; formal analysis, C.P.-O. and D.P.; investigation, C.P.-O. and D.P.; resources, C.P.-O.; data curation, C.P.-O.; writing—original draft preparation, C.P.-O. and D.P.; writing—review and editing, C.P.-O. and D.P.; visualization, C.P.-O. and D.P.; project administration, C.P.-O.; funding acquisition, C.P.-O. All authors have read and agreed to the published version of the manuscript.

**Funding:** This research and the corresponding APC have been funded by the SINTEF strategic project SEP 2023, through the Norwegian Research Council (NFR—Norges Forskningsråd): Ambient Noise Tomography and Passive Seismicity for $CO_2$ Imaging and Monitoring.

**Data Availability Statement:** This publication used open and public data, available at URL http://www.epos-no.org/ (accesed on 12 January 2023). For more details please follow the instructions detailed on Appendix A.

**Acknowledgments:** The authors thank the EPOS-N project for the open-source seismicity data available in the data portal http://www.epos-no.org/ (accesed on 12 January 2023). This work has been funded by the SINTEF strategic project SEP 2023, through the Norwegian Research Council

(NFR - Norges Forskningsråd): Ambient Noise Tomography and Passive Seismicity for $CO_2$ Imaging and Monitoring.

**Conflicts of Interest:** The authors declare no conflict of interest.

## Appendix A

The instructions to download the original catalogue as further processed to be used in this study are presented here.

After entering the EPOS-N portal, the SEARCH/FIND option must be selected (Figure A1).

**Figure A1.** Capture from https://epos$-$no.uib.no/eposn$-$data$-$portal/(accesed on 12 January 2023).

Then, a list of around 40 different data services will appear on the right-hand side of the screen. The option to download the NNSN seismological catalog is located at the end of the list (Figure A2). By pressing the ADD button (Figure A2), the data will be added to the project (Figure A3).

List of available services Num items: 40

| | | Category | Element Name | Service type |
|---|---|---|---|---|
| ADD | INFO | Geophysical data | Heat flow map | CSV file |
| LINK | INFO | Geophysical data | InSAR data | External Link |
| LINK | INFO | Geophysical data | Local geophysical and geochem | External Link |
| ADD | INFO | Geophysical data | NE Atlantic and Fennoscandia m | CSV file |
| LINK | INFO | Geophysical data | NORSAR-2D Model Builder | External Link |
| LINK | INFO | Geophysical data | World Stress Map | External Link |
| ADD... | INFO | Seismological data | ARCTIC earthquake parameters | Earthquake list |
| ADD | INFO | Seismological data | Explosion events ARCTIC 1960- | Earthquake list |
| ADD | INFO | Seismological data | Explosion events NNSN 1980-2( | Earthquake list |
| LINK | INFO | Seismological data | Interactive map of receiver functi | External Link |
| LINK | INFO | Seismological data | Macroseismic observations from | External Link |
| ADD | INFO | Seismological data | NNSN earthquake parameters | Earthquake list |
| ADD | INFO | Seismological data | NORSAR/UIB stations and array | Earthquake stations |
| LINK | INFO | Seismological data | SEISAN - Seismological Analysi | External Link |
| ADD | INFO | Seismological data | Seismic events ARCTIC 1960-2( | Earthquake list |
| ADD | INFO | Seismological data | Seismic events NNSN 1980-201 | Earthquake list |
| LINK | INFO | Seismological data | Seismological data portal for Noi | External Link |

**Figure A2.** Capture from https://epos$-$no.uib.no/eposn$-$data$-$portal/ (accesed on 12 January 2023).

Using the DISPLAY option, the data can be visualized in several formats. As shown in (Figure A3), the options include map, linear, and histogram plotting, among others. The catalog will also be available in .csv format to be downloaded and processed (Figure A3).

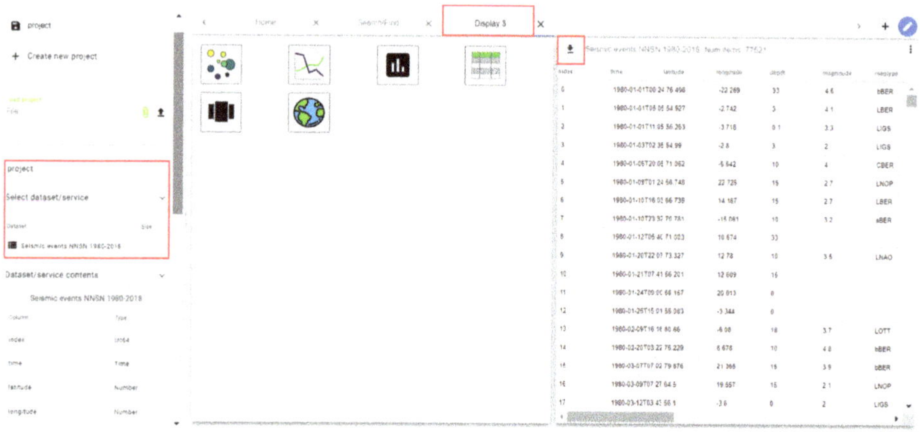


**Figure A3.** Capture from https://epos$-$no.uib.no/eposn$-$data$-$portal/ (accesed on 12 January 2023).

## References


1. Köhler, A.; Weidle, C.; Maupin, V. Directionality analysis and Rayleigh wave tomography of ambient noise in southern Norway. *Geophys. J. Int.* **2011**, *184*, 287–300. [CrossRef]
2. Köhler, A.; Weidle, C.; Maupin, V. Crustal and uppermost mantle structure of southern Norway: Results from surface wave analysis of ambient seismic noise and earthquake data. *Geophys. J. Int.* **2012**, *191*, 1441–1456. [CrossRef]
3. Loidl, B.; Behm, M.; Thybo, H.; Stratford, W. Three-dimensional seismic model of crustal structure in Southern Norway. *Geophys. J. Int.* **2014**, *196*, 1643–1656. [CrossRef]
4. Stratford, W.; Thybo, H. Crustal structure and composition of the Oslo graben, Norway. *Earth Planet. Sci. Lett.* **2011**, *304*, 431–442. [CrossRef]
5. Stratford, W.; Thybo, H. Seismic structure and composition of the crust beneath the southern Scandes, Norway. *Tectonophysics* **2011**, *502*, 364–382. [CrossRef]
6. Kolstrup, M.; Maupin, V. A Proterozoic boundary in Southern Norway revealed by joint-inversion of P-receiver functions and surface waves. *Precambrian Res.* **2013**, *238*, 186–198. [CrossRef]
7. Ottemöller, L.; Midzi, V. The crustal structure of Norway from inversion of teleseismic receiver functions. *J. Seismol.* **2003**, *7*, 35–48. [CrossRef]
8. Olesen, O.; Brönner, M.; Ebbing, J.; Gellein, J.; Gernigon, L.; Koziel, J.; Lauritsen, T.; Myklebust, R.; Sand, M.; Solheim, D.; et al. New aeromagnetic and gravity compilations from Norway and adjacent areas—Methods and applications. *Pet. Geol. Conf. Ser.* **2010**, *7*, 559–586. [CrossRef]
9. Olesen, O.; Ebbing, J.; Gellein, J.; Kihle, O.; Myklebust, R.; Sand, M.; Skilbrei, J.; Solheim, D.; Usov, S. Gravity Anomaly Map, Norway and Adjacent Areas. Geol. Surv. Nor. Map Publ. DOCID: 63478. 2010. Available online: https://openarchive.ngu.no/ngu-xmlui/bitstream/handle/11250/2661922/Mag_3_mill.pdf?sequence=1&isAllowed=y (accessed on 12 January 2023).
10. Olesen, O.; Gellein, J.; Gernigon, L.; Kihle, O.; Koziel, J.; Lauritsen, T.; Mogaard, J.; Myklebust, R.; Skilbrei, J.; Usov, S. Magnetic Anomaly Map, Norway and Adjacent Areas.Geol. Surv. Nor. Map Publ. DOCID: 63477. 2010. Available online: https://openarchive.ngu.no/ngu-xmlui/handle/11250/2661922?locale-attribute=en (accessed on 12 January 2023)
11. Grad, M.; Tiira, T; ESC Working Group. The Moho depth map of the European plate. *Geophys. J. Int.* **2009**, *176*, 279–292. [CrossRef]
12. Stratford, W.; Thybo, H.; Faleide, J.; Olesen, O.; Tryggvason, S. New Moho map for onshore southern Norway. *Geophy. J. Int.* **2009**, *178*, 1755–1765. [CrossRef]
13. Shiddiqi, H.A.; Ottemöller, L.; Rondenay, S.; Halpaap, F.; Gradmann, S.; Michálek, J. Crustal structure and intraplate seismicity in Nordland, Northern Norway: Insight from seismic tomography. *Geophy. J. Int.* **2022**, *230*, 813–830. [CrossRef]
14. Pavez, C.; Estay, R.; Brönner, M.; Ortiz, A.; Debarbieri, F.; Ibañez, J.; Guzmán, L. Frictional energy patterns related to the temperature increases due to intraplate seismicity, southern Norway, 2000–2019 catalogue. *Nor. J. Geol.* **2021**, *101*, 202105. [CrossRef]

15. Wahlström, R.; Grünthal, G. Probabilistic seismic hazard assessment (horizontal PGA) for Fennoscandia using the logic tree approach for regionalization and nonregionalization models. *Seismol. Res. Lett.* **2001**, *72*, 33–45. [CrossRef]
16. Johnsten, M. Seismic Hazard in Norway Due to Large Earthquakes. Master's Thesis, University of Bergen, Bergen, Norway, 2015; 133p.
17. Olesen, O.; Bungum, H.; Dehls, J.; Londholm, C.; Pascal, C.; Roberts, D. Neotectonics, seismicity and contemporary stress field in Norway, mechanisms and implications. In *Quaternary Geology of Norway, Geological Survey of Norway Special Publication, 13*; Olsen, L., Fredin, O., Olesen, O., Eds.; Geological Survey of Norway: Trondheim, Norway, 2013; Volume 13, pp. 145–174.
18. Keiding, M.; Kreemer, C.; Lindholm, C.; Gradmann, S.; Olesen, O.; Kierulf, H. A comparison of strain rates and seismicity for Fennoscandia: Depth dependency of deformation from glacial isostatic adjustment. *Geophys. J. Int.* **2015**, *202*, 1021–1028. [CrossRef]
19. Sørensen, M.; Ottemöller, L.; Havskov, J.; Atakan, K.; Hellevang, B.; Pedersen, R. Tectonic processes in the Jan Mayen Fracture Zone based on earthquake occurrence and bathymetry. *Bull. Seismol. Soc. Am.* **2007**, *97*, 772–779. [CrossRef]
20. Pastén, D.; Pavez-Orrego, C. Multifractal time evolution for intraplate earthquakes recorded in southern Norway during 1980–2021. *Chaos Solitons Fractals* **2023**, *167*, 113000. [CrossRef]
21. Paul, E.; Rényi, A. On the evolution of random graphs. *Publ. Math. Inst. Hung. Acad. Sci* **1960**, *5*, 17–60.
22. Watts, D.J.; Strogatz, S.H. Collective dynamics of 'small-world' networks. *Nature* **1998**, *393*, 440–442. [CrossRef]
23. Barabási, A.L.; Albert, R. Emergence of scaling in random networks. *Science* **1999**, *286*, 509–512. [CrossRef]
24. Thiery, J.; Sleeman, J. Complex networks orchestrate epithelial–mesenchymal transitions. *Nat. Rev. Mol. Cell. Biol.* **2006**, *7*, 131–142. [CrossRef]
25. Barabási, A.; Gulbahce, N.; Loscalzo, J. Network medicine: A network-based approach to human disease. *Nat. Rev. Genet.* **2011**, *12*, 56–68. [CrossRef] [PubMed]
26. Jeong, H.; Tombor, B.; Albert, R.; Oltvai, Z.; Barabási, A. The large-scale organization of metabolic network. *Nature* **2000**, *407*, 651–654. [CrossRef] [PubMed]
27. Moghadam, H.; Mohammadi, T.; Kashani, M.; Shakeri, A. Complex networks analysis in Iran stock market: The application of centrality. *Phys. A* **2019**, *531*, 121800. [CrossRef]
28. Yang, X.; Wu, Z.; Javaid, S. An Adaptive Hierarchical Network Model for Studying the Structure of Economic Network. *Entropy* **2022**, *24*, 702. [CrossRef]
29. Scabini, L.; Ribas, L.; Neiva, M.; Junior, A.; Farfán, A.; Bruno, O. Social interaction layers in complex networks for the dynamical epidemic modeling of COVID-19 in Brazil. *Phys. A* **2021**, *564*, 125498. [CrossRef]
30. Kertész, J.; Török, J.; Murase, Y.; Jo, H.H.; Kaski, K. Modeling the Complex Network of Social Interactions. In *Pathways Between Social Science and Computational Social Science*; Rudas, T., Péli, G., Eds.; Computational Social Sciences; Springer: Cham, Switzerland, 2021. [CrossRef]
31. Tenenbaum, J.; Havlin, S.; Stanley, H. Earthquake networks based on similar activity patterns. *Phys. Rev. E* **2012**, *86*, 046107. [CrossRef]
32. Varotsos, P.A.; Sarlis, N.V.; Skordas, E.S. Study of the temporal correlations in the magnitude time series before major earthquakes in Japan. *J. Geophys. Res.* **2014**, *119*, 9192–9206. [CrossRef]
33. Chorozoglou, D.; Kugiumtzis, D.; Papadimitriou, E. Testing the structure of earthquake networks from multivariate time series of successive main shocks in Greece. *Phys. A* **2018**, *499*, 28–39. [CrossRef]
34. Mintzelas, A.; Sarlis, N. Minima of the fluctuations of the order parameter of seismicity and earthquake networks based on similar activity patterns. *Phys. A* **2019**, *527*, 121293. [CrossRef]
35. Varotsos, P.; Perez-Oregon, J.; Skordas, E.; Sarlis, N. Estimating the Epicenter of an Impending Strong Earthquake by Combining the Seismicity Order Parameter Variability Analysis with Earthquake Networks and Nowcasting: Application in the Eastern Mediterranean. *Appl. Sci.* **2021**, *11*, 10093. [CrossRef]
36. Suyal, V.; Prasad, A.; Singh, H. Visibility-graph analysis of the solar wind velocity. *Sol. Phys.* **2014**, *289*, 379–389. [CrossRef]
37. Muñoz, V.; Garcés, N.E. Analysis of pulsating variable stars using the visibility graph algorithm. *PLoS ONE* **2021**, *16*, e0259735. [CrossRef]
38. Acosta-Tripailao, B.; Max-Moerbeck, W.; Pastén, D.; Moya, P. Assigning Degrees of Stochasticity to Blazar Light Curves in the Radio Band Using Complex Networks. *Entropy* **2022**, *24*, 1063. [CrossRef]
39. Abe, S.; Suzuki, N. Complex-network description of seismicity. *Nonlinear Proc. Geophys.* **2006**, **13**, 145–150. [CrossRef]
40. Pastén, D.; Torres, F.; Toledo, B.; Noz, V.M.; Rogan, J.; Valdivia, J.A. Time-Based Network Analysis Before and After the M w 8.3 Illapel Earthquake 2015 Chile. *Pure Appl. Geophys.* **2016**, *173*, 2267–2275. [CrossRef]
41. Martín, F.; Pastén, D. Complex Networks and the b-Value Relationship Using the Degree Probability Distribution: The Case of Three Mega-Earthquakes in Chile in the Last Decade. *Entropy* **2022**, *24*, 337. [CrossRef] [PubMed]
42. Abe, S.; Suzuki, N. Scale-free network of earthquakes. *Chin. Sci. Bull.* **2004**, *65*, 581. [CrossRef]
43. Abe, S.; Pastén, D.; Muñoz, V.; Suzuki, N. Universalities of earthquake-network characteristics. *Chin. Sci. Bull.* **2011**, *56*, 34. [CrossRef]
44. Pastén, D.; Czechowski, Z.; Toledo, B. Time series analysis in earthquake complex networks. *Chaos Interdiscip. J. Nonlinear Sci.* **2018**, *28*, 083128. [CrossRef] [PubMed]

45. Telesca, L.; Lovallo, M. Analysis of seismic sequences by using the method of visibility graph. *Europhys. Lett.* **2012**, *97*, 50002. [CrossRef]
46. Aguilar-San Juan, B.; Guzmán-Vargas, L. Earthquake magnitude time series: Scaling behavior of visibility networks. *Eur. Phys. J. B* **2013**, *86*, 454. [CrossRef]
47. Telesca, L.; Pastén, D.; Muñoz, V. Analysis of Time Dynamical Features in Intraplate Versus Interplate Seismicity: The Case Study of Iquique Area (Chile). *Pure Appl. Geophys.* **2020**, *177*, 4755–4773. [CrossRef]
48. Department of Earth Science. *Annual Report for the Norwegian National Seismic Network, Technical Report*; Institute of Solid Earth Physics, University of Bergen and Norwegian Oil and Gas Association: Bergen, Norway, 2018; Volume 1, pp. 1–62.
49. Bungum, H.; Lindholm, C.D.; Dahle, A.; Woo, G.; Nadim, F.; Holme, J.K.; Gudmestad, O.; Hagberg, T.; Karthigeyan, K. New seismic zoning maps for Norway, the North Sea, and the United Kingdom. *Seismol. Res. Lett.* **2000**, *71*, 687–697. [CrossRef]
50. Bannister, S.; Ruud, B.; Husebye, E. Tomographic estimates of sub-Moho seismic velocities in Fennoscandia and structural implications. *Tectonophysics* **1991**, *189*, 37–53. [CrossRef]
51. Kolstrup, M.; Hung, S.; Maupin, V. Multiscale finite frequency P and S tomography of the upper mantle in the south-western Fennoscandian Shield. *Geophys. J. Int.* **2015**, *202*, 190–218. [CrossRef]
52. Hejrani, B.; Balling, N.; Jacobsen, B.; England, R. Upper-mantle velocities below the Scandinavian Mountains from P- and S-wave traveltime tomography. *Geophys. J. Int.* **2017**, *208*, 177–192. [CrossRef]
53. Bøe, R.; Fossen, H.; Smelror, M. Mesozoic sediments and structures onshore Norway and in the coastal zone. *NGU Bull.* **2010**, *450*, 15–32.
54. Fjeldskaar, W.; Lindholm, C.; Dehls, J.; Fjeldskaar, I. Postglacial uplift, neotectonics and seismicity in Fennoscandia. *Quat. Sci. Rev.* **2000**, *19*, 1413–1422. [CrossRef]
55. Department of Earth Science. *Annual Report for the Norwegian National Seismic Network, Technical Report*; Institute of Solid Earth Physics, University of Bergen and Norwegian Oil and Gas Association: Bergen, Norway, 2019; Volume 1, pp. 1–58.
56. Varotsos, P.A.; Sarlis, N.V.; Skordas, E.S. Long-range correlations in the electric signals that precede rupture. *Phys. Rev. E* **2002**, *66*, 011902. [CrossRef]
57. Varotsos, P.; Sarlis, N.; Skordas, E. Seismic electric signals and seismicity: On a tentative interrelation between their spectral. *Acta Geophys. Pol.* **2002**, *50*, 337–354.
58. Varotsos, P.A.; Sarlis, N.V.; Tanaka, H.K.; Skordas, E.S. Similarity of fluctuations in correlated systems: The case of seismicity. *Phys. Rev. E* **2005**, *72*, 041103. [CrossRef] [PubMed]
59. Ro, H.; Faleide, J. A stretching model for the Oslo Rift. *Tectonophysics* **1992**, *208*, 19–36. [CrossRef]

**Disclaimer/Publisher's Note:** The statements, opinions and data contained in all publications are solely those of the individual author(s) and contributor(s) and not of MDPI and/or the editor(s). MDPI and/or the editor(s) disclaim responsibility for any injury to people or property resulting from any ideas, methods, instructions or products referred to in the content.

*entropy*

*Article*

# Solar Flare 1/f Fluctuations from Amplitude-Modulated Five-Minute Oscillation

**Masahiro Morikawa [1,*] and Akika Nakamichi [2]**

1 Department of Physics, Ochanomizu University, 2-1-1 Otsuka, Bunkyo, Tokyo 112-8610, Japan
2 General Education, Kyoto Sangyo University, Motoyama Kamigamo Kita-ku, Kyoto 603-8555, Japan; nakamichi@cc.kyoto-su.ac.jp
* Correspondence: hiro@phys.ocha.ac.jp

**Abstract:** We first report that the solar flare time sequence exhibits a fluctuation characterized by its power spectral density being inversely proportional to the signal frequency. This is the 1/f fluctuation, or pink noise, observed ubiquitously in nature. Using GOES16 data, we found that low-energy flares ($E \leq E_{mean}$) display 1/f fluctuations, whereas high-energy flares ($E > E_{mean}$) show a flat spectrum. Furthermore, we found that the timing sequence of the flares reveals clearer 1/f fluctuations. These observations suggest that the solar flare 1/f fluctuations are associated with low-energy phenomena. We investigated the origin of these 1/f fluctuations based on our recent hypothesis: 1/f fluctuations arise from amplitude modulation and demodulation. We propose that this amplitude modulation is encoded by the resonance with the solar five-minute oscillation (SFO) and demodulated by magnetic reconnections. We partially demonstrate this scenario by analyzing the SFO eigenmodes resolving the frequency degeneration in the azimuthal order number $m$ using the solar rotation and resonance. Given the robust nature of 1/f fluctuations, we speculated that the solar flare 1/f fluctuations may be inherited by the various phenomena around the Sun, such as the sunspot numbers and cosmic rays. In addition, we draw parallels between solar flares and earthquakes, both exhibiting 1/f fluctuations. Interestingly, the analysis applied to solar flares can also be adapted to earthquakes if we read the SFO as Earth's free oscillation and magnetic reconnections as fault ruptures. Moreover, we point out the possibility that the same analysis also applies to the activity of a black hole/disk system if we read the SFO as the quasi-periodic oscillation of a black hole.

**Keywords:** 1/f fluctuations; solar flare; solar five-minute oscillation; resonance; amplitude modulation

**Citation:** Morikawa, M.; Nakamichi, A. Solar Flare 1/f Fluctuations from Amplitude-Modulated Five-Minute Oscillation. *Entropy* **2023**, *25*, 1593. https://doi.org/10.3390/e25121593

Academic Editor: Georgios Michas

Received: 9 October 2023
Revised: 15 November 2023
Accepted: 18 November 2023
Published: 28 November 2023

**Copyright:** © 2023 by the authors. Licensee MDPI, Basel, Switzerland. This article is an open access article distributed under the terms and conditions of the Creative Commons Attribution (CC BY) license (https://creativecommons.org/licenses/by/4.0/).

## 1. Introduction

A solar flare is a sudden energy eruption in the solar atmosphere [1]. It is triggered by magnetic reconnections, and the enormous magnetic energy of $10^{17}$–$10^{26}$ J is converted into plasma particle acceleration, heating, and light emission. Solar flares are complex phenomena, and the statistical approach to predicting them is effective, as in the case of earthquakes, which are sudden energy eruptions of the Earth's crust.

It is well known that solar flares and earthquakes are similar to each other, and they show similar statistical properties. In particular, the scaling relation laws, such as the Gutenberg–Richter law [2] and the Omori law [3], are universal laws for both solar flares and earthquakes [4,5].

Here, in this study, we concentrated on solar flares and aimed to add one more universal scaling law in the ultra-low-frequency region of the power spectral density (PSD) for the long time sequence of solar flares. It was revealed that the solar flare time sequence shows a power law that is almost inversely proportional to the frequency in the PSD. This is often called the 1/f fluctuation or pink noise and appears in most fields of nature and human activities [6,7]. However, the origin of this fluctuation has not been clarified despite a large number of studies performed in the past century.

We recently proposed that the general origin of the 1/f fluctuation is amplitude modulation (AM), or the beat of many waves with accumulating frequencies [8]. In particular, this frequency accumulation is possible in resonance, where many eigenfrequencies are systematically concentrated in a narrow domain.

Applying this method in [9], we studied the 1/f fluctuations in seismic activity. The seismic-energy time sequence shows an apparent 1/f fluctuation in its PSD by more than three digits if giant earthquakes are excluded. Therefore, seismic 1/f fluctuations are considered to be associated with low-energy phenomena. In this case, perpetually exciting Earth's free oscillation (EFO) in the lithosphere results in resonance, which yields an AM or wave beats. A relatively low-energy EFO will sufficiently trigger the fault eruption and cause earthquakes.

In this study, applying this proposal to solar flares, we aimed to verify our proposal and determine the statistical properties of complex systems in general. When we naively used all the energy time series data, we obtained almost a flat PSD at low-frequency regions and discerned no clear pink noise. However, if we restricted the solar flare events to having their energy below the mean, we obtained clear pink noise. Therefore, we speculated that the 1/f fluctuation in solar flares is associated with low-energy phenomena. Interestingly, this situation is the same as for seismic activity, as explained above.

The resonance in the solar case would be characterized by solar five-minute oscillations (SFOs), which are perpetually excited by the turbulence in the solar convective region [10,11]. The eigenfrequencies have been precisely measured and calculated by many studies assuming appropriate solar models. Using these observational data, we constructed the superposition of many waves with these eigenoscillations, including the fine splitting structure using the solar rotation and the resonance effects. Then, we could obtain 1/f fluctuations from the thresholding of these data in the PSD. Thus, we can partially demonstrate the AM theory for 1/f fluctuations in solar flares.

Since pink noises often appear in various fields of science, we explored the neighboring phenomena to the solar flares. Then, we found 1/f fluctuations in several phenomena such as solar winds, sun spot numbers, and some terrestrial traces. These facts may indicate the robustness of pink noises.

This paper is organized as follows. In Section 2, we explore the GOES data and RHESSI data of solar flares and analyze the PSD. In Section 3, we explain the AM proposal for the origin of 1/f fluctuations from resonance. In Section 4, we superpose the eigenmodes of the SFO and obtain pink noise. In Section 5, we study another statistical characterization of solar flares using the Weibull distribution and compare it with 1/f characterization. In Section 6, we emphasize the robustness of 1/f fluctuations being the origin of a variety of pink noises. We point out that the 1/f fluctuation property is inherited by solar winds, sunspot events, and some traces on Earth. In Section 7, we conclude our work and briefly describe the possible future research, including the back hole.

## 2. Solar Flare Fluctuations

Solar flares are eruptive energy-release events in the solar atmosphere. Each event transforms enormous magnetic energy into plasma particle acceleration, visible light, X-ray emission, and more. Our focus was specifically directed towards the soft X-ray flux data acquired by the GOES16 satellite from February 2017 to September 2023, covering a span of 6.6 years [12]. We mainly used xrsa1_flux and xrsa2_flux xrsa_flux.

Initially, we used all the time sequence data of the soft X-ray energy flux in units of $W/m^2$, as shown in Figure 1 (left). The corresponding PSD appears almost flat and random in the low-frequency regions, as depicted in Figure 1 (right). However, this is consistent with the previous research [13], in which the authors partially extracted the 1/f fluctuations in the GOES6 data by superposing multiple PSDs. We adopted another approach to extract the entirety of the 1/f fluctuations by restricting the energy flux.

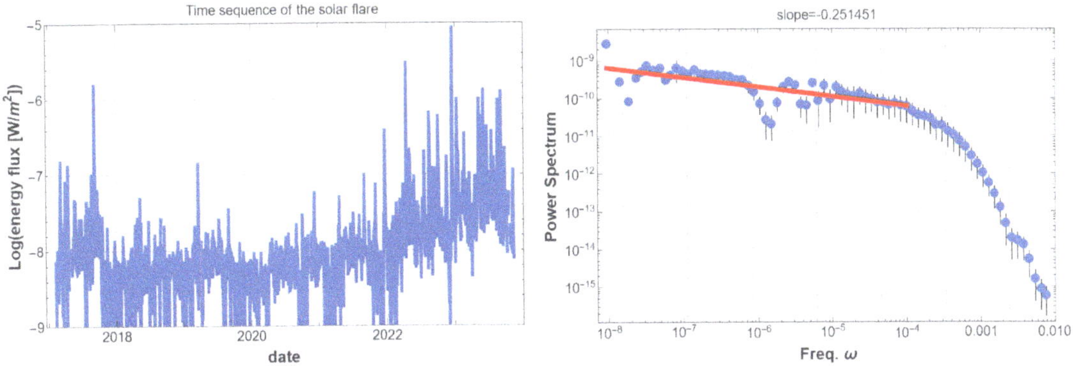


**Figure 1.** (**left**) Goes16 soft X-ray flux data from February 2017 to September 2023 (6.6 years) [12]. The vertical unit is $W/m^2$ in the logarithmic scale. Although the data show an apparent trend associated with the eleven-year solar activity cycle, we did not apply any artificial operation for our analysis, as extraction of the trend does not greatly affect the result. (**right**) Power spectral density (PSD) of the energy-flux time sequence of all the GOES16 solar flare data over 6.6 years. The time is measured in seconds, and the frequency unit is Hz. Since the time interval of the original data is not uniform, we redistributed the energy-flux data after making the time interval equal and assigned zero values for vacant intervals. Then, we applied the fast Fourier transformation to obtain the PSD. Various window functions were tested, but as they did not significantly alter the results, the data without any artificial arrangement were used. The Fourier-transformed data were averaged within the equal bins on a logarithmic scale. A red line in the low-frequency domain in the PSD graph represents the best fits of the data according to the least squares method within that domain. This procedure is the same for all PSD analyses below. This PSD illustrates that the energy time sequence is random, as indicated by the almost flat red line fitting the data points in the low-frequency domain.

Subsequently, we analyzed two datasets: the high-energy group, which include all events with energy exceeding the mean ($4.4 \times 10^{-8}$ $W/m^2$) (i.e., comprising part of class B, and classes C, M, and X), and the low-energy group, comprising events with energy below the mean (class A and part of class B). Figure 2 presents the PSDs for these two groups. The $1/f$ fluctuation is evident in Figure 2 (right), representing the low-energy data. Conversely, the high-energy group's PSD as in Figure 2 (left), does not exhibit $1/f$ fluctuations and resembles the pattern in Figure 1 (right); high-energy data disrupt the $1/f$ fluctuations in the solar flare. These observations suggest that solar flare $1/f$ fluctuations are associated with low-energy phenomena.

To confirm that the solar flare $1/f$ fluctuation is independent of energy, we removed the energy information from the data: all energy values in the time sequence were set to one. The PSD for the entire dataset then displayed a $1/f$ fluctuation with a power index of $-1.1$ as shown in Figure 3 (left). Similarly, the high-energy group's PSD, with energy information removed, exhibits a $1/f$ fluctuation with a power index of $-0.98$ as in Figure 3 (right).

All together, $1/f$ fluctuation with a power index of approximately $-0.9 \sim -1.1$ is observed within about five orders of frequencies, corresponding to timescales from about an hour ($10^{-3}$ Hz) to 6.6 years ($2 \times 10^{-8}$ Hz). This $1/f$ fluctuation appears when high-energy solar flare events are excluded or when energy information is completely removed.

These findings suggest that the solar flare $1/f$ fluctuation does not reflect the energy scaling structure typically caused by the self-organized criticality (SOC) formed by energy cascades from small to large, although SOC may be crucial for explaining popular scaling laws like the Gutenberg–Richter and Omori laws. On the contrary, solar flare $1/f$ fluctuation seems to be a low-energy phenomenon, probably triggered by a tiny energy source. This point will be further discussed in the next section.

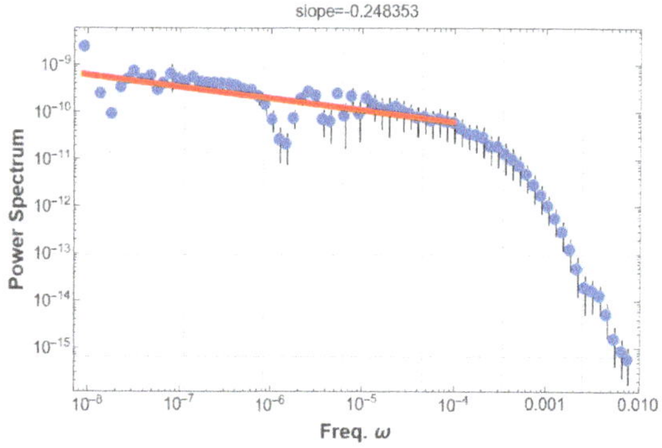


**Figure 2.** (**left**) PSD for the high-energy group, encompassing events with energy exceeding the mean. The PSD does not exhibit 1/f fluctuations, akin to Figure 1 (right). (**right**) PSD for the low-energy group, including events with energy below the mean. The PSD clearly shows 1/f fluctuation with an index of $-0.88$ over more than five orders of magnitude. Generally, 1/f fluctuation is defined by a power index of $-1 \pm 0.5$ in the PSD. A red line in the low-frequency domain in the PSD graph represents the best fits of the data according to the least squares method within that domain.

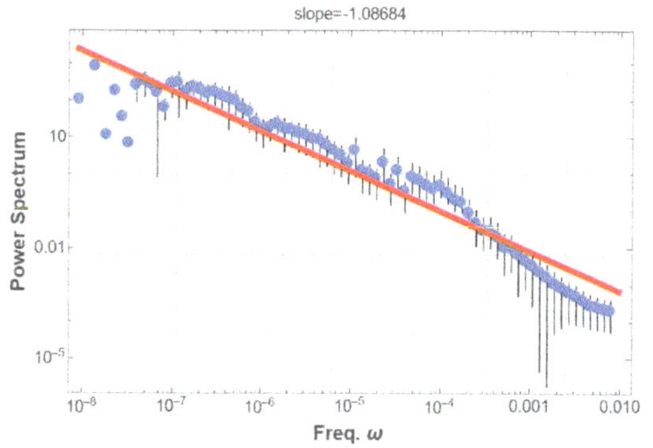


**Figure 3.** (**left**) Similar to Figure 1 (**right**), but with energy information entirely removed: following time-uniformization, finite energy values were reset to one, leaving the others at zero. This represents the time sequence of the flare occurrence, or the unweighted distribution. The PSD displays typical pink noise behavior with a power index of $-1.1$ in the frequency range $10^{-3}$ to $2 \times 10^{-8}$ Hz. (**right**) Corresponding PSD for the high-energy group. This clear 1/f fluctuation, with a power index $-0.98$, contrasts with Figure 2, where all energy information is included, resulting in a flat spectrum. A red line in the low-frequency domain in the PSD graph represents the best fits of the data according to the least squares method within that domain.

We also analyzed short-term GOES16 solar flare data for one week, chosen arbitrarily. The results are consistent, showing a 1/f fluctuation with an index of $-1.2$ across a frequency range of $2 \times 10^{-3}$ to $2 \times 10^{-6}$ Hz. This extends across the entire week, encompassing frequencies lower than the typical frequency of SFO.

So far, we have examined soft X-ray data from GOES observations, which provide one perspective of solar flares. We now turn to hard X-ray data from RHESSI observations for another perspective on solar flare.

Analysis of 16 years of RHESSI solar flare data from 2002 to 2018 [14] mostly aligns with the GOES findings. The RHESSI solar flare energy time sequence displays a marginal 1/f fluctuation with an index of $-0.59$ as shown in Figure 4 (left). In contrast, the occurrence time sequence exhibits a pronounced 1/f fluctuation with an index of $-0.90$ within the frequency range of $10^{-5}$ to $2 \times 10^{-9}$ Hz, as depicted in Figure 4 (right). This fluctuation spans from a day to the entire observation period, though data scattering at the lowest frequency domain is notable. Other flare indicators, such as total count, peak count rate, and duration times mean energy, were also examined. These indicators generally yielded a power index around $-0.40$, diverging significantly from 1/f fluctuations. However, the duration alone demonstrated a 1/f fluctuation with a power index of $-0.88$, with slightly less scattering than Figure 4 (right).

Finally, in our analysis of solar flares in this section, we acknowledge certain limitations. Each solar flare event has been characterized as instantaneous in our dataset. However, realistically, a solar flare involves substantial energy transfer from the magnetic field to plasma and particles over a finite duration; large flares may have prolonged durations, while smaller flares are more instantaneous. We have excluded the extended profiles, such as the decay phases of large flares. This decision is based on the fact that, in the 16-year RHESSI dataset, the average time interval between flares is 4865 s, while the average duration of a flare is 642 s, with a maximum of 4500 s. Consequently, we believe this approach does not significantly alter our conclusion that solar flare 1/f fluctuation is predominantly a low-energy phenomenon. This does not discount the high-energy core component of a solar flare. The relationship between the low-energy and high-energy components of solar flares, particularly in the context of SOC, will be further discussed in the next section.

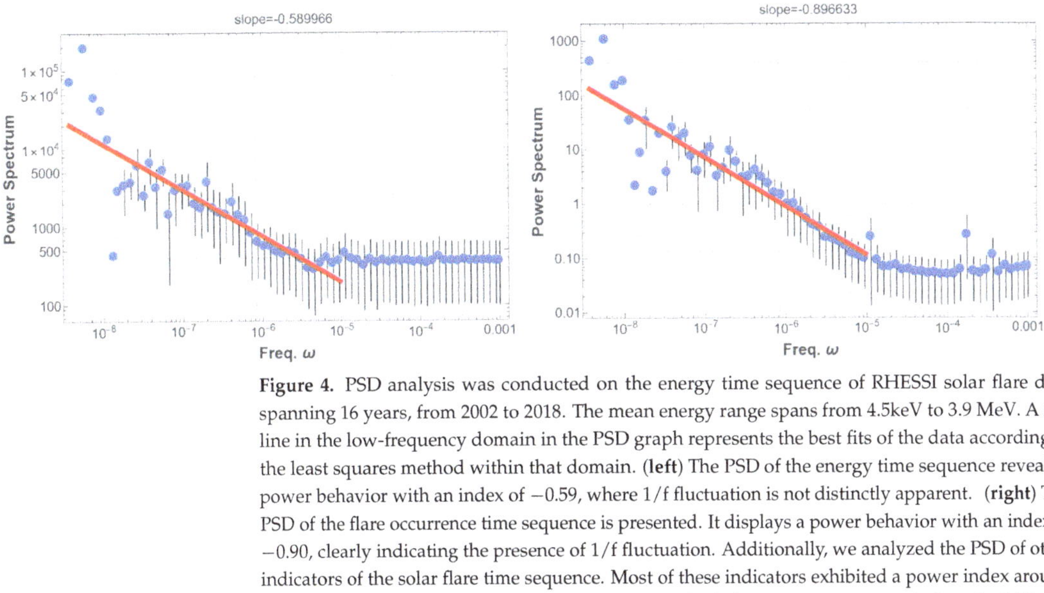


**Figure 4.** PSD analysis was conducted on the energy time sequence of RHESSI solar flare data spanning 16 years, from 2002 to 2018. The mean energy range spans from 4.5keV to 3.9 MeV. A red line in the low-frequency domain in the PSD graph represents the best fits of the data according to the least squares method within that domain. (**left**) The PSD of the energy time sequence reveals a power behavior with an index of $-0.59$, where 1/f fluctuation is not distinctly apparent. (**right**) The PSD of the flare occurrence time sequence is presented. It displays a power behavior with an index of $-0.90$, clearly indicating the presence of 1/f fluctuation. Additionally, we analyzed the PSD of other indicators of the solar flare time sequence. Most of these indicators exhibited a power index around $-0.40$, with the notable exception of the duration, which demonstrates a power index of $-0.88$.

What mechanism then gives rise to this universal 1/f fluctuation at the low-energy regime of various solar flare datasets?

## 3. Amplitude Modulation from Resonance

We recently proposed a potential origin for 1/f fluctuation, attributing it to wave beat or amplitude modulation [8]. Given the generality of this mechanism, our intention is to extend its application to the context of solar flare 1/f fluctuation within the scope of this paper.

The foundation of this theory rests on the observation that waves with accumulating frequencies yield robust low-frequency signals. In cases where this accumulation is systematic, such as in instances of resonance, synchronization, or infrared divergence, the wave beats consistently manifest a power law with an index approximately equal to −1. The ubiquity of this phenomenon in nature stems from the prevalence of simple physics: wave beat or amplitude modulation.

Consider an example of a beat: waves with frequencies of 440 Hz and 441 Hz yield a beat. In musical sounds, this beat is 'audible' as a sinusoidal amplitude oscillation with a frequency of 1 Hz. However, a Fourier transform of the original signal does not yield this 1 Hz signal, only the original two frequencies. To extract this encoded 1 Hz signal, one simple method is to square the data and then apply a Fourier transform. This approach allows for the extraction of the encoded low-frequency signal at 2 Hz, though it is twice the original frequency. Decoding is not limited to squaring but can also involve absolute value, rectification, fourth-order power, thresholding, or other methods, resulting in a variety of pink noise patterns.

Another example is AM radio, where amplitude modulation is employed. Using high-frequency radio waves between 526.5 kHz and 1606.5 kHz, a low-frequency audible signal is encoded. However, the encoded sound cannot be directly heard, as the rapidly oscillating positive and negative parts in the wave cancel each other out, leaving no audible signal. Demodulation is achieved by rectifying the radio wave signal, traditionally through germanium diodes or vacuum tubes. This process is indispensable for extracting the encoded low-frequency signal, such as 1/f fluctuation [8].

In the context of solar flares, we hypothesize that the resonant mode crucial for the manifestation of 1/f fluctuations is the solar five-minute oscillation (SFO), a phenomenon consistently activated within the solar atmosphere through turbulent convection [10,11]. Specifically, pressure modes of SFO exhibit accumulating eigenfrequencies, particularly converging towards lower angular indices $l$. We aim to examine whether this frequency accumulation effectively produces a 1/f power spectral density.

If SFO induces amplitude modulation, demodulation becomes imperative for observing 1/f fluctuation [8]. This necessity arises due to the cancellation of positive and negative components within the relatively high-frequency wave, encompassing 1/f modulation. In the context of solar flares, the demodulation process is envisioned to be facilitated by the threshold established through magnetic reconnection. The tiny energy required for magnetic reconnection may exhibit the 1/f fluctuation characteristic, aligning with the tiny energy associated with SFO that can trigger solar flares.

The energy associated with SFO may be significantly small compared with the total energy of a flare. Despite this, we believe that the SFO determines the 1/f fluctuation property of flares.

In general, a solar flare appears to involve two consecutive stages: (a) the gradual accrual of core magnetic energy manifested as accumulating intensity of magnetic fields with opposite polarities and (b) a subsequent tiny trigger that reconnects a very local strained magnetic fields initiating a sudden discharge of energy. From the preceding discussion, it is apparent that the 1/f fluctuation is closely associated with the second stage, (b).

The first stage, (a), is characterized by the progressive buildup of magnetic strain energy within many local domains of concentrated magnetic fields. This buildup phenomenon may be well described by applying the theory of self-organized criticality (SOC).

Regarding the second stage, (b), the final trigger, though minor, introduces sufficient energy to cause the reconnection of magnetic fields in a local area, overcoming the energy threshold and resulting in the burst release of magnetic energy. This minor final trigger, possibly linked to ongoing fluctuations of SFO on the solar surface, determines the timing of a solar flare; the characteristics of 1/f fluctuation in SFO may be inherited by the flares.

Moving forward, our analysis will focus more closely on this latter stage, (b), by applying the amplitude modulation theory to the analysis of 1/f fluctuations in solar flares.

## 4. Resonating Solar Five-Minute Oscillation

We delve into the potentiality of the solar five-minute oscillation (SFO) as a catalyst for 1/f fluctuation in solar flare activity. Specifically, our focus centers on elucidating how SFO eigenmodes contribute to the accumulation of frequencies, thereby generating low-frequency signals through amplitude modulation mechanisms.

The small displacement, denoted as $u(t, r, \theta, \phi)$, of the solar atmosphere from its equilibrium position follows the Poisson equation

$$\rho\ddot{u} = \kappa\triangle u - \rho\nabla\phi_g, \; \Delta\phi_g = 4\pi G\rho \tag{1}$$

where $\rho, \kappa, G, \phi_g$ represent the mass density, bulk modulus, gravitational constant, and gravitational potential, respectively.

The stationary solution $u(t, r, \theta, \phi) = v(r, \theta, \phi)e^{-i\omega t}$ leads to the eigenvalue equation. Utilizing the variable separation method in the spherical coordinate system, we obtain a solution of the form

$$u(t, r, \theta, \phi) = R_{n,l,m}(r)Y_l^m(\theta, \phi)e^{-i\omega_{n,l,m}t}, \tag{2}$$

where $Y_l^m(\theta, \phi)$ represents spherical harmonics. Modes are characterized by quantum numbers $n = 0, 1, 2, \ldots, l = 0, 1, 2, \ldots,$ and $-l \leqq m \leqq l$. These modes are further categorized as pressure and gravitational modes. All parameters are uncertain depending on the detail of the solar interior, making the solution of the eigenvalue equation a complex task. Numerous numerical calculations and observational studies have been conducted on these eigenmode equations.

We utilize observational data pertaining to eigenmodes of solar oscillations obtained through helioseismology [15]. This dataset provides valuable information on numerous observed frequencies, disregarding the degeneration in the azimuthal order number parameter $m$ $(-l \leq m \leq l)$. A distinctive characteristic of these modes is the accumulation of frequencies towards smaller values of $l$ for each $n$ parameter, typically around $3 \times 10^{-3}$ Hz. This property is pivotal for the emergence of 1/f fluctuation through the amplitude modulation mechanism [8].

To simulate the phenomenon, we randomly superimposed all sinusoidal waves with frequencies ranging from the lowest at 848.241 μHz up to 4669.16 μHz. The wave mode superposition is expressed as

$$\Phi(t) = \sum_{k=1}^{N} \xi_k sin(2\pi\Omega_k t), \tag{3}$$

where $\xi_k$ is a random variable within the range $[0, 1]$, and $N = 2247$ represents the total number of eigenfrequencies in the dataset. Subsequently, we conducted Fourier analysis (FFT) on the power spectral density (PSD) for the time series of the absolute value $|\Phi(t)|$. Notably, calculating the PSD for the bare $\Phi(t)$ yields no signal in the low-frequency domain. As the 1/f fluctuation signal is modulated in our model, a demodulation process is imperative; taking the absolute value is a typical demodulation method, essential for extracting 1/f fluctuations in the PSD analysis. The specifics of the demodulation process will be explored further in the subsequent discussion.

The PSD analysis results in a power-law with an index of approximately−0.42 within the low-frequency range of $2 \times 10^{-5}$–$5 \times 10^{-3}$ Hz, as illustrated in Figure 4 (left). However, it is noteworthy that the observed solar flare 1/f fluctuation occurs in the range of $2 \times 10^{-8} \sim 10^{-3}$ Hz, considerably lower than our analysis. Further, the power index is definitely larger than in the case of the observed solar flare. These discrepancies suggest the need for a more nuanced consideration of realistic fine structures of the eigenstates and additional resonances, a facet we will delve into in the subsequent analysis.

Our analysis of resonance is currently incomplete, with several aspects requiring further exploration. Firstly, (a) each eigenmode is associated with a resonance curve, and numerous frequency-accumulating modes are linked to each mode. Secondly, (b) the degeneration in the azimuthal order number $m$ should give rise to a fine structure around each principal frequency characterized by $n$ and $l$. This degeneration in $m$ is resolved by the solar non-spherical symmetry or the solar rotation. In this paper, we examine representative modes for both cases (a) and (b) to illustrate how the fine structure contributes to the emergence of $1/f$ fluctuation. A comprehensive analysis encompassing these aspects will be presented in our forthcoming publications.

To refine the power spectral density (PSD), we introduce the following effects: (a) each eigenfrequency labeled by $n$ and $l$ possesses a finite width, and (b) the degeneration in $m$ is resolved by the solar rotation.

(a) Resonant modes are typically modeled by the Lorentzian distribution,

$$R[\omega] = \frac{1}{\left(\frac{\kappa}{2}\right)^2 + (\omega - \Omega)^2}, \tag{4}$$

where $\Omega$ is the fiducial resonance frequency, and $\kappa$ characterizes the sharpness of the resonance. This function represents the frequency distribution density associated with the fiducial frequency $\Omega$. The inverse function (tangent) of the cumulative distribution function (hyperbolic tangent) generates this distribution from the Poisson random field.

(b) Solar rotation resolves the degeneration in $m$ by breaking the spherical symmetry of the system. Although the details are intricate, a rough estimate is provided by the resolved frequency [16,17] in the lowest perturbation in $\Omega/\omega$ ($\ll 1$),

$$\omega_{nlm} = \omega_{nl} + \frac{m}{l(l+1)}\Omega, \tag{5}$$

where $\omega_{nl}$ is the degenerate eigenfrequency, and $\Omega = 4.3 \times 10^{-7}$ Hz is the frequency associated with the solar rotation. The coefficient of $\Omega$ is chosen approximately according to [16,17].

These effects are implemented through a specific process. Initially, we construct wave data by superposing $N$ sinusoidal waves with eigenfrequencies after eliminating the degeneration in $m$. Additionally, we superimpose $M$ resonant waves with frequencies proximate to the fiducial frequency, following the distribution in Equation (4). The fully superposed wave is defined as:

$$\Phi(t) = \sum_{n=1}^{N} \sum_{i=1}^{M} \sin(2\pi(1 + c\tan(\xi_i))\Omega_n t), \tag{6}$$

where the parameter $c = \kappa/\Omega$ represents the relative line width for each eigenfrequency. The random variable $\xi_i$, ranging in $[0, \pi/2]$, generates the frequency distribution through $R(\omega)$ in Equation (4). While $c$ actually depends on each $n$, for simplicity, we use $c = 0.01$. We utilized data [15], limiting $M$ to 100 and $N$ to 100.

As before, the power spectral density (PSD) of the bare $\Phi(t)$ exhibits no signal in the low-frequency region. However, taking the absolute value $|\Phi(t)|$ or applying arbitrarily set threshold data produces $1/f$ fluctuations (details in the caption of Figure 5). These operations essentially function as a demodulation of the original signal. Consequently, the $1/f$ fluctuation becomes evident only after demodulation and proves to be quite robust. Figure 5 (right) illustrates the PSD of the thresholded data, demonstrating an approximate $1/f$ fluctuation with a power index of $-1.1$, covering a frequency range extended down to $2 \times 10^{-7}$ Hz. This range partially coincides with the observed range below $10^{-3}$ Hz. In our future study, further refinement of the PSD analysis is intended, incorporating finer structures of eigenfrequencies, decay times, and deviations of the Sun from spherical symmetry. The introduction of gravitational modes, alongside pressure modes, which operate in much lower frequency domains, is also of interest.

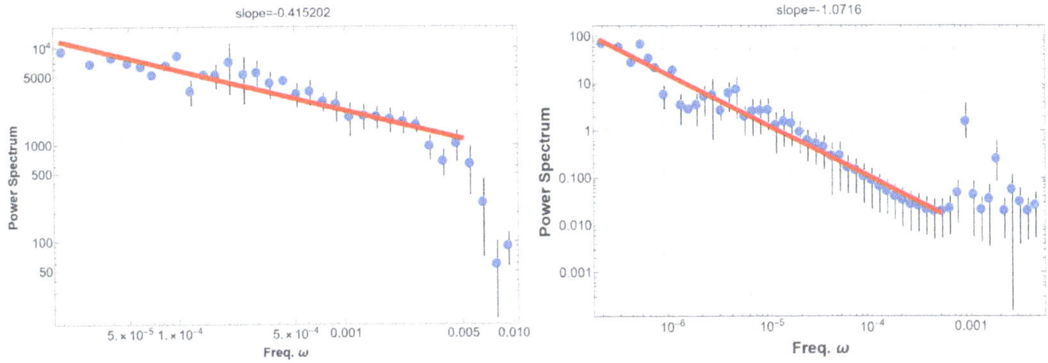


**Figure 5. left**: On the left side, the graph depicts the power spectral density (PSD) of the absolute value of the time sequence given by Equation (3), denoted as $|\Phi(t)|$. Here, $\Phi(t)$ represents the superposition of sinusoidal waves with the $N = 2247$ eigenfrequencies of the solar five-minute oscillation (SFO), each with a random amplitude. Each mode is identified by the parameters $n, l$, and the azimuthal order number $m$, which is degenerate. Despite exhibiting a power law, this presentation barely demonstrates the characteristics of pink noise. **right**: On the right side, analogous to the left graph, this graph includes resonant modes and fine eigenmodes after resolving the degeneration in $m$. In constructing the data $\Phi(t)$, we superimpose sinusoidal waves with 100 frequencies from the lowest and introduce $N = 100$ Lorentzian-distributed modes. The latter are randomly generated following Equation (4). The graph displays the PSD of the thresholded timing sequence of Equation (6), $\Phi(t)$. The threshold is set to select data points $\Phi(t)$ that surpass the mean although is moderately insensitive to the demodulation method. This presentation reveals nearly 1/f fluctuation with an index of $-1.1$ spanning over four digits. Notably, variations in thresholds and sample sizes in the PSD analysis consistently yield similar 1/f fluctuations. A red line in the low-frequency domain in the PSD graph represents the best fits of the data according to the least squares method within that domain.

In the preceding discussion, we superimposed the eigenmodes of the solar five-minute oscillation (SFO) to obtain amplitude modulation and pink noise, explaining the observed 1/f fluctuation in solar flares. However, a more direct examination of the bare data of SFO before decomposition into eigenmodes is warranted. This implies that the resonance of SFO directly yields pink noise. Initially, we utilize SOHO-GOLF data on the fluctuations of the time elapsed $T(t)$, for the acoustic waves to travel through the solar center [18] measured in the unit second. Details are expounded in [19], with the data spanning about 16.5 years and an interval of 80 s, albeit with some data-missing periods.

We commenced by calculating the power spectral density (PSD) of the original data $T(t)$. The result is depicted in the left graph of Figure 6. A prominent peak appears around $3 \times 10^{-3}$ Hz, corresponding to the typical five-minute mode of solar oscillation. From there, a partial power-law behavior is observed toward $6 \times 10^{-6}$ Hz with an index of about $-1$, followed by a flat behavior at lower frequencies. This partial 1/f fluctuation may stem from instrumental origins [20], potentially not reflecting genuine solar properties.

Conversely, when taking the absolute value of the data, the 1/f fluctuation region in the PSD extends toward the lowest frequency limit, as illustrated in Figure 6 (right). The behavior of the SOHO-GOLF data $T(t)$ is precisely the same as the sound data of orchestra music [21] or the sound of a big bell. While the sound wave amplitude time-sequence data do not exhibit pink noise, the square of the amplitude showcases apparent pink noise. These operations of squaring or taking the absolute value naturally correspond to the demodulation process, revealing the encoded pink noise.

This forms part of the rationale behind our belief that solar eigenoscillations contribute to the 1/f fluctuation observed in solar flares. To extract 1/f fluctuation from the wave $T(t)$, we required operational demodulation, such as taking the absolute value, due to the low-frequency 1/f fluctuation being encoded as amplitude modulation. In contrast, operational demodulation was not necessary in the case of solar flares, as the demodulation process is considered intrinsic to the Sun and is already facilitated by magnetic reconnection.

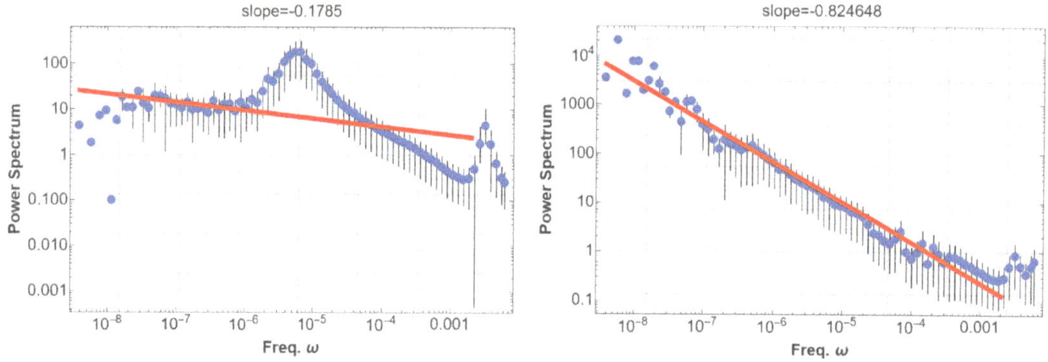


**Figure 6.** (**left**) The power spectral density (PSD) is presented for the SOHO-GOLF 16.5-year data, focusing on the time elapsed T(t) for the waves circumnavigating the Sun [18]. The rightmost peak corresponds to the typical five-minute mode of the solar five-minute oscillation (SFO) at $3 \times 10^{-3}$ Hz. (**right**) The PSD is shown for the absolute value of the original data, revealing $1/f$ fluctuation with a slope of $-0.82$ over approximately six digits. In both figures, the frequency region corresponding to SFO is deliberately excluded from the fitting region of $1/f$ fluctuations. This exclusion is based on our consideration that the $1/f$ fluctuation originates from the wave beat produced by the resonating SFO; such a beat typically manifests in a lower frequency domain than the resonator itself. Practically, including the SFO frequency region in the analysis is not expected to influence the results significantly. A red line in the low-frequency domain in the PSD graph represents the best fits of the data according to the least squares method within that domain.

An alternative analysis involves ground-based data from BiSON [22,23], measuring radial velocity from January 1985 to January 2023 (all sites, optimized for quality). The PSD of the original bare data does not exhibit pink noise. However, taking the absolute value of the original data results in clear $1/f$ fluctuation with a slope of $-0.71$, although this index is slightly larger than the right side of Figure 6. This discrepancy is likely influenced by artificial peaks corresponding to the periods of the Earth's spin and rotation at $1.1 \times 10^{-5}$ Hz and $3 \times 10^{-8}$ Hz.

A variety of demodulation methods have been considered in association with this BiSON data manipulation. In our previous analysis, we focused on the absolute value of the data, but it is noteworthy that other manipulations can also be employed to extract pink noise. For instance, the squared data exhibit pink noise, while the $1/f$ fluctuation disappears in the cubed data. However, when the data is raised to the fourth power, $1/f$ fluctuation reappears. These findings strongly indicate that the amplitude-modulated $1/f$ fluctuation emerges after specific demodulation processes.

Similar phenomena are often observed in sound systems. For example, we examined sound data collected at the water-harp cave (Suikinkutsu) at HosenIn Temple in Kyoto [24]. The sound is generated by the perpetual impact of water drops on the water surface in the two-meter Mino-yaki pot underground [25]. Although the original sound data barely show pink noise, the squared data from this sound source clearly display $1/f$ fluctuation with an index of $-0.80$ for four digits. The resonator, in this case, is presumed to be the Mino-yaki pot.

## 5. Timing Statistics

In our analysis of Section 2, we identified $1/f$ fluctuation in the time series of solar flare timing and speculated that this characteristic might be indicative of the low-energy trigger for solar flares. Similar pink noise behavior is observed in seismic activity [9], where time series of earthquake occurrences is often described by the Weibull distribution function [26,27]:

$$f(x) = \frac{\alpha}{\beta}\left(\frac{x}{\beta}\right)^{\alpha-1} \exp\left(-\left(\frac{x}{\beta}\right)^{\alpha}\right). \tag{7}$$

This leads us to explore whether the solar flare timing sequence is characterized by the Weibull distribution and its potential connection to pink noise.

Upon examination of the GOES data [12] used in our analysis, we discover that, the logarithm of the time intervals between solar flares follows the Weibull distribution, as illustrated in Figure 7 (left). The best-fit parameters are $\alpha = 5.3$ and $\beta = 9.9$. Therefore, similar to seismic activity, the statistical distribution of solar flare timing can be effectively characterized by the Weibull distribution. The question then arises: to what extent does this Weibull distribution characterize the $1/f$ fluctuation property?

We have checked that a time sequence simply following the Weibull distribution does not exhibit pink noise; the power spectral density (PSD) becomes flat in the low-frequency range, as depicted in Figure 7 (right). Consequently, the $1/f$ fluctuation observed in solar flare timing is independent of the Weibull distribution, as the behavior observed in seismic cases. This property is readily understood; the time sequence, constructed with randomly chosen intervals according to Weibull distribution statistics, lacks long correlation times. In contrast, $1/f$ fluctuation inherently possesses long correlation times, thereby manifesting as a low-frequency property.

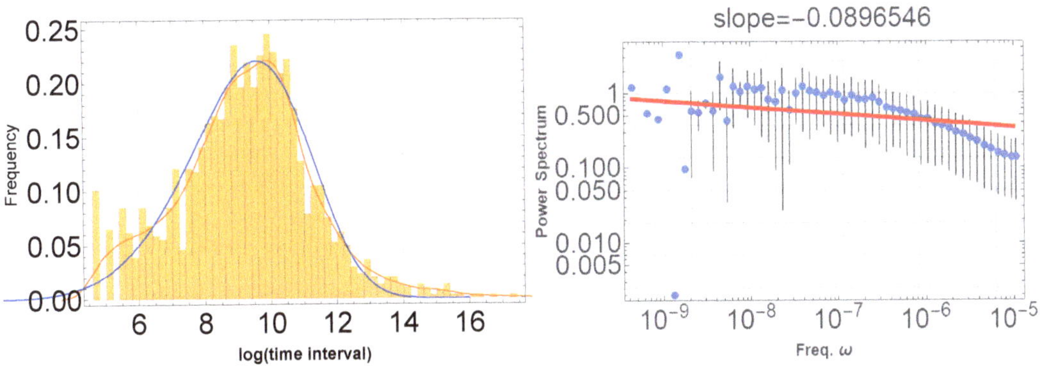


**Figure 7.** (**left**) Orange bars indicate the frequency distribution of the logarithm of the time intervals between solar flare occurrences, while the thin red line represents the smoothed version of the fitting curve. The blue line denotes the Weibull distribution with the parameters $\alpha = 5.3$ and $\beta = 9.9$, best fitting the solar flare data. (**right**) The PSD is presented for artificial data with random time intervals generated by the best-fit Weibull distribution. It is evident that the PSD of these artificial data is flat, indicative of a random distribution. A red line in the low-frequency domain in the PSD graph represents the best fits of the data according to the least squares method within that domain.

## 6. Robustness and Inherited 1/f Fluctuation

We have delved into the origin of $1/f$ fluctuation in solar flares, applying the overarching concept that the accumulation of frequencies in numerous waves leads to beat or amplitude modulation. Frequency accumulation is achieved through resonance, and we have successfully explored this concept using the solar five-minute oscillation (SFO), which are consistently resonating. It is plausible that magnetic reconnection serves as the demodulation (DM) of this amplitude modulation (AM), resulting in $1/f$ fluctuation in solar flares. If this holds true, the combination of SFO (as AM) and magnetic reconnection (as DM) may generate $1/f$ fluctuation beyond solar flare in other extended regions of the solar neighborhood.

For instance, if magnetic reconnection induces solar wind through jetlets [28], triggered by SFO, then the solar wind [29] may also exhibit pink noise. Indeed, 1/f fluctuation in the solar wind has been observed over many years [30–32]. SFO has been detected in the solar corona [33], suggesting the potential observation of 1/f fluctuation in that context as well.

Furthermore, the solar wind may interact with the Earth's atmosphere, inducing chemical reactions leading to the production of the $NO_3^-$ isotope. This isotope could then become embedded in Antarctic ice cubes. Research in this area is ongoing. If solar wind and solar flares influence the Earth's surface, then sea surface temperature may also exhibit 1/f fluctuation, similar to the case of seismic activity.

In seismic activity, which also displays pink noise, seismic events inherit a 1/f fluctuation pattern, potentially resulting from amplitude modulation due to resonance with Earth free oscillation (EFO) in the lithosphere [9]. This EFO may further contribute to the 1/f fluctuation observed in the time sequences of volcano eruptions and the fluctuation of the Earth's rotation axes.

Including the above robustness of pink noise, we compare solar flares and earthquakes in Table 1. This table is preliminary and will be finalized in our forthcoming study.

**Table 1.** Similarity of solar flares and earthquakes from the view point of pink noise. This is a tentative table, and the detail will be reported soon by the authors.

| | **Solar Flare** | **Earthquakes** |
|---|---|---|
| AM (resonator) | Solar five-minute oscillation (SFO) | Earth free oscillation (EFO) |
| DM | magnetic reconnection | fault rupture |
| PSD total data | flat (GOES16, RHESSI) | flat (USGS) |
| PSD low-energy | pink (GOES16, RHESSI) | pink (USGS) |
| PSD timing | pink (GOES16, RHESSI) | pink (USGS) |
| PSD superposed eigenmodes | pink (JSOC) | pink (T. G. Masters, R. Widmer ) |
| Weibull distribution | yes: $\alpha = 5.3$ and $\beta = 9.9$ (GOES16) | yes: $\alpha = 6.3$ and $\beta = 7.63$ (USGS) |
| PSD resonator | pink (SOHO-GOLF, BiSON) | ? |
| inherent phenomena | solar wind, sunspot number, nitrate, SST, cosmic ray | volcano eruption, rotation axes |

## 7. Conclusions and Prospects

In conclusion, our investigation has identified 1/f fluctuation in the solar flare time series, and we have partially elucidated its origin by applying our proposed mechanism: 1/f fluctuation emerges from the resonance of the solar five-minute oscillations via amplitude modulation and demodulation.

Our GOES data analysis of the solar flare time sequence has revealed distinct low-frequency properties. Specifically, the power spectral density of low-energy flares ($E \leq E_{mean}$) exhibited 1/f fluctuations, while high-energy flares ($E > E_{mean}$) displayed a flat spectrum. Notably, the time sequence of flare occurrences demonstrated clearer 1/f fluctuations, indicating that low-energy characteristics play a pivotal role in triggering the observed 1/f fluctuations in solar flares. Building on our recent proposal that 1/f noise arises from amplitude modulation and demodulation, we postulated that this modulation is encoded through resonance with the solar five-minute oscillation (SFO) and demodulated via magnetic reconnection.

To test this hypothesis, we constructed a dataset by superposing sinusoidal waves with 2247 eigenfrequencies of SFO. The absolute value of this time sequence marginally exhibited 1/f fluctuations with a power index of $-0.42$ down to $2 \times 10^{-5}$ Hz. Further refinement of the data, considering resonance effects and finer structures labeled by $m$ induced by solar rotation, involved adding 100 extra modes generated by resonant Lorentzian distributions for the first 100 eigenfrequencies of SFO after resolving the degeneration in m. The absolute value of this refined time sequence clearly displayed 1/f fluctuations with a power index of $-1.1$ down to $10^{-7}$ Hz, largely overlapping with the observed range of solar flare 1/f fluctuations (power index $-1.0$ from about $2 \times 10^{-3}$ Hz down to2 $\times 10^{-8}$ Hz). Thus, our analysis provided partial verification that SFO triggers seismic 1/f fluctuations.

Further investigation into SOHO-GOLF data and BiSON data for velocity fluctuations in the solar atmosphere revealed that the original time sequence of these data barely exhibited 1/f fluctuations, while the absolute values of the time sequence did display clear 1/f fluctuations. This lends additional support to our proposition: SFO is the origin of solar flare 1/f fluctuations.

Additionally, our examination of the time sequence of solar flare occurrences revealed adherence to a Weibull distribution. However, an artificial time sequence composed from the Weibull distribution barely exhibited 1/f fluctuations, suggesting that the Weibull distribution does not fully characterize solar 1/f fluctuations.

Lastly, our comparison of 1/f fluctuations in solar flares and earthquakes, as demonstrated above, has revealed remarkable similarities between them [34,35]. The underlying commonality in these phenomena is the generation of 1/f fluctuations through amplitude modulation by spherical resonators in the universe, such as stars and planets. This concept can naturally extend to other celestial objects, including black holes and neutron stars. Specifically, we have conducted a preliminary analysis of six years of MAXI X-ray data from Cyg X-1, obtaining 1/f fluctuations with a power index of $-0.8$ over four orders of magnitude. We hypothesize that the resonator in this case is the quasi-periodic oscillation of a black hole. Additionally, the similarity of repeated Fast Radio Burst (FRB) data with solar flares and earthquakes is noteworthy. It would be intriguing to explore whether this signal is related to the eigenoscillations of neutron stars or magnetars. These analyses will be detailed in our forthcoming separate papers. This broader perspective underscores the potential universality of the proposed amplitude modulation and demodulation mechanism across diverse astrophysical phenomena.

**Author Contributions:** Contributions: Conceptualization, M.M. and A.N.; methodology, A.N.; software, M.M.; validation, A.N. and M.M.; formal analysis, M.M.; investigation, A.N.; resources, M.M. and A.N.; data curation, A.N.; writing—original draft preparation, M.M.; writing—review and editing, A.N.; visualization, M.M.; supervision, A.N.; project administration, M.M. and A.N.; funding acquisition, M.M. All authors have read and agreed to the published version of the manuscript.

**Funding:** This research was funded by Japan Society for the Promotion of Science (JSPS) grant number 18K18765.

**Data Availability Statement:** The data presented in this study are available on request from the corresponding author.

**Acknowledgments:** We extend our heartfelt gratitude to the members of the Lunch-Time Remote Meeting and the Department of Physics at Ochanomizu University, as well as Izumi Uesaka and Manaya Matsui at Kyoto Sangyo University, for their engaging discussions and unwavering support. Special appreciation goes to Tim Larson and Takashi Sekii for their invaluable insights into solar data. Additionally, we express our sincere thanks to Satoru Ueno for enlightening discussions on the topic of solar flare pink noise. Their contributions have enriched our work and made this research journey a truly collaborative and meaningful experience.

**Conflicts of Interest:** The authors declare no conflict of interest. The funders had no role in the design of the study; in the collection, analyses, or interpretation of data; in the writing of the manuscript; or in the decision to publish the results.

## References

1. Benz, A.O. Flare Observations. *Living Rev. Sol. Phys.* **2017**, *14*, 2. [CrossRef]
2. Gutenberg, B.; Richter, C.F. Frequency of Earthquakes in California. *Bull. Seismol. Soc. Am.* **1944**, *34*, 185–188. [CrossRef]
3. Omori, F. On the Aftershocks of Earthquakes. *J. Coll. Sci. Imp. Univ. Tokyo* **1894**, *7*, 111.
4. de Arcangelis, L.; Godano, C.; Lippiello, E.; Nicodemi, M. Universality in Solar Flare and Earthquake Occurrence. *Phys. Rev. Lett.* **2006**, *96*, 051102. [CrossRef] [PubMed]

5. Najafi, A. The Modified Form of the Gutenberg-Richter Law in Solar Flare Complex Network: Approach of Genetic Algorithm on the Thresholded Power-Law Behavior. *Iran. Astron. Astrophys.* **2020**, *7*, 47–56.
6. Raychaudhuri, A.K. Measurement of 1/f noise and its application in materials science. *Curr. Opin. Solid State Mater. Sci.* **2002**, *6*, 67–85. [CrossRef]
7. Milott, E. 1/f noise: A pedagogical review. *arXiv* **2002**, arXiv:physics/0204033.
8. Morikawa, M.; Nakamichi, A. A simple model for pink noise from amplitude modulations. *Sci. Rep.* **2023**, *13*, 8364. [CrossRef]
9. Nakamichi, A.; Matsui, M.; Morikawa, M. Seismic 1/f Fluctuations from Amplitude Modulated Earth's Free Oscillation. *arXiv* **2023**, arXiv:2307.03192.
10. Leighton, R.B.; Noyes, R.W.; Simon, G.W. Velocity Fields in the Solar Atmosphere. I. Preliminary Report. *Astrophys. J.* **1962**, *135*, 474. [CrossRef]
11. Evans, J.W.; Michard, R. Observational Study of Macroscopic Inhomogeneities in the Solar Atmosphere. III. Vertical Oscillatory Motions in the Solar Photosphere. *Astrophys. J.* **1962**, *136*, 493. [CrossRef]
12. NOAA. National Centers for Environmental Information. Available online: https://data.ngdc.noaa.gov/platforms/solar-space-observing-satellites/goes/goes16/l2/data/xrsf-l2-avg1m_science/ (accessed on 1 October 2023).
13. Ueno, S.; Mineshige, S.; Negoro, H.; Shibata, K.; Hudson, H.S. Statistics of Fluctuations in the Solar Soft X-ray Emission. *Astrophys. J.* **1997**, *484*, 920–926. [CrossRef]
14. The Reuven Ramaty High Energy Solar Spectroscopic Imager (RHESSI). Available online: https://en.wikipedia.org/wiki/Reuven_Ramaty_High_Energy_Solar_Spectroscopic_Imager (accessed on 14 September 2022).
15. Joint Science Operations Center (JSOC). Available online: http://jsoc.stanford.edu/ (accessed on 1 October 2023).
16. Duvall, T.L., Jr.; Harvey, J.W. Rotational frequency splitting of solar oscillations. *Nature* **1984**, *310*, 19–22. [CrossRef]
17. Gilbert, F.; Backus, G. The Rotational Splitting of the Free Oscillations of the Earth, 2. *Rev. Geophys.* **1965**, *3*, 1–9. [CrossRef]
18. GOLF Patrick Boumier. Available online: https://www.ias.u-psud.fr/golf/templates/access.html (accessed on 1 October 2023).
19. Fossat, E.; Boumier, P.; Corbard, T.; Provost, J.; Salabert, D.; Schmider, F.X.; Gabriel, A.H.; Grec, G.; Renaud, C.; Robillot, J.M.; et al. Asymptotic g modes: Evidence for a rapid rotation of the solar core. *Astron. Astrophys.* **2017**, *604*, A40. [CrossRef]
20. Gabriel, A.H.; Grec, G.; Charra, J.; Robillot, J.-M.; Roca Cortés, T.; Turck-Chièze, S.; Bocchia, R.; Boumier, P.; Cantin, M.; Cespédes, E.; et al. Global Oscillations at Low Frequency from the SOHO Mission (GOLF). *SOHO Mission* **1995**, *162*, 61–99.
21. Morikawa, M. Low-Frequency Characterization of Music Sounds—Ultra-Bass Richness from the Sound Wave Beats. *arXiv* 2021, arXiv:2104.08872.
22. Davies, G.R.; Chaplin, W.J.; Elsworth, Y.; Hale, S.J. BiSON data preparation: A correction for differential extinction and the weighted averaging of contemporaneous data. *Mon. Not. R. Astron. Soc.* **2014**, *441*, 3009–3017. [CrossRef]
23. Hale, S.J.; Howe, R.; Chaplin, W.J.; Davies, G.R.; Elsworth, Y.P. Performance of the Birmingham Solar-Oscillations Network (BiSON). *Sol. Phys.* **2016**, *291*, 1–28. [CrossRef]
24. HosenIn at Kyoto. Available online: http://www.hosenin.net/ (accessed on 1 October 2023).
25. Suikinkutsu. Available online: http://www.suikinkutsu.com/ (accessed on 1 October 2023).
26. Tanaka, H.; Aizawa, Y. Detailed Analysis of the Interoccurrence Time Statistics in Seismic Activity. *J. Phys. Soc. Jpn.* **2017**, *86*, 024004. [CrossRef]
27. Hatano, T. The Third Law of Earthquake Statistics? *JPSJ News Comments* **2017**, *14*, 03. [CrossRef]
28. Raouafi1, N.E.; Stenborg, G.; Seaton, D.B.; Wang, H.; Wang, J.; DeForest, C.E.; Bale, S.D.; Drake, J.F.; Uritsky, V.M.; Karpen, J.T.; et al. Magnetic Reconnection as the Driver of the Solar Wind. *Astrophys. J.* **2023**, *945*, 28. [CrossRef]
29. Parker, E.N. Dynamics of the Interplanetary Gas and Magnetic Fields. *Astrophys. J.* **1958**, *128*, 664. [CrossRef]
30. Verdini, A.; Grappin, R.; Pinto, R.; Velli, M. On the origin of the 1/f spectrum in the solar wind magnetic field. *Astrophys. J. Lett.* **2012**, *750*, L33. [CrossRef]
31. Matteini, L.; Chen, C.H.; Stansby, D.; Horbury, T.S.; Perrone, D.; Tenerani, A.; Velli, M.; Bale, S.; Pulupa, M.; Malaspina, D.; et al. Large scale 1/f magnetic field spectrum in the solar wind close to the Sun: Comparison between 0.15 and 0.3AU. In Proceedings of the American Geophysical Union Fall Meeting 2019, San Francisco, CA, USA, 9–13 December 2019; abstract SH21C-3329 2019.
32. Huang, Z.; Sioulas, N.; Shi, C.; Velli, M.; Bowen, T.; Davis, N.; Chandran, B.D.G.; Matteini, L.; Kang, N.; Shi, X.; et al. New Observations of Solar Wind 1/f Turbulence Spectrum from Parker Solar Probe. *Astrophys. J. Lett.* **2023**, *950*, L8. [CrossRef]
33. Didkovsky1, L.; Kosovichev, A.; Judge1, D.; Wieman, S.; Woods, T. Variability of Solar Five-Minute Oscillations in the Corona as Observed by the Extreme Ultraviolet Spectrophotometer (ESP) on the Solar Dynamics Observatory Extreme Ultraviolet Variability Experiment (SDO/EVE). *arXiv* **2012**, arXiv:1211.0711.
34. Kossobokov, V.G.; Lepreti, F.; Carbone, V. Complexity in Sequences of Solar Flares and Earthquakes. *Pure Appl. Geophys.* **2008**, *165*, 761–775. [CrossRef]
35. de Arcangelis1, L.; Lippiello, E.; Godano, C.; Nicodemi, M. Statistical properties and universality in earthquake and solar flare occurrence. *Eur. Phys. J. B* **2008**, *64*, 551–555. [CrossRef]


**Disclaimer/Publisher's Note:** The statements, opinions and data contained in all publications are solely those of the individual author(s) and contributor(s) and not of MDPI and/or the editor(s). MDPI and/or the editor(s) disclaim responsibility for any injury to people or property resulting from any ideas, methods, instructions or products referred to in the content.

MDPI
St. Alban-Anlage 66
4052 Basel
Switzerland
www.mdpi.com

*Entropy* Editorial Office
E-mail: entropy@mdpi.com
www.mdpi.com/journal/entropy

Disclaimer/Publisher's Note: The statements, opinions and data contained in all publications are solely those of the individual author(s) and contributor(s) and not of MDPI and/or the editor(s). MDPI and/or the editor(s) disclaim responsibility for any injury to people or property resulting from any ideas, methods, instructions or products referred to in the content.

www.ingramcontent.com/pod-product-compliance
Lightning Source LLC
Chambersburg PA
CBHW061606210926
51850CB00009B/2229
*9783725802050*